Alcohol, Cell Membranes, and Signal Transduction in Brain

Alcohol, Cell Membranes, and Signal Transduction in Brain

Edited by

Christer Alling
University of Lund
Lund, Sweden

Ivan Diamond
The Ernest Gallo Clinic and Research Center
University of California
San Francisco, California

Steven W. Leslie
The University of Texas at Austin
Austin, Texas

Grace Y. Sun
University of Missouri
Columbia, Missouri

and

W. Gibson Wood
Veterans Affairs Medical Center
and University of Minnesota School of Medicine
Minneapolis, Minnesota

Springer Science+Business Media, LLC

Library of Congress Cataloging-in-Publication Data

Alcohol, cell membranes, and signal transduction in brain / edited by
   Christer Alling ... [et al.].
        p.   cm.
      "Proceedings of the Marcus Wallenberg Symposium on Alcohol, Cell
   Membranes, and Signal Tansduction in Brian, held June 28-July 1,
   1993, in Lund, Sweden"--Copr. p.
      Includes bibliographical references and index.
      ISBN 978-0-306-44583-5   ISBN 978-1-4615-2470-0 (eBook)
      DOI 10.1007/978-1-4615-2470-0
      1. Alcohol--Physiological effect--Congresses. 2. Brain--Effect of
   drugs on--Congresses. 3. Neurochemistry--Congresses. 4. Cell
   membranes--Congresses. 5. Cellular signal transduction--Congresses.
   I. Alling, Christer. II. Marcus Wallenberg Symposium on Alcohol,
   Cell Membranes, and Signal Transduction in Brain (1993 : Lund,
   Sweden)
   QP801.A3A427 1993
   615'.7828--dc20   93-23593
                                                                CIP
                                                                r93

Proceedings of the Marcus Wallenberg Symposium on Alcohol, Cell Membranes, and Signal
Transduction in Brain, held June 28–July 1, 1993, in Lund, Sweden

ISBN 978-0-306-44583-5

©1993 Springer Science+Business Media New York
Originally published by Plenum Press, New York in 1993

This book is dedicated to **Dora B. Goldstein, M.D.**, Professor of Pharmacology, Stanford University School of Medicine, whose pioneering work on how alcohol acts on brain membranes was a major contribution and stimulus to understanding the cellular mechanisms involved in intoxication and tolerance.

ACKNOWLEDGEMENTS

This book is based in part on the Marcus Wallenberg Symposium entitled "Alcohol, Cell Membranes, and Signal Transduction in Brain" held in Lund, Sweden, June 28 - July 1, 1992. The major sponsor of the symposium was the Marcus Wallenberg Foundation for International Scientific Exchange. Cosponsors were the National Institute on Alcohol Abuse and Alcoholism, USA, and the Swedish Medical Research Council. Additional support was provided by Bristol-Myers AB, Ciba-Geigy AB, Janssen Pharma AB, Kabi-Pharmacia Diagnostics Sverige AB, Labdesign AB and Scandinavian Diagnostic Services. A special note of appreciation is extended to the Local Organizing Committee (Lena Gustavsson, Christer Larsson, Christofer Lundqvist and F. David Rodríguez) and the conference secretaries (Helena Elding and Marie Elmér) for their administrative and technical support that contributed significantly to the overall success of the symposium.

PREFACE

Alcohol abuse and alcoholism are international problems whose costs economically, psychologically and medically have been well documented. Alcohol is a unique drug in that the effects of excessive use can have a deleterious effect on most if not all organs of the body. The brain is one of the organs most affected by excessive alcohol consumption. Effects on the brain can be seen in cognitive function, brain structure and neurochemistry. Over the past few years, there have been significant advances made in understanding how alcohol affects brain neurochemistry. This book examines four major areas, *i.e.*, membrane lipids, receptors and ion channels, second messengers, and gene expression, where significant advancements have been made.

The book is divided into four sections based on the four major areas. In each section, data are examined that cover a range of approaches from *in vitro* to *in vivo* studies. The section on membrane lipids includes recent developments in how ethanol affects membrane cholesterol domains, polyunsaturated fatty acids, the cause and consequences of phosphatidylethanol formation, and the modulation of membrane protein function by lipid-protein interaction. The second section comprises chapters on NMDA and 5-HT$_3$ receptors, including new aspects on alcohol neurotoxicity and the molecular heterogeneity that may underlie differences in alcohol sensitivity as well as chapters on GABA-gated chloride flux, and calcium channels. The third section on second messengers includes chapters on ethanol effects on nucleoside transport and the induction of heterologous desensitization, G-proteins, phospholipases generating second messengers, and discussions on the selectivity of ethanol actions at different levels of the signal transduction. Chapters in the fourth section on gene expression describe the regulation of neuronal gene transcription by ethanol and the effects of acute and chronic administration of ethanol on *c-fos* expression.

Alcohol has a ubiquitous effect on the neurochemistry of the brain. It is clear that no single approach or system can explain alcohol-induced phenomena such as intoxication, tolerance, and dependence. Ethanol seems to be able to induce sequential effects or parallel effects as well as direct effects on defined molecules in the cellular machinery. Some systems may be more affected by acute alcohol than others and chronic effects of alcohol include changes in gene expression.

The organization of the book follows the effects of ethanol from actions on membrane lipids, ion channels and receptors, followed by effects on signal transduction and intracellular proteins to the regulation of the genes. It is our hope that this volume will stimulate new ideas and approaches to the understanding of the actions of such a simple molecule as ethanol on nerve cells and integrated brain function.

Christer Alling
W. Gibson Wood

CONTENTS

I. LIPID STRUCTURE, DOMAINS AND PROTEINS

II. RECEPTORS AND ION CHANNELS

IV. GENE EXPRESSION

MOLECULAR MECHANISMS OF ETHANOL AND ANESTHETIC ACTIONS: LIPID- AND PROTEIN-BASED THEORIES

Christopher D. Stubbs and Emanuel Rubin

Department of Pathology and Cell Biology
Thomas Jefferson University
Philadelphia, PA 19107, USA

INTRODUCTION

The use of ethanol goes back to the dawn of civilization, but in spite of intense efforts its mode of action at the molecular level has remained elusive. Ethanol is classed as a weak anesthetic. Apart from its metabolic fate, it is generally believed that the mechanisms of ethanol intoxication and general anesthesia share important features (in this review the term anesthetic therefore includes ethanol). The precise location of molecular action has continued to be a subject of controversy to the present day. The current consensus is that anesthetics have a site(s) of action, located within the cell membrane, presumably at the synapse. However, debate has continued as to whether the site is located exclusively in the lipid bilayer, at a hydrophobic site on a protein or at the membrane protein-lipid interface.

THE MEYER-OVERTON RULE

A landmark in anesthesia research were the observations by Meyer (1889; 1901) and Overton (1901) that the potencies of general anesthetics increase in proportion to their solubility in olive oil (commonly referred to as the Meyer-Overton rule of anesthesia). These early observations were followed by a series of refinements, in which olive oil was replaced by octanol-water partition coefficients and the physical properties of the lipid bilayer studied.

LIPID THEORIES

A number of theories propose various perturbations of bulk physical properties of the lipids of cell membranes as the primary event leading to inebriation or anesthesia. One of the major lipid bilayer properties that has been proposed to be relevant to anesthesia is membrane fluidity, which describes the motional and structural properties of the phospholipid acyl chains. Thus the activities of a number of membrane enzymes have been correlated with spectroscopic parameters that are modified by ethanol and anesthetics, observations that suggest a route by which these compounds could modulate membrane protein function (e.g. see Richards et al., 1978). Membrane proteins are also sensitive to changes in membrane dimensions, such as the thickness of the lipid bilayer, which is also influenced by anesthetics (reviewed by Franks and Lieb, 1982; Dluzewski et al., 1983). In the case of membrane fluidity, it is worth noting that the effects varied considerably according to the lipid composition of the membrane (see below under protein theories).

Alcohol, Cell Membranes, and Signal Transduction in Brain
Edited by C. Alling *et al.*, Plenum Press, New York, 1993

CRITICISMS OF LIPID THEORIES

By the early 1980's it had become apparent that lipid theories may not adequately explain the mechanism of action of anesthetics. The main criticism leveled against lipid theories is that perturbations of the lipid bilayer at clinically relevant concentrations of anesthetics are too small (Franks and Lieb, 1982) to account for anesthesia. As a corollary, the concentrations used in lipid theory studies tend to be unrealistically high.

The second argument that has been leveled against lipid theories is that the magnitude of the change in the physical parameter is comparable to, or less than that, achieved by less than one degree change in temperature. Since such a change in temperature does not lead to anesthesia or intoxication, the altered physical parameter cannot have been responsible for the anesthesia. As a result recent efforts to find alternative sites in proteins have intensified. However, to completely abandon the idea that perturbation of lipid bilayer properties plays a role in anesthesia and ethanol action may be premature (see below).

PROTEIN THEORIES

It is axiomatic that a protein site for anesthetic action implies that a conformational change is involved. It was stated by Claude Bernard in 1875 that "...the way in which anesthetics act on nerves...consists of semi-coagulation of the nerve cells, a coagulation which would not be permanent, the anatomical unit being able to return to its normal state after the toxic agent is eliminated." This *reversible-coagulation* is another way of stating "protein conformational change" (Seeman, 1972).

For a protein to be a feasible site for anesthetic binding one would anticipate that a hydrophobic pocket is required. For any measurable anesthetic effect to be considered a crucial element in anesthesia, its magnitude should follow the Meyer-Overton rule, although the effects of long term exposure to ethanol do not have this restriction. Several examples of proteins with hydrophobic sites for anesthetic molecules have now been described, including myoglobin, bovine serum albumin, B-lactalbumin, adenylate kinase and firefly luciferase (reviewed by Franks and Lieb, 1982; Dluzewski et al., 1983; Miller, 1985). The most studied protein is firefly luciferase (e.g. see Halsey and Smith, 1970; Ueda and Kameya, 1973; Franks and Lieb, 1984; 1985). Using it as a model, Franks and Lieb (1985) have shown that it obeys the Meyer-Overton rule and that it displays the "cut off" effect (see below). From this they argued that the primary anesthetic site could be a hydrophobic pocket on a protein and even independent of membrane lipids. However, to date, no specific protein hydrophobic site critical for anesthesia or ethanol intoxication has been identified.

IS THERE A TARGET FOR ETHANOL AND ANESTHETICS IN THE BRAIN ?

There is general agreement that some effect on membrane ion channels at the synapse must ultimately be involved in anesthesia and ethanol intoxication, and that these membrane proteins offer potential sites for anesthetic interaction. A membrane protein is a metastable entity maintained or "stored" in an optimum conformation by the surrounding lipid matrix (LoGrasso et al., 1988). This is easily seen since if the lipids are removed, since without detergent replacement the secondary structure of the protein irreversibly collapses. The physical properties of the lipids also dictate protein conformation, and therefore their ability to respond to appropriate physiological stimuli. Thus one problem in demonstrating hydrophobic sites on a membrane protein is the difficulty in distinguishing effects on the proteins from indirect effects due to perturbed lipids. A range of functional effects of anesthetics on ion channels are now known, as recently reviewed (Rubin et al., 1991). Although hydrophobic sites of action have been rarely implicated (Wood et al., 1991), the quest to identify such regions has now intensified.

Recently studies with *C. elegans* mutants suggest that distinct anesthetic sites are expressed by different genes (Morgan et al., 1991). Other evidence for potentially multiple, overlapping target molecules for anesthetics also emerged from studies with

Drosophila mutants (Nash et al., 1991), in which single locus mutations that alter the neuromuscular system of the fruit-fly have been shown to modify the response to halothane. Other studies from our laboratory have shown that ethanol selectively blocks a non-inactivating potassium current encoded by a *Drosophila Shaw2,* but not four other structurally homologous channels (Covarrubias, 1993). These studies all support the thesis of multiple anesthetic sites of action.

Recent studies in our laboratory have shown that protein kinase C (PKC) is inhibited by anesthetics, including ethanol (Slater et al., 1993a). PKC has a central role in regulating synaptic function (reviewed by Nishizuka, 1988; Bell and Burns, 1991), affecting ion channels as well as receptors and enzymes. For PKC to become active it first translocates from the cytosol to the membrane bilayer surface. This process is triggered by an increase in intracellular free calcium, which induces the binding of PKC to anionic phospholipids, especially phosphatidylserine (PS). Once associated with the membrane, the conformation of the enzyme is altered, thereby exposing a hydrophobic site to the membrane interior that binds diacylglycerol (DAG). While normally requiring a lipid co-factor, PKC can also be artificially activated in the absence of lipids, thereby providing a test for the lipid dependence of anesthetic effects not easily accomplished with intrinsic transmembrane proteins.

PKC purified from rat brain, and activated in a lipid free assay system, was inhibited by members of the homologous series of aliphatic *n*-alkanols (Slater et al., 1993a and see Fig. 1). In agreement with the Meyer-Overton rule, the concentration

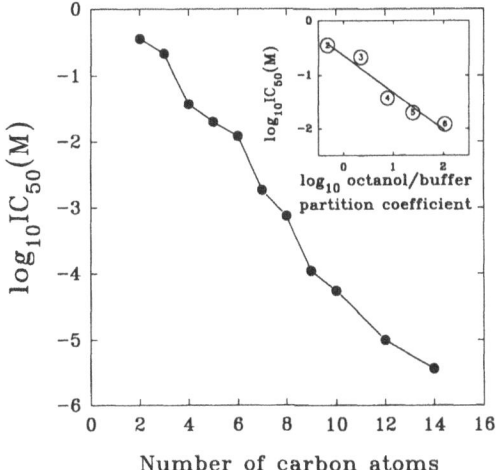

Figure 1. The inhibition of protein kinase C follows the Meyer-Overton rule. The dependence of the potency of PKC inhibition (IC_{50}) by a homologous series of *n*-alkanols in a lipid free assay system. IC_{50} values were calculated from PKC activities determined over a range of alcohol concentrations. **INSET:** the potency of inhibition by alcohols (chain length inside symbols) as a function of the buffer/octanol partition coefficients (from Slater et al., 1993a, with permission).

required for a 50% inhibition of the enzyme (IC_{50}) increased as a linear function of the alcohol chain length of the alcohol, and was also a linear function of the octanol/water partition coefficient. Enflurane and halothane were also found to inhibit at clinically relevant concentrations. The fact that the inhibition was obtainable in the absence of lipids suggests a mechanism involving a direct interaction with the protein, at a hydrophobic site, rather than via an indirect perturbation of the lipids in the membrane.

This result supports the contention of Franks and Lieb, (1984;1985) that a (membrane) protein may be a direct target for anesthetic interactions.

When activation of PKC was achieved in a lipid -dependent assay involving insertion into the lipid bilayer, it was found that the *potency* of inhibition now showed a dependence on lipid composition. The potency increased in parallel with decreasing strength of interlipid hydrogen bonding, as induced by an increased level of membrane phospholipid *sn*-2 unsaturation or by the presence of bilayer destabilizing lipids such as phosphatidylethanolamine (Slater et al., 1993a). Whereas anesthetics also weaken interlipid hydrogen bonding and space the lipid head groups, thereby tending to activate PKC, this effect is apparently overcome by the much stronger direct, inhibitory interaction with the enzyme.

The sensitivity of inhibition PKC by anesthetics to lipid composition establishes the point that even in the protein-site model of anesthesia, lipids may still play an important role, in this case setting the potency of inhibition. Details of the exact location of the site of anesthetic interaction with PKC has yet to be ascertained, but the catalytic domain of PKC, conserved to a high degree among kinases in general, and obtainable by proteolytic cleavage from the regulatory domain of PKC, was not inhibited. This finding suggests that the site of inhibition may be on the (lipid) regulatory subunit, which is unique to protein kinase C. The inhibitory effect may be specific to this kinase.

THE "CUT OFF EFFECT"

For homologous series of anesthetic agents, the Meyer-Overton rule breaks down, and there is a fall off in potency, known as the "cut-off effect". For example, with the alkanol series, using the loss of the righting reflex of tadpoles as a measure of anesthesia, there is a decline in potency that occurs beyond dodecanol (Alifimoff, et al., 1989). A possible explanation for the effect is that as the chain length of the anesthetic approaches that of the membrane phospholipid acyl chains, then the anesthetic ceases to become a perturbent, an explanation compatible with lipid theories (Alifimoff et al., 1989). Another possibility is that hydrophobic pocket(s) in a critical protein must be of finite size, so that with increasing size an anesthetic eventually no longer fits thereby causing a cut-off in potency. The fact that the "cut-off" tends to vary according to the homologous series in question tends to argue for latter, and thus favors a protein theory. Indeed, a "cut-off" has been demonstrated for fire-fly luciferase (Franks and Lieb, 1985). The potency of inhibition of PKC by the alcohol series also begins to show a deviation from linearity after dodecanol (Slater et al., 1993a), an observation that is again consistent with a "cut off effect".

RE-EVALUATION OF THE ROLE OF LIPIDS: RECENT ADVANCES

The need to redefine lipid bilayer properties

There is a possibility that the problem with membrane fluidity as a key component in the anesthetic process may lie with the theory of membrane fluidity itself, rather than with its application to the problem of anesthesia. Thus, in parallel with revising ideas of the mechanism of anesthesia, our understanding of the physical parameters that describe membrane properties has also been advancing (reviewed by Stubbs and Williams, 1992). Do current ideas of lipid bilayer properties provide improved prospects for lipid theories?

The term "membrane fluidity" is often misused. It arose from a combination of spectroscopic studies, the realization that a membrane can be regarded as a two dimensional fluid, and the drive to obtain a simple single physical parameter that would describe the property. The difficulty with the membrane fluidity concept is that any physical parameter chosen will be a property of the spectroscopic method employed, specifically its particular time window (from $\sim 10^{-5}$ sec. for NMR to $\sim 10^{-9}$ sec. for fluorescence and ESR) and the properties of the probe (shape, charge, location etc) (Stubbs and Williams, 1992). It also depends on the assumption that the hydrophobic region of cell membranes is structurally and dynamically homogeneous, an assumption that is now under serious challenge. Thus while it may be true to state that bulk or average spectroscopic properties of cell membranes may not be useful in building a hypothesis for the mechanism of anesthesia, local properties pertaining to domains or in the immediate environment of a membrane protein may be very relevant.

Another argument against lipid theories, namely that the concentrations to produce anesthetic effects are too high, also has to be carefully evaluated. First, many, although not all, effects of anesthetics on membrane protein function also tend to require concentrations that are on the high side. A counter argument may be made that extrapolation can be made back to lower and clinically relevant levels where effects, albeit small, are still observable. We cannot say at present whether these small effects may indeed be sensed, perhaps by a membrane protein such as an ion channel, leading to a cascade of events that lead eventually to anesthesia. Indeed recent studies from this laboratory (Janes et al., 1992; Wang et al., 1993) have established the principle that low enthalpy modifications in membranes, such as domain formation, could potentially be triggered by clinically relevant anesthetic concentrations.

Effects on hydration

Water has a primary role in the formation and maintenance of cell membrane architecture (reviewed by Crowe and Crowe, 1984), and there is considerable potential in allosteric regulation by a so called "perturbed water layer" found around proteins and lipid bilayers (e.g. see review by Rand and Parsegian, 1989). It has been stated that hydration in membranes "provides a fast and inexpensive regulatory mechanism for lipid membranes to adapt their characteristics, at least locally and transiently, to new requirements" (Cevc, 1987). Early theories involving hydration, involving ice-like or clathrate structures, promoted by Pauling (1961) and Miller (1961), are now discarded, owing to the poor correlation of potency with hydrate dissociation pressure, compared to that of potency with the oil/gas partition coefficient of anesthetics. However, the recognition that hydrogen bonds and hydrogen-bonded water play such an important role in membrane structure requires that these factors now be taken more seriously in the search for a mechanism of anesthesia.

Anesthetics partition into the membrane surface itself, into a region termed the "hydration layer", which is a hydrogen-bonded network of water molecules extending between phospholipid head groups. This hydration layer arises from a "bound" hydration shell of 9-20 waters per phospholipid. Direct inter-lipid hydrogen bonding will also occur between some lipids, for example, between, the $-NH_3$ of phosphatidylethanolamine and a phosphate oxygen of another lipid. The contribution of these types of hydrogen bonding to membrane surface lipid-lipid interactions and membrane stability is well recognized (reviewed by Boggs, 1987; Rand and Parsegian, 1989). There is now clear evidence that (halogenated) inhalation anesthetics compete with water that is hydrogen bonded to the surface of lipid bilayers (reviewed by Ueda, 1991; Curatola et al., 1991; Urry and Sandorfy, 1991). Although so far little has appeared on ethanol, some studies have been reported (e.g. see Slater et al., 1993b and references cited therein).

We have recently shown that the hydrogen-bonded network of water molecules at the membrane surface extends into the protein-lipid interface deep in the membrane core (Ho and Stubbs, 1992). Perturbance of this region by anesthetics could provide a mechanism whereby protein conformation is influenced by anesthetics partitioning into the lipid bilayer.

When anesthetics partition into membranes, they intercalate between lipids into the hydrophobic portion of the lipid bilayer, concentrating at the glycerol backbone region. This expands the membrane, increases the cross sectional areas of the hydrocarbon chains and induces a more disordered chain motion that thins the bilayer, owing to chain contraction. Although the effect is small this allows an increase in interstitial water. The extent of water penetration can be mapped according to the values of the dielectric constant across the bilayer, which is 78 for bulk water and ~5 at the bilayer center. Thus there is a dielectric constant gradient (or water penetration gradient) across the bilayer, which is steeper at the head group region (Zannoni et al., 1983). At the steep part of the dielectric constant gradient, at the glycerol backbone, shifting the gradient by only a small degree could mean the dielectric constant may change by a large amount. This could be sensed by an adjacent portion of a membrane protein and be translated into a functional effect.

Disruption of hydrogen bonding by ethanol was investigated in our laboratory, based on measurements of the rates of lipid desorption (Slater et al., 1993b). For phosphatidylcholine (PC) bilayers, interlipid hydrogen bonding can only occur via the hydrogen bonded network of water molecules. The desorption rate of a phospholipid reflects various inter-lipid forces (e.g. hydrogen bonding, van der Waals interactions etc). However, it was found that the desorption rate of a fluorophore labelled PC from PC "donor" vesicles to "acceptor" vesicles (or bovine serum albumin) was dominated by interlipid surface hydrogen bonding involving the hydrogen bonded network of water molecules, and not by other forces or lipid order. With increased phospholipid unsaturation the rate of desorption of PC increased (because wider head group spacing leads to a weakened inter-lipid hydrogen-bonded water molecule network. Ethanol (and other alcohols and anesthetics) also increased the rate of desorption, the effect increasing in relative potency with increased phospholipid unsaturation (see Fig. 2.). Again this effect was found to be due largely to a weakening of the inter-lipid hydrogen-bonded water network, rather than its relatively much

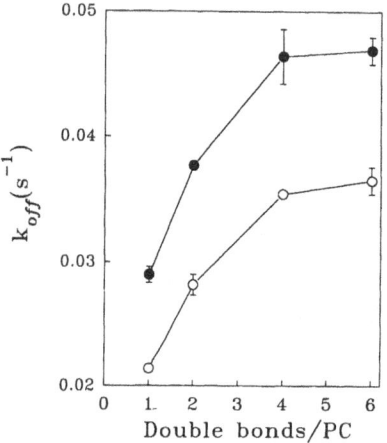

Figure 2. Ethanol weakens inter-lipid hydrogen bonding involving water with increasing potency for greater levels of phospholipid unsaturation. The effect of increasing the *sn*-2 unsaturation of PC donor vesicles on the rate of desorption (k_{off}) of PC labelled with a fluorophore (*1*-palmitoyl-*2*-*N*-[7-nitrobenzo-2-oxa-1, 3-diazole-4-yl]aminohexanoyl-PC) from donor vesicles to bovine serum albumin, in the absence (open circles), and presence (filled circles) of ethanol. Donor vesicles were composed of PC where the *sn*-1 chain was palmitoate (saturated, no double bonds), and the *sn*-2 chain was either oleate, arachidonate or docosahexaenoate, with one, four and six *cis*- double bonds/ chain respectively, also with egg-PC with an average of two double bonds (from Slater et al., 1993b, with permission).

weaker effects on lipid order, which had little influence on desorption. The results indicate that surface hydrogen bonding involving water is a target for ethanol perturbation in membranes. Similar hydrogen bonded water at the protein-lipid interface (see above) would also be a target, and its perturbation could be sensed by the protein, leading to a conformational change that could modify function. The results also suggest that changes in phospholipid unsaturation, known to occur after chronic ethanol ingestion (see below), and domain formation could alter local sensitivity to ethanol perturbation.

Chronic ethanol ingestion and membrane tolerance

A particularly intriguing effect of ethanol is that chronic exposure leads to the development of "membrane tolerance". Thus membranes from animals adapted to chronic ethanol are resistant to lipid disordering by ethanol itself (reviewed by Goldstein, 1984; Sun and Sun, 1985, Taraschi and Rubin, 1985). This phenomenon has now been demonstrated in many tissues and is also manifest as a reduced membrane partitioning of hydrophobic compounds, including ethanol (Rottenberg et al., 1981; Nie et al., 1989), and a decreased susceptibility to phospholipid hydrolysis by exogenous phospholipase A_2 (Stubbs et al., 1988; 1991). Membrane tolerance appears to be due to a modification in anionic phospholipids (Taraschi et al., 1986; Ellingson et al., 1988; Stubbs et al., 1991; and see review by Hoek and Taraschi, 1988). Another important observation is that the effect of anesthetics is also reduced, a phenomenon known as "cross tolerance" (Rottenberg et al., 1981). Thus, a study and understanding of the membrane tolerance may lead to increased knowledge of the location of anesthetic action.

To date studies of the membrane tolerance phenomenon have been largely confined to effects on lipids. For membrane tolerance effects to have a physiological impact they must be expressed at the membrane protein-lipid interface, irrespective of whether the perturbation is local or even mediated via bulk lipid modulation, as in the "lipid theories".

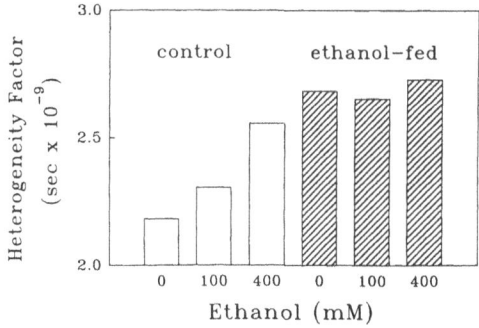

Figure 3. Membrane tolerance is manifest at the protein-lipid interface. Analysis of the fluorescence decay of the fluorophore DPH tethered to PC in liver microsomes shows an increase in the heterogeneity factor after chronic ethanol-treatment. Addition of ethanol only affects the heterogeneity factor in untreated controls, whereas membranes from chronic ethanol-fed animals are resistant even to 400 mM ethanol.

Recently we have found evidence for membrane tolerance at the protein-lipid interface (Ho and Stubbs, 1993). The effects were revealed by probing this region with the fluorophore 1,6-Diphenyl-1,3,5-hexatriene (DPH) attached to the sn-2 position of the glycerol backbone of PC. The fluorescence decay of DPH-PC is homogeneous in vesicles of microsomal lipids, but heterogeneous in intact membranes (Ho et al., 1992). This means the fluorophore experiences many different environments while in the excited state (i.e. at the protein-lipid interface), and there is a range of different decay rates. The *degree* of heterogeneity or diversity of decay rates (reflecting the diversity of the fluorophore environments) is measured by modeling the decay as a continuous lifetime distribution. This is usually chosen to be a Lorentzian distribution (Williams and Stubbs, 1988; Williams et al., 1990; Stubbs and Williams, 1992) in which the broader the width at half-maximum peak height the greater the heterogeneity. This "heterogeneity factor" thus refers to a property of the protein-lipid interface in terms of structural diversity. Chronic ethanol treatment results in a marked increase in the heterogeneity factor of rat liver microsomes (see Fig 3). Furthermore, whereas ethanol itself also increased the value of the heterogeneity factor, this effect was not found for membranes prepared from animals chronically fed ethanol which were resistant even up to 400 mM ethanol. These effects were also found for apocytochrome C reconstituted into vesicles made from microsomal lipids extracted from microsomes of ethanol treated and control animals. Thus these fluorescence spectroscopic studies reveal that membrane tolerance is expressed at the protein-lipid interface and that the effects are due to lipid compositional changes.

Effect on membrane lipid organization

The composition of the cytoplasmic and external sides of a cell membrane distinctly differ. This "lipid bilayer asymmetry" has been shown to be modified as a result of chronic ethanol ingestion (reviewed by Wood and Schroeder, 1988) and may be influenced by cholesterol. Lateral heterogeneity may also play a role in the mechanism of anesthesia. Altered membrane domain organization could provide a plausible mechanism for anesthetic action, provided membrane proteins are sensitive to different lipid environments. It is important to note here that domain organizational changes could have profound effects on proteins even *without* overall compositional changes.

Ethanol consumption has a profound effect on the lipid composition of membranes (reviewed e.g. Sun and Sun 1985). It is important to recognize that responses are tissue, or even organelle, specific. However, one of the few general effects is on the level of polyunsaturation, and on the arachidonate (omega-3) to mainly docosahexaenoate (omega-6) ratio. Membrane tolerance itself has been shown to be associated with modifications to anionic phospholipids (see above). In liver microsomes the docosahexaenoate (22:6, 22-carbons with 6 *cis*-double bonds) increases at the expense of arachidonate (20-carbons with 4 *cis*-double bonds) (Stubbs et al., 1991). By contrast, 22:6 levels in synaptosomes, which are higher than in liver cell membranes, *decrease* as a result of chronic ethanol-treatment (Sun and Sun 1985). 22:6 has *cis*-double bonds that are distributed almost to the terminal methyl. This causes a helical, bulky and relatively inflexible structure (reviewed by Stubbs, 1989), which is likely to accommodate higher levels of interstitial water compared to 20:4. In addition phospholipids high in 22:6, such as PS, associate with proteins, so that altered levels of 22:6 could modify the level of water at the protein-lipid interface.

The protein-lipid interface: a hybrid theory

The lipid perturbation by anesthetics, as sensed by the protein, could be bulk or average lipid properties, as implied in lipid theories. It is also possible that the critical perturbation could be within a local domain or confined solely to the first shell of lipids round the protein. Whichever is the case any effects on the protein have to be felt at or transmitted via the protein-lipid interface if they are to be relevant to functional modulation of membrane proteins and hence to the cell functioning as a whole.

So far studies focused on ethanol alone are few. However, if one includes anesthetics, there are several relevant studies that provide evidence for effects on structures other than purely lipids (reviewed by Miller, 1985). It was stated as long ago as 1978 that general anesthetics "..by primarily labilizing lipid-protein interactions on biomembranes, induce changes in protein conformation that become translated into functional alterations", and that the effect is "...primarily evident in synaptic membranes where concentrations promoting labilization of lipid-protein interactions are close to those abolishing synaptic transmission" (Curatola et al., 1978). Theories supporting these effects were lacking, and it was not until 1985 (Miller, 1985) that the theory began to be seriously considered. Recent observations include effects of anesthetics on the acetylcholine receptor (Fraser et al., 1990), Ca^{2+} ATP'ase (Lopes and Louro, 1991), the $Na^+/K^+/Cl^+$ cotransporter in glial type cells (Tas et al., 1987), a Ca^{2+} activated K^+ channel (Tas et al., 1990) and Na^+, K^+-ATP'ase (Marques and Guerri, 1988). There are also several studies showing that protein conformation is affected by anesthetics (Grog and Belagyi, 1983; Almeida et al., 1986; Kaminoh et al.,

1992). Studies involving activation energies of membrane enzymes (arrhenius plots) also yield evidence for a role of protein-lipid interactions (e.g. Harris and Schroeder, 1981; Swann, 1984).

CONCLUSIONS

There have recently been advances in our understanding of membrane receptor function and cell membrane structure, dynamics and organization. It may be anticipated that this should allow a testable mechanism of anesthesia to emerge. A membrane protein, or a number of proteins, may be found to be a key site of interaction, but it seems likely that membrane lipids will also be found to play an important role.

Acknowledgements

This work was supported by U.S. Public Health Service Grants AA08022, AA07186 and AA07215.

REFERENCES

Alifimoff, J. K., Firestone, L. L., and Miller, K. W. 1989, Anaesthetic potencies of primary alkanols: implications for the molecular dimensions of the anaesthetic site.*Br. J. Pharmacol.* 96: 9-16.

Almeida, L. M., Vaz, W. L. C. Stumpel, J., and Madiera, V. M. C. 1986, Effect of short-chain primary alcohols on fluidity and activity of sarcoplasmic reticulum membranes. *Biochemistry* 25: 4832-4839.

Bell, R. M., and Burns, D. J. 1991, Lipid activation of protein kinase C. *J. Biol. Chem.* 266: 4661-4664.

Bernard, C. 1875 Lecons sur les anesthesiques et sur l'asphyxie. 37a: pp363.

Boggs, J. M. 1987, Lipid interactions and molecular hydrogen bonding. *Biochim. Biophys. Acta* 906: 353-404.

Cevc, G. 1987, How membrane chain melting properties are regulated by the polar surface of the lipid bilayer. *Biochemistry* 26: 6305-6310.

Covarrubias, M. 1993, Ethanol selectively bblocks a non-activating K+ current encoded by *Drosophila Shaw2. Biophys. J.* 64: A227.

Crowe, J. H., and Crowe, L. M. 1984, Effects of dehydration on membranes. *Biological Membranes* 5: 58-103.

Curatola, G., Lenaz, G., and Zolese, G. 1991 Anesthetic-membrane interactions, *in*: "Membrane Fluidity" (vol 5) R. C. Aloia, C. C. Curtain, and L. M. Gordon, eds., Wiley, NY., pp 35-70.

Curatola, G., Mazzanti, L., Bertoli, E., and Lenaz, G. 1978, The action of general anesthetics on lipid-protein interactions. *Bull. Mol. Biol. Med.* 3: 123-135.

Dluzewski, A. R., Halsey, M. J., and Simmonds, A. C. 1983, Membrane interactions with general and local anesthetics: a review of molecular hypotheses of anesthesia. *Molec. Aspects Med.* 6: 459-573.

Ellingson, J. S., Janes, N., Taraschi, T. F., and Rubin, E. 1991, The effect of chronic ethanol consumption of the fatty acid composition of phosphatidylinositol in rat liver microsomes as determined by gas chromatography and ^{1}H-NMR. *Biochim. Biophys. Acta* 1062: 199-205.

Ellingson, J. S., Taraschi, T. F., Wu, A., Zimmerman, R., and Rubin, E. 1988, Cardiolipin from ethanol-fed rats confers tolerance to ethanol in liver mitochondrial membranes. *Proc. Natl. Acad. Science (USA)* 85: 3353-3357.

Franks, N. P., and Lieb, W. R. 1982, Moleculr mechanism of general anesthesia. *Nature* 300: 487-493.

Franks, N. P., and Lieb, W. R. 1984, Do general anesthetics act by competetive binding to specific receptors? *Nature* 310: 599-601.

Franks, N. P., and Lieb, W. R. 1985, Mapping of general anesthetic target sites provides a molecular basis for cut-off effects. *Nature* 316: 349-351.

Fraser, D. M., Louro, S. R. W., Horvath, L. I., Miller, K. W., and Watts, A. 1990, A study of the effec of general anesthetics on protein-lipid interactions in acetylcholine receptor enriched membranes. *Biochemistry* 29: 2664-2669.

Goldstein, D. B. 1984, Effect of drugs on membrane fluidity. *Ann. Rev. Pharm. Toxicol.* 24: 43-64.

Grog, P., and Belagyi, J. 1983, The effect of anesthetics on protein conformation in membranes as studied by the spin-labelling technique. *Biochim. Biophys. Acta* 734: 3198-328.

Halsey, M. J., and Smith, E. B. 1970, Effects of anesthetics on luminous bacteria. *Nature* 227: 1363-1365.

Harris, R. A. and Schroeder, F. 1981, Ethanol and the physical properties of brain membranes. *Mol. Pharm.* 20: 128-137.

Ho, C., and Stubbs, C. D. 1992, Hydration at the hydrophobic membrane protein-lipid interface. *Biophys. J.* 63: 897-902.

Ho, C., Kelly, M. B., and Stubbs, C. D. 1993, Fluorescence lifetime heterogeneity as a probe of the membrane protein-lipid interface: Effects of increased phospholipid unsaturation and perturbation by ethanol. *Biophy. J.* (submitted).

Ho, C., Williams, B. W., and Stubbs, C. D. 1992, Analysis of cell membrane micro-heterogeneity using the fluorescence lifetime of DPH-type fluorophores. *Biochim. Biophys. Acta* 1104: 273-282..

Hoek, J, B., and Taraschi, T. F. 1988, Cellular adaptation to ethanol. *Trends in Biochemical Sciences* 13: 269-274.

Janes, N., Hsu, J. W., Rubin, E., and Taraschi, T. F. 1992, Nature of alcohol and anesthetic action on cooperative membrane equilibria. *Biochemistry* 31: 9467-9472.

Kaminoh, Y., Nishimura, S., and Ueda, I. 1992, Alcohol interaction with high entropy states of macromolecules: critical temperature hypothesis for anesthesia cut-off. *Biochim. Biophys. Acta* 1106: 335-343.

Klemm, W. R., and Yurttas, L. 1992 The dehydration theory of alcohol intoxication, *in*: "Treatment of Drug and Alcohol Abuse," R. R. Watson, ed., Humana Press, Clifton, NJ. pp 1-15.

LoGrasso, P. V., F. Moll, and T. A. Cross. (1988,) Solvent history dependence of gramicidin a conformations in hydrated lipid bilayers. *Biophysical J.* 54: 259-267.

LoGrasso, P. V., F. Moll, and T. A. Cross. 1988, Solvent history dependence of gramicidin a conformations in hydrated lipid bilayers. *Biophysical J.* 54: 259-267.

Lopes, C. M. B., and Louro, S. R. W. 1991, The effects of n-alkanols on the lipid/protein interface of Ca2+-ATPase of sarcoplasmic reticulum vesicles. *Biochim. Biophys. Acta* 1070: 467-473.

Marques, A., and Guerri, C. 1988, Effects of ethanol on rat brain Na+, K+-ATPase from native and delipidized synaptic membranes. *Biochem. Pharm.* 37: 601-606.

Meyer, H. 1889, Eigenschaft der anathetica bedingt ihre narkitische wirkung? *Arch Exp. Path. Pharmak. (Naunyn-Schmiedebergs)* 42: 109-118.

Meyer, H. 1901, Der einfluss wechselnder temperatur auf wirkungsstarke und theilungscoefficient der narcotica. *Arch Exp. Membr. Path. Pharmak. (Naunyn-Schmiedebergs)* 46: 338-346.

Miller, K. W. 1985, Nature of the site of general anesthesia. *Int. Rev. Neurobiol.* 27: 1-61.

Miller, S. L. 1961, A theory of gaseous anesthetics. *Proc. Natl. Acad. Sci.* 47: 1515-1524.

Morgan, P. G., Sedensky, M. M., and Meneely, P. M. 1991, The genetics of response to volatile anesthetics in *C. Elegans. Ann. NY Acad. Sci.* 625: 524-531.

Nash, H. A., Campbell, D. B., and Krishnan, K. S. 1991, New Mutants of *Drosophila* that are resistant to the anesthetic effects of halothane. *Ann. NY Acad. Sci.* 625: 540-544.

Nie, Y., Stubbs, C. D., Williams, B. W., and Rubin, E. 1989, Ethanol causes decreased partitioning into biological membranes without changes in lipid order. *Archives Biochem. Biophys.* 268: 349-359.

Nishizuka, Y. 1988, Protein kinase C. *Nature* 334: 661-665.

Overton, E. Studien uber die Narkose (Fischer, Jena, 1901).

Pauling, L. 1961, A molecular theory of general anesthesia. *Science* 134: 15-21.

Rand, R. P., and Parsegian, V. A. 1989, Hydration forces between phospholipid bilayers. *Biochim. Biophys. Acta* 988: 351-375.

Richards, C. D., Martin, K., Gregory, S., Keightley, C. A., Hesketh, T. R., Smith, G. Warren, G. B., and Metcalfe, J. C. 1978, Degenerate perturbations of protein structure as the mechanism of anesthetic action. *Nature* 276: 775-779.

Rottenberg, H., Waring, A., and Rubin, E. 1981, Tolerance and cross tolerance in chronic alcoholics: reduced membrane binding of ethanol and other drugs. *Science* 213: 583-585.

Rubin, E., Miller, K. W., and Roth, S. H. (eds) Molecular and Cellular Mechanisms of Alcohol and Anesthetics. *Ann. NY Acad. Sci.* vol 625, (1991).

Seeman, P. 1972, The membrane actions of anesthetics and tranquilizers. *Pharamocol. Reviews* 24: 583-655.

Slater, S. J., Cox, K. A. J., Lombardi, J. V., Ho, C., Kelly, M. B., Rubin, E., and Stubbs, C. D. 1993a, Inhibition of protein kinase C by alcohols and anesthetics. Nature (in press).

Slater: S. J., Ho, C., Taddeo, F. J. Kelly, M., and Stubbs, C. D. 1993b, The contribution of hydrogen bonding to lipid-lipid interactions in membranes and the role of lipid order: Effects of cholesterol, increased phospholipid unsaturation and ethanol. *Biochemistry* (in press).

Stubbs, C. D. 1989, Physico-chemical responses of cell membranes to dietary manipulation. *Colloque Inserm* 195: 125-134.

Stubbs, C. D. 1983, Membrane fluidity. Structure and dynamics in membrane lipids. *Essays in Biochemistry* 19: 1-39.

Stubbs, C. D., and Williams, B. W. (1992) Fluorescence in membranes, *in:* "Fluorescence Spectroscopy in Biochemistry" (vol III), J. R. Lakowicz, ed., Plenum NY, pp231-263.

Stubbs, C. D., Kisielewski, A., and Rubin, E. 1991, Chronic ethanol ingestion modifies liver microsomal phosphatidylserine inducing resistance to hydrolysis by exogeneous phospholipase A_2. *Biochim. Biophys. Acta* 1070: 349-354.

Stubbs, C. D., Williams, B. W., and Ho, C 1990, Fluorophore lifetime distributions as a probe of lipid bilayer organization. *Time-Resolved Laser Spectroscopy in Biochemistry II.* (Lakowicz, J. R. ed.) Proc. SPIE 1204: 448-455.

Stubbs, C. D., Williams, B. W., Pryor, C. L., and Rubin, E. 1988, Ethanol induced modifications to membrane lipid structure: Effect on phospholipase A_2 - membrane interactions. *Archives Biochem. Biophys.* 262: 560-573.

Sun, G. Y., and Sun. A. Y. 1985, Ethanol and membrane lipids. *Alcohol Clin. Exp. Res.* 9: 164-180.

Swann, A. C. 1984, Chronic ethanol and Na-K ATP'ase. *J. Pharm. Exp. Ther.* 232: 475-479.

Taraschi, T. F., and Rubin, E. 1985, Effect of ethanol on the chemical and stuctural properties of membranes. *Lab. Invest.* 52: 120-131.

Taraschi, T. F., Ellingson, J. S., Wu, A., Zimmerman, R., and Rubin, E. 1986,. Phosphatidylinositol from ethanol-fed rats confers membrane tolerance to ethanol. *Proc. Natl. Acad. Science (USA)* 83: 9398-9402.

Tas, P. W. L., Kress, H. G., and Koschel, K. 1987, General anesthetics can competitively interfere with sensitive membrane proteins. *Proc. Natl. Acad. Sci. USA* 84: 5972-5975.

Tas, P. W. L., Kress, H. G., and Koschel, K. 1990, Lipid solubility is not the sole criterion for the inhibition of a Ca2+ activated K+ channel by alcohols. *Biochim. Biophys. Acta* 1023: 436-440.

Ueda, I. (1991) Interfacial effects of anesthetics on membrane fluidity, *in:* "Membrane Fluidity" (vol 5) R. C. Aloia, C. C. Curtain, and L. M. Gordon, eds., Wiley, NY., pp 91-131.

Ueda, I., and Kameya, H. 1973, Kinetic and thermodynamic aspects of the mechanism of general anesthesia in a model system of firefly luminescence *in vitro. Anesthesiology* 38: 425-436.

Urry, D. W., and Sandorfy, C. (1991) Chemical Modification of transmembrane protein structure and function, *in:* "Membrane Fluidity" (vol 5) R. C. Aloia, C. C. Curtain, and L. M. Gordon, eds., Wiley, NY., pp 91-131.

Wang, D., Taraschi, T. F., Rubin, E., and Janes, N. 1993, Configurational entropy is the driving force of ethanol action in membrane architecture. *Biochim. Biophys. Acta* 1145: 141-148.

Williams, B. W., and Stubbs, C. D. 1988, Properties influencing fluorescence lifetime distributions in membranes. *Biochemistry* 27: 7994-7999.

Williams, B. W., Scotto, A. W., and Stubbs, C. D. 1990, The effect of proteins on fluorophore lifetime heterogeneity in lipid bilayers. *Biochemistry* 29: 3248-3255..

Wood, S. C., Forman, S. A., and Miller, K. W. 1991, Short chain and long chain alkanols have different sites of action on nicotinic acetylcholine receptor channels from Torpedo. *Mol. Pharmacol.* 39: 332-338.

Wood, W. G., and Schroeder, F. 1988, Membrane Effects of ethanol: Bulk lipid versus lipid domains.*Life Science* 43: 467-475.

Zannoni, C., Arcioni, A., and Cavatorta, P. 1983, Fluorescence depolarization in liquid crystals and membrane bilayers. *Chem. Phys. Lipids* 32: 179-250.

MEMBRANE CHOLESTEROL AND ETHANOL: DOMAINS, KINETICS, AND PROTEIN FUNCTION

W. Gibson Wood[1], A. Muralikrishna Rao[1], Friedhelm Schroeder[2], and Urule Igbavboa[1]

[1]Geriatric Research, Education and Clinical Center
VA Medical Center, and Department of Pharmacology
University of Minnesota School of Medicine
Minneapolis, Minnesota 55417 and [2]Department of Pharmacology and
Cell Biophysics, University of Cincinnati, Cincinnati, OH 45267-004

INTRODUCTION

Approximately 15 years ago, Chin and Goldstein (1977) published a study that was the stimulus for a major effort towards understanding the mechanisms involved in cellular tolerance to ethanol. They had shown that membranes of ethanol-tolerant mice were resistant to the fluidizing effects of ethanol *in vitro*. This finding has been replicated in several different laboratories using different techniques and different types of membranes (Crews et al.,1983; Harris et al.,1984; Waring et al.,1982; Rottenberg et al.,1981; Ponnappa et al.,1982; Kelly-Murphy et al.,1983; Taraschi et al.,1986; Taraschi et al.,1990; Wood et al.,1989b). Resistance to ethanol-induced fluidization has also been reported in erythrocyte membranes of alcoholics patients (Beauge et al.,1985; Wood et al.,1987).

Some element of the membrane had changed in response to chronic ethanol consumption, that conferred resistance to effects of ethanol *in vitro*. It had been widely known that different lipids could affect the physical properties of the membrane. Hill and Bangham (1975) had hypothesized that alterations in membrane lipids may be involved in drug-induced cellular tolerance and dependency. There is now an extensive literature on effects of chronic ethanol consumption on membrane lipid composition and many of those studies have been reviewed previously (Sun and Sun, 1985; Wood and Schroeder, 1988; Wood et al.,1991b; Wood and Schroeder, 1992; Deitrich et al.,1989). Changes in the total amount of individual phospholipids and acyl group composition have not adequately accounted for ethanol-induced changes in membrane structure and protein function.

Cholesterol is another membrane lipid that has received a large amount of attention in alcohol studies. It has been proposed that ethanol may increase the total amount of membrane cholesterol and that increase may account for cellular tolerance to ethanol (Chin and Goldstein, 1984; Daniels and Goldstein, 1982; Parsons et al.,1982; Chin et al.,1978; Chin and Goldstein, 1981). Interest in cholesterol derives from the fact that cholesterol has a rigidifying effect on the membrane above the phase transition temperature of the membrane lipid (Curtain et al.,1988). In addition, membrane cholesterol is one of the major lipids of plasma membranes and is important with respect to membrane structure and activity of specific membrane proteins (Yang et al.,1990; Berstein et al.,1989; Kanner and Shouffani, 1990; Artigues et al.,1989; Maguire and Druse, 1989b; McMurchie and Patten, 1988; Maguire and Druse, 1989a; Michelangeli et al.,1990; Zhou et al.,1991; Incerpi et al.,1992).

Most of the studies on effects of ethanol on membrane cholesterol examined the total amount of cholesterol in membranes. Changes in the total amount of membrane cholesterol in response to chronic ethanol consumption have been variable. However, as shown in Table 1, membrane cholesterol can be described in other ways, in addition to the total or bulk amount of cholesterol in the membrane. Cholesterol is not equally dispersed throughout the

Table 1. Different Aspects of Cholesterol.

1. Total membrane content.
2. Transbilayer distribution.
3. Lateral exchangeable and non-exchangeable cholesterol pools.
4. Kinetics of cholesterol exchange between membranes.
5. Sterol carrier proteins.
6. Regulation of protein structure and function.

membrane but consists of both vertical and lateral domains (Schroeder et al.,1989; Schroeder et al.,1991a; Wood and Schroeder, 1992). Kinetics of cholesterol domains can be described as well as factors that alter the movement of cholesterol in and out of membranes (Phillips et al.,1987; Schroeder et al.,1991a). The association between cholesterol and proteins has been demonstrated with respect to proteins that bind cholesterol (Veerkamp et al.,1991) and the involvement of cholesterol in the optimal functioning of certain proteins (Yeagle, 1989).

The purpose of this chapter will be to discuss the different structural and functional characteristics of membrane cholesterol described in Table 1, and to examine the association between those aspects of cholesterol and ethanol. The emphasis of this chapter will be on brain cholesterol, however, where appropriate, data will be included on non-neuronal membranes because of the paucity of studies on the different structural and functional aspects of brain cholesterol. It is our general hypothesis that chronic ethanol consumption modifies specific aspects of membrane cholesterol and that such specific effects contribute to ethanol-induced changes in membrane structure and the activity of particular membrane proteins.

DIFFERENT ASPECTS OF MEMBRANE CHOLESTEROL

Total Membrane Cholesterol

As was discussed in the **Introduction** to this chapter, there have been several studies on effects of chronic ethanol consumption on the total amount of membrane cholesterol in chronic ethanol-treated animals and humans. Much of that work has been reviewed previously (Hunt, 1985; Wood and Schroeder, 1988; Sun and Sun, 1985). Generally, effects of chronic ethanol consumption on total membrane cholesterol have not been consistently observed. It has been reported in some studies that the total amount of membrane cholesterol increased in membranes of ethanol-treated animals, whereas, other studies have found that the total amount of cholesterol was reduced in membranes of ethanol consuming individuals. Several studies have reported that the total amount of cholesterol did not differ between membranes of ethanol-treated and control groups. Obviously, some of the variability in effects of ethanol on total membrane cholesterol might be explained in terms of species, membrane type, method of ethanol administration (e.g., liquid diet, intubation, inhalation) and duration of ethanol exposure.

In the studies that have reported that the total amount of cholesterol was increased in membranes of chronic ethanol consuming individuals, it would be of interest to determine the origin of the additional cholesterol. One possibility would be that the synthesis of cholesterol might be stimulated as a result of chronic ethanol consumption. The major site of cholesterol synthesis is the liver but there are also data that indicate that synthesis does occur in brain (Sastry, 1985). To our knowledge, effects of ethanol on the synthesis of brain cholesterol have not been examined. Another possibility to account for ethanol-induced changes in the total amount of cholesterol is that ethanol may affect the influx and efflux of membrane cholesterol. This possibility has been examined and is discussed elsewhere in this chapter (see **Kinetics of Cholesterol Exchange**).

Transbilayer Distribution of Cholesterol

Plasma membranes are a bilayer that actually consist of two monolayers or membrane leaflets (Op den Kamp, 1979). These leaflets differ in electrical charge, fluidity, protein distribution and function, lipid distribution, and do not appear to be coupled (Curtain et al.,1988). The outer or exofacial leaflet of the membrane is zwitterionic or neutral and the inner or cytofacial leaflet is considered to be negatively charged. The two leaflets are asymmetric in fluidity. The exofacial leaflet of synaptic plasma membranes (SPM), LM fibroblasts and erythrocytes have been reported to be significantly more fluid as compared to the cytofacial leaflet (Schachter et al.,1983; Wood et al.,1989b; Schroeder et al.,1988; Sweet et al.,1987; Incerpi et al.,1992). In SPM, the limiting anisotropy of diphenylhexatriene was 0.236 ± 0.001 in the exofacial leaflet and 0.271 ± 0.001 in the cytofacial leaflet (Schroeder et al.,1988). The difference in fluidity between the two leaflets is similar to the effects of a 3 to 4°C change in temperature on the membrane.

The exofacial and cytofacial membrane leaflets also differ in the distribution of lipids. Most of the data on lipid distribution were derived from studies using erythrocytes (Roelofsen, 1982). The exofacial leaflet contained primarily phosphatidylcholine and sphingomyelin and the cytofacial leaflet contained primarily phosphatidylethanolamine and the two negatively charged phospholipids, phosphatidylserine and phosphatidylinositol. There have been few studies that have examined the transbilayer distribution of phospholipids in neuronal membranes. Phosphatidylethanolamine and phosphatidylserine in SPM were found to be concentrated in the cytofacial leaflet (Fontaine et al.,1980). Sphingomyelin in SPM was reported to be exclusively in the exofacial leaflet (Rao et al.,1993). It has been shown that phosphatidylcholine was predominantly located in the exofacial leaflet of neuronal plasma membranes from the electroplax cell of electric eel (Supernovich et al.,1991a; Supernovich et al.,1991b).

Cholesterol is asymmetrically distributed in the membrane exofacial and cytofacial leaflets (Schroeder et al.,1991a; Wood and Schroeder, 1992). Table 2 summarizes data on SPM, erythrocytes, and LM fibroblasts. It was found in each type of membrane that there was substantially more cholesterol in the cytofacial leaflet as compared to the exofacial leaflet. In SPM, there was 7 times as much cholesterol in the cytofacial leaflet as was observed in the exofacial leaflet (Wood et al.,1990).

Table 2. Transbilayer distribution of cholesterol in plasma membranes.

Membrane	Exofacial Leaflet (%)	Cytofacial Leaflet (%)
Synaptic plasma membrane[1]	12 ± 2	88 ± 2
Erythrocyte[2]	26 ± 3	75 ± 5
Erythrocyte[3]	13	87
LM fibroblasts[4]	20 ± 3	80 ± 3

[1]Wood et al., 1990; [2]Schroeder et al., 1991b; [3]Brasaemle et al., 1988; [4]Kier et al., 1986.

We had previously reported that fluidity of the SPM exofacial leaflet was more affected by ethanol *in vitro* as compared to the cytofacial leaflet (Schroeder et al.,1988). In a subsequent study, we found that chronic ethanol consumption had a greater effect on fluidity of the exofacial leaflet than the cytofacial leaflet (Wood et al.,1989b). The SPM exofacial leaflet of the chronic ethanol group was also resistant to perturbation by ethanol *in vitro*. The susceptibility of the exofacial leaflet to effects of ethanol *in vitro* and effects of chronic ethanol consumption has been shown to be associated with the transbilayer distribution of cholesterol in the exofacial and cytofacial leaflets (Wood et al.,1990). We examined the distribution of cholesterol in the SPM exofacial and cytofacial leaflets of mice that were administered an ethanol liquid diet for 8 weeks. Cholesterol distribution was estimated using the fluorescent sterol, dehydroergosterol. There was approximately a two-fold increase in exofacial cholesterol in SPM of the ethanol group (Figure 1). The large change in SPM transbilayer cholesterol distribution occurred in the absence of a change in the total amount of SPM cholesterol. Cholesterol was reorganized in the membrane without changing the total amount of membrane cholesterol.

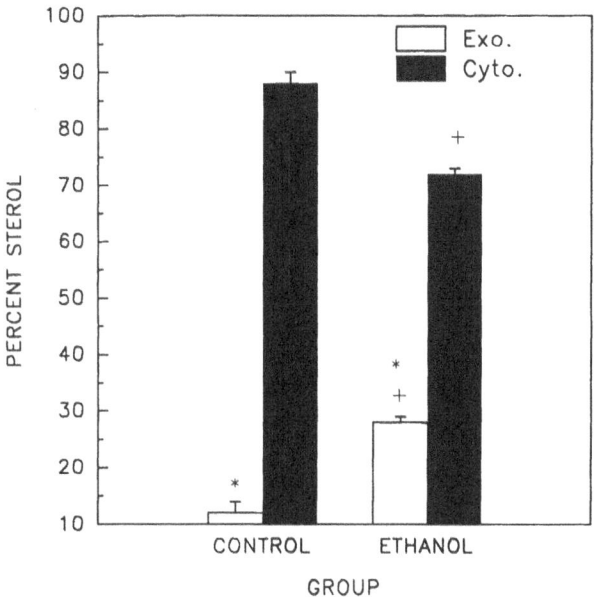

Figure 1. Transbilayer distribution of dehydroergosterol (cholesterol analogue) in SPM of control and chronic ethanol groups. Synaptosomes were treated with plus or minus trinitrobenzenesulfonic acid (TNBS) after which SPM were isolated. Values represent the mean ± SEM of the percent sterol as determined from fluorescence intensity/mg protein (n=4 SPM preparations per group). *p <0.01 compared with the cytofacial leaflet and +p <0.01 compared with the exofacial or cytofacial leaflet of the control group. Data are from Wood et al.,(1990).

The mechanism that regulates the asymmetric distribution of cholesterol in the exofacial and cytofacial membrane leaflets has not been identified. While there is evidence that a protein may mediate the transbilayer distribution of phosphatidylserine and phosphatidylethanolamine (Devaux, 1988; Devaux, 1991), whether there is a protein regulating the transbilayer distribution of cholesterol is unclear. There have been proteins identified that bind cholesterol (see **Sterol Carrier Proteins** in this chapter), but their involvement in the transbilayer distribution of membrane cholesterol has not been well-characterized. Chronic ethanol consumption may alter a sterol binding protein, that in turn, modifies the transbilayer distribution of cholesterol in the membrane.

Exchangeable and Non-Exchangeable Pools. Membrane cholesterol can be described with respect to distribution across the lateral or horizontal plane of the membrane. These lateral domains have also been referred to as exchangeable and non-exchangeable cholesterol pools (Schroeder et al.,1991a; Wood and Schroeder, 1992). The exchangeable pool of cholesterol has been attributed to cholesterol monomers and cholesterol-phospholipid complexes (Nemecz et al.,1988). The non-exchangeable cholesterol pool may result from cholesterol-cholesterol complexes. Most of the studies examining lateral domains in biological membranes have been with erythrocytes and LM fibroblasts. In erythrocytes, the size of the exchangeable pool has been reported to be approximately 70 percent of total membrane cholesterol (Poznansky and Lange, 1976). One study, however, found that the size of the exchangeable pool of cholesterol was over 90 percent of total membrane cholesterol (Bloj and Zilversmit, 1977). A similar value has been reported for LM fibroblasts membranes (Schroeder et al.,1991a).

We have recently shown in the synaptosomal membrane, that the exchangeable pool of cholesterol was approximately 50% of total membrane cholesterol when measured at 37°C (Table 3) (Rao et al.,1993). Lowering the incubation temperature to 25°C had a significant effect on the exchangeable and non-exchangeable pools. It can be seen in Table 3 that only 27% of the cholesterol was exchangeable.

The size of the cholesterol exchangeable pool could be affected by temperature. We next examined whether chronic ethanol consumption would have an effect on the

Table 3. Cholesterol exchange between small unilamellar vesicles and synaptosomes at 25° and 37°C.

T°C	Rate Constant (h^{-1})	t$_{1/2}$ (hr)	Ex. Pool (%)
25°C	0.021 ± 0.001	33.51 ± 1.95	27 ± 1
37°C	0.079 ± 0.008[*]	8.79 ± 0.90[*]	51 ± 1[*]

Cholesterol exchange was between radiolabeled SUV and synaptosomes. SUV contained POPC/cholesterol (58:42) and trace amounts of [1,2-^{3}H]cholesterol and [1-^{14}C]cholesteryl oleate. At timed intervals, over a 24 hour period, SUV and synaptosomes were separated by centrifugation at 40,000 x g for 20 min and radioactivity in the supernatant containing the SUV was counted. [1-^{14}C]cholesteryl oleate was used as a nonexchangeable marker to determine fusion between SUV and synaptosomes. Fusion was less than 3%. The incubation temperatures were 25°C and 37°C. Kinetic data in all experiments were analyzed by an iterative nonlinear least-squares analysis (ISI Software, Philadelphia, PA) to determine rate of exchange and amount of exchange (Nemecz et al.,1988). Values are means $\pm$ SEM of three different experiments. [*] P < 0.01 as compared to 25°C using two-tailed t-tests. Data are taken from Rao et al.,1993.

exchangeable and non-exchangeable cholesterol pools in synaptosomal membranes (Wood et al.,1993). Liquid diets containing ethanol or sucrose were administered to mice for 12 weeks. The size of the cholesterol exchangeable pool in the pair-fed control group was 47% (Table 4) and similar to what we had reported previously in synaptosomal membranes of untreated mice (Table 3). Chronic ethanol consumption did not modify the size of the cholesterol exchangeable pool. Approximately 51% of the cholesterol was exchangeable in synaptosomal membranes of chronic ethanol-treated mice and was not significantly different from the pair-fed control group (Table 4).

Table 4. Kinetics and Size of Cholesterol Lateral Domains in Synaptosomes of Pair-Fed Control and Ethanol-Treated Mice.

Group	Rate Constant (h^{-1})	$t_{1/2}$ (hr)	Ex. Pool (%)
Control	0.065 ± 0.001	10.70 ± 0.25	47 ± 3.51
Ethanol	$0.053 \pm 0.001^*$	$13.33 \pm 0.58^*$	51 ± 4.04

Ethanol and control diets were administered for 12 weeks. Cholesterol exchange was between radiolabeled SUV and synaptosomes as described in Table 3. Data are means ±SEM of three different membrane preparations (n=4 mice/preparation). Data are from Wood et al.,1993.

Chronic ethanol consumption altered the transbilayer distribution of cholesterol but had no effect on the cholesterol lateral pools. A possible explanation for the absence of an effect on the lateral pools was that the data represented the total amount of exchangeable cholesterol. The amount of cholesterol that is exchangeable may differ in the exofacial and cytofacial leaflets. There was a two-fold increase in the amount of cholesterol in the exofacial leaflet of SPM of chronic ethanol-treated mice (Figure 1). Such an increase may alter the exchangeable and non-exchangeable pools within each leaflet.

Kinetics of Cholesterol Exchange. The movement of cholesterol into and out of membranes has been well-characterized in model membrane systems and non-neuronal membranes (Phillips et al.,1987). Experiments have examined cholesterol exchange, which involves a one for one exchange of cholesterol molecules between a donor and acceptor pair. In cholesterol exchange experiments, the rate of cholesterol exchange could be affected by factors such as temperature (McLean and Phillips, 1982), mol% cholesterol (Nemecz et al.,1988), and various phospholipids (Thomas and Poznansky, 1988; Bar et al.,1987; Gold and Phillips, 1990).

The kinetics of cholesterol exchange have been reported recently in synaptosomal membranes (Rao et al.,1993). Table 3 contains kinetic data on cholesterol exchange between small unilamellar vesicles (SUV) and synaptosomal membranes. The $t_{1/2}$ for cholesterol exchange was approximately 10 hours. This value was similar to some studies that have used erythrocytes (Gold and Phillips, 1990; Phillips et al.,1987), but was longer when compared to studies using brush-border membranes (Bittman et al.,1990) and *Mycoplasma capricolum* membranes (Thurnhofer and Hauser, 1990). The rate of cholesterol exchange was significantly slowed by reducing the incubation temperature (Table 3).

It has been proposed that sphingomyelin may be involved in regulation of membrane cholesterol (Clejan and Bittman, 1984). Depletion of sphingomyelin has been reported to reduce the uptake of cholesterol into LM fibroblasts membranes, modify transbilayer distribution of cholesterol and increase intracellular cholesterol (Slotte et al.,1989; Porn et al.,1991; Slotte and Bierman, 1988; Jefferson et al.,1991). Sphingomyelin was depleted using sphingomyelinase C (*Staphylococcus aureus*) in those studies. Data in Table 5 show that sphingomyelinase treatment of synaptosomes significantly increased the $t_{1/2}$ of cholesterol exchange by approximately 35%.

Table 5. Cholesterol exchange between small unilamellar vesicles and synaptosomes treated with and without sphingomyelinase

Treatment	Rate Constant (h^{-1})	$t_{1/2}$ (hr)	Ex. Pool (%)
Synaptosomes	0.065 ± 0.002	10.59 ± 0.29	53 ± 4
Sphingomyelinase-Treated Synaptosomes	$0.048 \pm 0.001^*$	$14.26 \pm 0.32^{**}$	51 ± 4

Synaptosomes were incubated with buffer or sphingomyelinase for 5 min. Following incubation, cholesterol exchange procedures were used to determine the kinetics and size of cholesterol lateral pools at 37°C as described in Table 3. Values are the means ± SEM of three different experiments per treatment group. *P <0.01, **P <0.001 as compared to untreated synaptosomes. Data are taken from Rao et al.,1993.

Neither the exchangeable pool of cholesterol nor the total amount of membrane cholesterol was affected by sphingomyelinase treatment. It has been suggested that sphingomyelin may facilitate the incorporation of cholesterol into membranes (Slotte and Bierman, 1988). Sphingomyelinase treated fibroblasts showed less cholesterol taken up into the membranes as compared to untreated membranes. The data on synaptosomal membranes were consistent with the hypothesis that sphingomyelin increases the capacity of the membrane to solubilize cholesterol. The mechanism for such an effect has not been identified.

The rate of cholesterol exchange between membranes could be altered by temperature and hydrolysis of sphingomyelin. The kinetics of cholesterol exchange between SUV and synaptosomes were also significantly affected by chronic ethanol consumption in mice (Wood et al.,1993). The $t_{1/2}$ of cholesterol exchange in the pair-fed control group was 10.7 (hr) (Table 4) and was similar to what we have observed in synaptosomal membranes of mice on regular laboratory diets (Table 3). There was a 25% increase in the $t_{1/2}$ of cholesterol exchange in synaptosomal membranes of the chronic ethanol group (Table 4).

The slower exchange of cholesterol observed in synaptosomal membranes of the ethanol group may have resulted from the two-fold increase in cholesterol content of the SPM exofacial leaflet (Figure 1). There were no differences between the two groups in the total amount of cholesterol, phospholipid phosphorus or the molar ratio of cholesterol to phospholipid.

Previously it was shown that treatment of synaptosomes with sphingomyelinase significantly slowed the rate of cholesterol exchange (Table 5). Effects of sphingomyelin hydrolysis by sphingomyelinase were examined in synaptosomal membranes of pair-fed controls and chronic ethanol-treated mice. Sphingomyelinase treatment significantly increased the $t_{1/2}$ of cholesterol exchange in membranes of both the pair-fed control and ethanol-treated mice (Table 6). However, it can be seen in Table 6 and Figure 2, that the effects of sphingomyelinase treatment were significantly greater in the pair-fed control group as compared to the chronic ethanol-treated group. One explanation for such an effect may have been that the membrane content of sphingomyelin or the amount hydrolyzed differed between the two groups. It has been previously reported that hydrolysis of phospholipids by phospholipase A_2 was altered by chronic ethanol consumption in liver microsomes and erythrocytes (Stubbs et al.,1988; Wood et al.,1991a). No significant differences were observed in the total amount of sphingomyelin or the amount hydrolyzed in comparisons between the pair-fed control and the chronic ethanol-treated group. Another explanation for the differences between the two groups in effects of sphingomyelinase on cholesterol exchange may be due in part to ethanol-induced changes in the transbilayer distribution of cholesterol as were discussed earlier in this chapter.

Table 6. Effects of Sphingomyelinase (SMASE) on the Kinetics and Size of Cholesterol Lateral Domains in Synaptosomes of Control and Ethanol Groups

Group	Rate Constant (h^{-1})	$t_{1/2}$ (hr)	Ex. Pool (%)
Control			
0	0.0653 ± 0.0004	10.61 ± 0.07	52 ± 0.9
50mU SMASE	$0.0466 \pm 0.0001^{**}$	$14.90 \pm 0.05^{**}$	54 ± 1.9
Ethanol			
0	$0.0588 \pm 0.0013^{+}$	$11.80 \pm 0.27^{+}$	53 ± 3.3
50mU SMASE	$0.0504 \pm 0.0011^{+*}$	$13.76 \pm 0.28^{+*}$	54 ± 2.3

Ethanol and control liquid diets were administered to mice for 12 weeks. Synaptosomes were incubated with sphingomyelinase (50mU/mg membrane protein) for 5 min, after which time the reaction was stopped and cholesterol exchange between SUV and synaptosomes performed as described in Table 3. Data are the means ± SEM of three different membrane preparations (n=4 mice/preparation). Data are from Wood et al.,1993.

We found that chronic ethanol consumption slowed the movement of cholesterol in synaptosomal membranes. In another study, it was reported that the uptake of cholesterol was greater in erythrocytes of chronic ethanol-treated squirrel monkeys (Doyle et al.,1986). In a subsequent study, that group found that cholesterol efflux was slower in erythrocyte membranes of ethanol-treated monkeys (Doyle et al.,1988). In both studies, membrane cholesterol was higher and the cholesterol/phospholipid ratio was larger in the ethanol as compared to the control group. Generally, it has been shown in model membrane studies and studies using biological membranes, that increasing the amount of membrane cholesterol slows cholesterol transport.

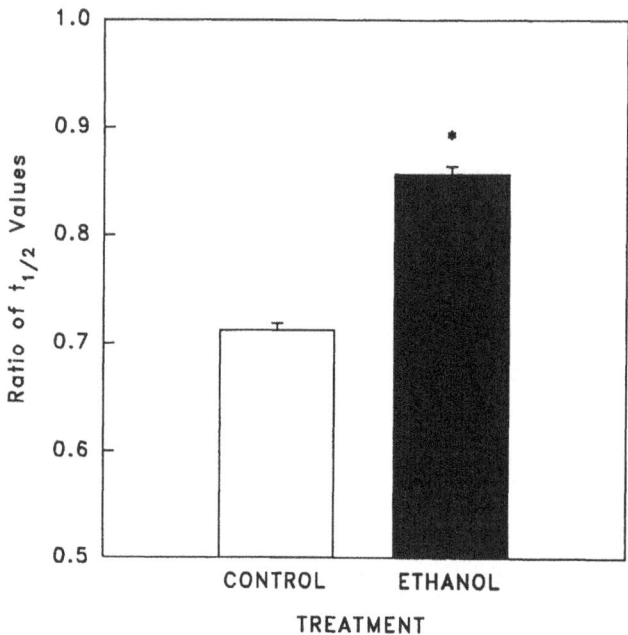

Figure 2. The ratio for the $t_{1/2}s$ of synaptosomes not treated with sphingomyelinase to synaptosomes treated with sphingomyelinase for the pair-fed control and chronic ethanol groups. Values are the means $\pm$ SEM of three different membrane preparations for each group. [*]p <0.01 compared to the control group. Data are from Wood et al., (1993).

An explanation for the contradictory findings between the erythrocyte studies on cholesterol influx and the synaptosomal membrane studies may be due to differences in species, membrane type, uptake procedures, and length of ethanol administration. Another possible explanation is that the phospholipid composition of the erythrocyte is very different as compared to the synaptosomal membrane, particularly with respect to sphingomyelin (Wood et al.,1989a; Wood et al.,1991a). Sphingomyelin composition in the erythrocyte is approximately 5-fold more as compared to the synaptosomal membrane. Sphingomyelin can affect cholesterol transport as previously discussed.

Sterol Carrier Proteins

There have been several sterol carrier proteins identified and studies on such proteins have been recently reviewed (Veerkamp et al.,1991). For purposes of this discussion, we will focus on two proteins that have been shown to bind cholesterol. Sterol carrier protein (SCP) (also called liver-fatty acid binding protein) sterol carrier protein-2 (SCP-2) (also called nonspecific lipid transfer protein) bind cholesterol in a 1:1 molar ratio (Schroeder et al.,1991a). SCP also binds fatty acids and SCP-2 binds phospholipids. Both of these proteins have been found to increase cholesterol exchange in L-cell fibroblast plasma membranes (Schroeder et al.,1991a). We have reported recently that the transbilayer distribution of cholesterol in plasma membrane of L-cell fibroblasts transfected with cDNA encoding for rat SCP was altered as compared to membranes of control L-cells that contain very little SCP (Incerpi et al.,1992). The exofacial leaflet of membranes from the transfected cells contained 34% of the total cholesterol as compared to 16% for the control cells. There is very little known about the occurrence of brain proteins that bind cholesterol. A fatty acid binding protein (similar to SCP) from bovine brain has been identified and found to bind fatty acids (Schoentgen et al.,1989). Binding of cholesterol was not examined. It has been reported that addition of exogenous fatty acid binding proteins stimulated synaptosomal Na^+-dependent amino acid transport (Bass et al.,1984). Effects of the binding proteins on amino acid transport could be due to binding of fatty acids or by acting on specific properties of membrane cholesterol. There has been only one study that examined effects of chronic ethanol consumption on SCP (Pignon et al.,1987). It was reported that SCP was increased in the liver of chronic ethanol animals. The maximal binding capacity of the protein for fatty acids was increased but the affinity for fatty acids was reported to be decreased. Effects on cholesterol binding or other properties of cholesterol were not examined. The increase in SCP may also have altered cholesterol domains and would be consistent with our findings with the transfected cells discussed above.

Regulation of Protein Activity and Cell Function

It has been well-established that lipids are involved in regulating the activity of certain membrane proteins (Yeagle, 1989). For example, several studies have shown an association between cholesterol and regulation of different membrane proteins. $Na^+ + K^+$-ATPase activity was modified by varying the total amount of cholesterol in membranes (Yeagle, 1983). Cholesterol enrichment and depletion of the erythrocyte exofacial leaflet affected membrane sulfhydryl group exposure and antigen exposure (Schachter et al.,1983). Dopamine uptake was inhibited by increasing the cholesterol/phospholipid ratio in synaptosomes (Maguire and Druse, 1989a). It was also reported that increasing the cholesterol/phospholipid ratio inhibited the activity of dopamine-stimulated adenylate cyclase (Maguire and Druse, 1989b). It is reasonable to predict that changing the cholesterol/phospholipid ratio would modify different membrane cholesterol domains.

Membrane cholesterol is involved in regulation of cell calcium (Locher et al.,1984; Zhou et al.,1991; Madden et al.,1981; Aepfelbacher et al.,1991). Increasing cholesterol in erythrocytes and in bovine arterial smooth cells increased intracellular calcium (Locher et al.,1984; Zhou et al.,1991). Effects on calcium were attributed to cholesterol acting on a calcium channel and $Ca^{2+} + Mg^{2+}$-ATPase. Increasing cholesterol in sarcoplasmic reticulum membranes decreased the activity of $Ca^{2+} + Mg^{2+}$-ATPase (Madden et al.,1981). Incorporation of cholesterol into synaptosomes altered the membrane location of a series of calcium channel drugs (Moring et al.,1990). Incorporation of phosphatidylserine into rat synaptosomes increased spontaneous calcium uptake, and Na^+/Ca^{2+} exchange but did not affect KCl-depolarization-induced calcium uptake at an equivalent amount of phosphatidylserine (Floreani et al.,1991). The significance of phosphatidylserine with respect to cholesterol is that it has been demonstrated in model membranes that phosphatidylserine increases the rate of cholesterol transport between membranes and modified the exchangeable cholesterol pool (Hapala et al.,1990; Butko et al.,1990). One interpretation of the effects of phosphatidylserine on synaptosomal calcium uptake is that the addition of phosphatidylserine produced a change in cholesterol domains, resulting in the increased calcium uptake.

There have not been any studies that have examined effects of ethanol, either *in vitro* or chronic administration, on cholesterol domains and protein structure and function. However, data from recent experiments with sphingomyelinase treated synaptosomes and intracellular calcium may be instructive. As discussed elsewhere in this chapter, sphingomyelinase treatment of LM fibroblast membranes altered the transbilayer distribution of cholesterol, decreased cholesterol in the membrane and reduced the rate of cholesterol exchange in synaptosomal membranes. Preliminary data indicated that intracellular calcium as measured by Fura-2 was increased in synaptosomes incubated with sphingomyelinase. Similar effects on intracellular calcium were observed when ethanol was added *in vitro* and were consistent with earlier studies (Daniell and Harris, 1988; Daniel et al.,1987). Sphingomyelinase treatment alters transbilayer cholesterol domains and kinetics of cholesterol exchange. The addition of ethanol *in vitro* may modify those aspects of membrane cholesterol that may contribute to the regulation of intracellular calcium (e.g., $Ca^{2+} + Mg^{2+}$-ATPase, Na^+/Ca^{2+} exchange).

CONCLUSIONS

It was the purpose of this chapter to discuss the multi-faceted aspects of membrane cholesterol and how chronic ethanol consumption modifies some but not all aspects of membrane cholesterol. Ethanol-induced changes in the total amount of membrane cholesterol have not been consistently reported. It has been shown that the transbilayer distribution of cholesterol in the exofacial and cytofacial leaflets were altered as a result of chronic ethanol consumption. On the other hand, the exchangeable and non-exchangeable cholesterol pools were not affected in membranes of chronic ethanol-treated animals. The kinetics of cholesterol exchange were slower and less affected by hydrolysis of sphingomyelin in membranes of chronic ethanol consuming animals as compared to membranes of pair-fed controls. It would appear that chronic ethanol consumption increases the amount of a protein that binds fatty acids and cholesterol. Finally, it is clear that changes in cholesterol content and in particular cholesterol domains can alter the behavior of certain proteins. The different aspects of membrane cholesterol are important factors that contribute to the structural and functional organizational of the membrane and are differentially affected by chronic ethanol consumption.

ACKNOWLEDGEMENTS

This work was supported in part, by grants from the National Institute on Alcohol Abuse and Alcoholism (#07292, W.G.W.) and the Medical Research Service, and the Geriatric Research, Education, and Clinical Center of the Department of Veterans Affairs.

REFERENCES

Aepfelbacher, M., N. Hrboticky, I. Lux and P.C. Weber, 1991, Cholesterol modulates PAF-stimulated Ca^{2+}-mobilization in monocytic U937 cells, Biochim Biophys Acta 1074, 125.

Artigues, A., M.T. Villar, A.M. Fernandez, J.A. Ferragut and J.M. Gonzalaez-Ros, 1989, Cholesterol stabilizes the structure of the nicotinic acetylcholine receptor reconstituted in lipid vesicles, Biochim Biophys Acta 985, 325.

Bar, L.K., Y. Barenholz and T.E. Thompson, 1987, Dependence on phospholipid composition of the fraction of cholesterol undergoing spontaneous exchange between small unilamellar vesicles, Biochem. 26, 5460.

Bass, N.M., D.E. Raghupathy, D.E. Rhoads, J.A. Manning and R.K. Ockner, 1984, Partial purification of molecular weight 12000 fatty acid binding proteins from rat brain and their effect on synaptosomal Na^+-dependent amino acid uptake, Biochem. 23, 6539.

Beauge, F., H. Stibler and S. Borg, 1985, Abnormal fluidity and surface carbohydrate content of erythrocyte membrane in alcoholic patients, Alcohol Clin Exp Res 9, 322.

Berstein, G., T. Haga and A. Ichiyama, 1989, Effect of the lipid environment on the differential affinity of purified cerebral and atrial muscarinic acetylcholine receptors for pirenzepine, Mol. Pharmacol. 36, 601.

Bittman, R., S. Clejan and S.W. Hui, 1990, Increased rates of lipid exchange between *Mycoplasma capricolum* membranes and vesicles in relation to the propensity of forming nonbilayer lipid structures, J. Biol. Chem. 265(25), 15110.

Bloj, B. and D.B. Zilversmit, 1977, Complete exchangeability of cholesterol in phosphatidylcholine/cholesterol vesicles of different degrees of unsaturation, Biochem. 16, 3943.

Brasaemle, D.L., A.D. Robertson and A.D. Attie, 1988, Transbilayer movement of cholesterol in the human erythrocyte membrane, J. Lipid Res. 29, 481.

Butko, P., I. Hapala, T.J. Scallen and F. Schroeder, 1990, Acidic phospholipids strikingly potentiate sterol carrier protein 2 mediated intermembrane sterol transfer, Biochem. 29, 4070.

Chin, J.H., L.M. Parsons and D.B. Goldstein, 1978, Increased cholesterol content of erythrocyte and brain membranes in ethanol-tolerant mice, Biochim Biophys Acta 513, 358.

Chin, J.H. and D.B. Goldstein, 1977, Drug tolerance in biomembranes: a spin label study of the effects of ethanol, Science 196, 684.

Chin, J.H. and D.B. Goldstein, 1981, Membrane-disordering action of ethanol. Variation with membrane cholesterol content and depth of the spin label probe, Mol. Pharmacol. 19, 425.

Chin, J.H. and D.B. Goldstein, 1984, Cholesterol blocks the disordering effects of ethanol in biomembranes, Lipids 19(12), 929.

Clejan, S. and R. Bittman, 1984, Decreases in rates of lipid exchange between Mycoplasma gallisepticum cells and unilamellar vesicles by incorporation of sphingomyelin, J. Biol. Chem. 259, 10823.

Crews, F.T., E. Majchrowicz and R. Meeks, 1983, Changes in cortical synaptosomal plasma membrane fluidity and composition in ethanol dependent rats, Psychopharm 81, 208.

Curtain, C.C., L.M. Gordon and R.C. Aloia, 1988, Lipid domains in biological membranes: Conceptual Development and Significance, in: Lipid Domains and the Relationship to Membrane Function, eds. R.C. Aloia, C.C. Curtain and L.M. Gordon (Alan R. Liss, New York) p. 1.

Daniel, L.C., E.P. Brass and R.A. Harris, 1987, Effect of ethanol on intracellular ionized calcium concentrations in synaptosomes and hepatocytes, Mol. Pharmacol. 32, 831.

Daniell, L.C. and R.A. Harris, 1988, Effect of chronic ethanol treatment and selective breeding for hypnotic sensitivity to ethanol on intracellular ionized calcium concentrations in synaptosomes, Alcohol Clin Exp Res 12, 179.

Daniels, C.K. and D.B. Goldstein, 1982, Movement of free cholesterol from lipoproteins or lipid vesicles into erythrocytes, Mol. Pharmacol. 21, 694.

Deitrich, R.A., T.V. Dunwiddie, R.A. Harris and V.G. Erwin, 1989, Mechanism of action of ethanol: Initial central nervous system actions, Pharmacol Rev 41, 489.

Devaux, P.F., 1988, Phospholipid flippases, FEBS Lett 234(1), 8.

Devaux, P.F., 1991, Static and dynamic lipid asymmetry in cell membranes, Biochem. 30, 1163.

Doyle, K., J. Cluette-Brown, F. Igoe and J. Hojnacki, 1986, Ethanol induced alterations in erythrocyte membrane lipids due to enhanced cholesterol influx, Res Comm Sub Abuse 7, 133.

Doyle, K., J. Cluette-Brown, N. Rencricca and J. Hojnacki, 1988, Ethanol induced alterations in red blood cell membrane lipid composition due to decreased cholesterol efflux, Res Comm Sub Abuse 9, 157.

Floreani, M., P. Debetto and F. Carpenedo, 1991, Phosphatidylserine vesicles increase Ca^{2+} uptake by rat synaptosomes, Arch. Biochem. Biophys. 285, 116.

Fontaine, R.N., R.A. Harris and F. Schroeder, 1980, Aminophospholipid asymmetry in murine synaptosomal plasma membrane, J. Neurochem 34, 269.

Gold, J.C. and M.C. Phillips, 1990, Effects of membrane lipid composition on the kinetics of cholesterol exchange between lipoproteins and different species of red blood cells, Biochim Biophys Acta 1027, 85.

Hapala, I., P. Butko and F. Schroeder, 1990, Role of acidic phospholipids in intermembrane sterol transfer, Chem Phys Lipids 56, 37.

Harris, R.A., D.M. Baxter, M.A. Mitchell and R.J. Hitzemann, 1984, Physical properties and lipid composition of brain membranes from ethanol tolerant-dependent mice, Mol. Pharmacol. 25, 401.

Hill, M.W. and A.D. Bangham, 1975, General depressant drug dependency: A biophysical hypothesis, Adv. Exp. Med. Biol. 59, 1.

Hunt, W.A., 1985, Alcohol and Biological Membranes (The Guilford Press, New York).

Incerpi, S., J.R. Jefferson, W.G. Wood, W.J. Ball and F. Schroeder, 1992, Na pump and plasma membrane structure in L-cell fibroblasts expressing rat liver fatty acid binding protein, Arch. Biochem. Biophys. 298, 35.

Jefferson, J.R., J.P. Slotte, G. Nemecz, A. Pastuszyn, T.J. Scallen and F. Schroeder, 1991, Intracellular sterol distribution in transfected mouse L-cell fibroblasts expressing rat liver fatty acid-binding protein, J. Biol. Chem. 266, 5486.

Kanner, B.I. and A. Shouffani, 1990, Cholesterol is required for reconstitution of the sodium- and chloride-coupled, GAMMA-aminobutyric acid trnsporter from rat brain, J. Biol. Chem. 265(11), 6002.

Kelly-Murphy, S., A.J. Waring, H. Rottenberg and E. Rubin, 1983, Effects of chronic ethanol consumption on the partition of lipophilic compounds into erythrocyte membranes, Lab. Invest. 50, 174.

Kier, A.B., W.D. Sweet, M.S. Cowlen and F. Schroeder, 1986, Regulation of transbilayer distribution of a fluorescent sterol in tumor cell plasma membranes, Biochim Biophys Acta 861, 287.

Locher, R., L. Neyses, M. Stimpel, B. Kuffer and W. Vetter, 1984, The cholesterol content of the human erythrocyte influences calcium influx through the channel, Biochem. Biophys. Res. Commun. 124, 822.

Madden, T.D., M.D. King and P.J. Quinn, 1981, The modulation of Ca^{2+}-ATPase activity of sarcoplasmic reticulim by membrane cholesterol: The effect of enzyme coupling, Biochim Biophys Acta 641, 265.

Maguire, P.A. and M.J. Druse, 1989a, The influence of cholesterol on synaptic fluidity, dopamine D1 binding and dopamine-stimulated adenylate cyclase, Brain Res. Bull. 23, 69.

Maguire, P.A. and M.J. Druse, 1989b, The influences of cholesterol on synaptic fluidity and dopamine uptake, Brain Res. Bull. 22, 431.

McLean, L. and M.C. Phillips, 1982, Cholesterol desorption from clusters of phosphatidylcholine and cholesterol in unilamellar vesicle bilayers during lipid transfer or exchange, Biochem. 21, 4053.

McMurchie, E.J. and G.S. Patten, 1988, Dietary cholesterol influences cardiac BETA-adrenergic receptor adenylate cyclase activity in the marmoset monkey by changes in membrane cholesterol status, Biochim Biophys Acta 942, 324.

Michelangeli, F., J.M. East and A.G. Lee, 1990, Structural effects on the interaction of sterols with the $(Ca^{2+} + Mg^{2+})$-ATPase, Biochim Biophys Acta 1025, 99.

Moring, J., W.J. Shoemaker, V. Skita, R.P. Mason, H.C. Hayden, R.M. Salomon and L.G. Herbette, 1990, Rat cerebral cortical synaptoneurosomal membranes structure and interactions with imidazobenzodiazepine and 1,4-dihydropyridine calcium channel drugs, Biophys J 58, 513.

Nemecz, G., R.N. Fontaine and F. Schroeder, 1988, A fluorescence and radiolabel study of sterol exchange between membranes, Biochim Biophys Acta 943, 511.

Op den Kamp, J.A.F., 1979, Lipid asymmetry in membranes, Annu. Rev. Biochem. 48, 47.

Parsons, L.M., E.J. Gallaher and D.B. Goldstein, 1982, Rapidly developing functional tolerance to ethanol is accompanied by increased erythrocyte cholesterol in mice, J Pharmacol Exp Ther 223(2), 472.

Phillips, M.C., W.J. Johnson and G.H. Rothblat, 1987, Mechanisms and consequences of cellular cholesterol exchange and transfer, Biochim Biophys Acta 906, 223.

Pignon, J., N.C. Bailey, E. Baraona and C.S. Lieber, 1987, Fatty acid-binding protein: a major contributor to the ethanol-induced increase in liver cytosolic proteins in the rat, Hepatology 7(5), 865.

Ponnappa, B.C., A.J. Waring, J.B. Hoek, H. Rottenberg and E. Rubin, 1982, Chronic ethanol ingestion increases calcium uptake and resistance to molecular disordering by ethanol in liver microsomes, J. Biol. Chem. 257, 10141.

Porn, M.I., J. Tenhunen and J.P. Slotte, 1991, Increased steroid hormone secretion in mouse Leydig tumor cells after induction of cholesterol translocation by sphingomyelin degradation, Biochim Biophys Acta 1093, 7.

Poznansky, M.J. and Y. Lange, 1976, Transbilayer movement of cholesterol in dipalmitoyl-lecithin-cholesterol vesicles, Nature 259, 420.

Rao, A.M., I. Igbavboa, M. Semotuk, F. Schroeder and W.G. Wood, 1993, Kinetics and size of cholesterol lateral domains in synaptosomal membranes: Modification by sphingomyelinase and effects on membrane enzyme activity, Neurochem. Int., in press.

Roelofsen, B., 1982, Phospholipases as tools to study the localization of phospholipids in biological membranes. A critical review, J. Toxicol. 1(1), 87.

Rottenberg, H., A. Waring and E. Rubin, 1981, Tolerance and cross-tolerance in chronic alcoholics: Reduced membrane binding of ethanol and other drugs, Science 213, 583.

Sastry, P.S., 1985, Lipids of nervous tissue: composition and metabolism, Prog. Lipid Res. 24, 69.

Schachter, D., R.E. Abbott, U. Cogan and M. Flamm, 1983, Lipid fluidity of the individual hemileaflets of human erythrocyte membranes, Ann NY Acad Sci 414, 19.

Schoentgen, F., G. Pignede, L.M. Bonanno and P. Jolles, 1989, Fatty acid-binding protein from bovine brain, Eur. J. Biochem. 185, 35.

Schroeder, F., W.J. Morrison, C. Gorka and W.G. Wood, 1988, Transbilayer effects of ethanol on fluidity of brain membrane leaflets, Biochim Biophys Acta 946, 85.

Schroeder, F., W.G. Wood, W.J. Morrison, R.N. Fontaine and A.B. Keir, 1989, Synaptosomal plasma membrane lipid and structural asymmetry, in: Neurochemical Aspects of Phospholipid Metabolism, eds. L. Freysz, J.N. Hawthorne and G. Toffano (Liviana Press, Padova,Italy) p. 17.

Schroeder, F., J.R. Jefferson, A.B. Kier, J. Knittel, T.J. Scallen, W.G. Wood and I. Hapala, 1991a, Membrane cholesterol dynamics: Cholesterol domains and kinetic pools, Proc. Soc. Exp. Bio. Med. 196, 235.

Schroeder, F., G. Nemecz, W.G. Wood, G. Morrot, M. Ayraut-Jarrier and P.F. Devaux, 1991b, Transmembrane distribution of sterol in the human erythrocyte, Biochim Biophys Acta 1066, 183.

Slotte, J.P., G. Hedstrom, S. Rannstrom and S. Ekman, 1989, Effects of sphingomyelin degradation on cell cholesterol oxidizability and steady-state distribution between the cell surface and the cell interior, Biochim Biophys Acta 985, 90.

Slotte, J.P. and E.L. Bierman, 1988, Depletion of plasma-membrane sphingomyelin rapidly alters the distribution of cholesterol between plasma membranes and intracellular cholesterol pools in cultured fibroblasts, Biochem J 250, 653.

Stubbs, C.D., B.W. Williams, C.L. Pryor and E. Rubin, 1988, Ethanol-induced modifications to membrane lipid structure: effect of phospholipid A_2-membrane interactions, Arch. Biochem. Biophys. 262, 560.

Sun, G.Y. and A.Y. Sun, 1985, Ethanol and membrane lipids, Alcohol Clin Exp Res 9(2), 164.
Supernovich, C., R. Crain and P. Rosenberg, 1991a, Phosphatidylcholine asymmetry in electroplax from the electric eel: Use of a phosphatidylcholine exchange protein, J. Neurochem 57, 575.

Supernovich, C., R. Crain and P. Rosenberg, 1991b, Effect of soman and sarin on phosphatidylcholine asymmetry in the electroplax from the electric eel, J. Neurochem 57, 585.

Sweet, W.D., W.G. Wood and F. Schroeder, 1987, Charged anesthetics selectively alter plasma membrane order, Biochem. 26, 2828.

Taraschi, T.F., J.S. Ellingson, A. Wu, R. Zimmerman and E. Rubin, 1986, Phosphatidylinositol from ethanol-fed rats confers membrane tolerance to ethanol, Proc. Natl. Acad. Sci. U. S. A. 83, 9398.

Taraschi, T.F., J.S. Ellingson, A. Wu-Sun and E. Rubin, 1990, Rats withdrawn from ethanol rapidly re-acquire membrane tolerance after resumption of ethanol feeding, Biochim Biophys Acta 1021, 51.

Thomas, P.D. and M.J. Poznansky, 1988, Cholesterol transfer between lipid vesicles. effect of phospholipids and gangliosides, Biochem J 251, 55.

Thurnhofer, H. and H. Hauser, 1990, Uptake of cholesterol by small intestinal brush border membrane is protein-mediated, Biochem. 29, 2142.

Veerkamp, J.H., R.A. Peeters and R.G.H.J. Maatman, 1991, Structural and functional features of different types of cytoplasmic fatty-acid-binding proteins, Biochim Biophys Acta 1081, 1.

Waring, A.J., H. Rottenberg, T. Ohnishi and E. Rubin, 1982, The effect of chronic ethanol consumption on temperature-dependent physical properties of liver mitochondrial membranes, Arch. Biochem. Biophys. 216, 51.

Wood, W.G., S. Lahiri, C. Gorka, H.J. Armbrecht and R. Strong, 1987, In vitro effects of ethanol on erythrocyte membrane fluidity of alcoholic patients: An electron spin resonance study, Alcohol Clin Exp Res 11, 332.

Wood, W.G., M. Cornwell and L.S. Williamson, 1989a, High performance thin-layer chromatography and densitometry of synaptic plasma membrane lipids, J. Lipid Res. 30, 775.

Wood, W.G., C. Gorka and F. Schroeder, 1989b, Acute and chronic effects of ethanol on transbilayer membrane domains, J. Neurochem 52, 1925.

Wood, W.G., F. Schroeder, L. Hogy, A.M. Rao and G. Nemecz, 1990, Asymmetric distribution of a fluorescent sterol in synaptic plasma membranes: Effects of chronic ethanol consumption, Biochim Biophys Acta 1025, 243.

Wood, W.G., C. Gorka, J.A. Johnson, G.Y. Sun, A.Y. Sun and F. Schroeder, 1991a, Chronic ethanol consumption alters transbilayer distribution of phosphatidylcholine in erythrocytes of Sinclair (S-1) Miniature Swine, Alcohol 8, 395.

Wood, W.G., F. Schroeder and A.M. Rao, 1991b, Significance of ethanol-induced changes in membrane lipid domains, Alc. & Alcohol. Suppl. 1, 221.

Wood, W.G., A.M. Rao, U. Igbavboa and M. Semotuk, 1993, Cholesterol exchange and lateral cholesterol pools in synaptosomal membranes of pair-fed control and chronic ethanol-treated mice, Alcohol Clin Exp Res, 17, 345.

Wood, W.G. and F. Schroeder, 1988, Membrane effects of ethanol: Bulk lipid versus lipid domains, Life Sci 43, 467.

Wood, W.G. and F. Schroeder, 1992, Membrane exofacial and cytofacial leaflets: A new approach to understanding how ethanol alters brain membranes, in: Alcohol and Neurobiology: Receptors, Membranes, and Channels, ed. R.R. Watson (CRC Press, Boca Raton, FL) p. 161.

Yang, J., G.L. Anderle and R. Mendelsohn, 1990, Effects of cholesterol on the interaction of Ca^{2+}-ATPase with 1-palmitoyl-2-oleoylphosphatidylethanolamine. An FTIR study, Biochim Biophys Acta 1021, 27.

Yeagle, P.L., 1983, Cholesterol modulation of $(Na^{+}+K^{+})$-ATPase ATP hydrolyzing activity in the human erythrocyte, Biochim Biophys Acta 729, 39.

Yeagle, P.L., 1989, Lipid regulation of cell membrane structure and function, FASEB J. 3, 1833.

Zhou, Q., S. Jimi, T.L. Smith and F.A. Kummerow, 1991, The effect of cholesterol on the accumulation of intracellular calcium, Biochim Biophys Acta 1085, 1.

THE EFFECTS OF ETHANOL ON POLYUNSATURATED FATTY ACID COMPOSITION

Norman Salem, Jr., and Glenn Ward

Laboratory of Membrane Biochemistry and Biophysics, DICBR
National Institute on Alcohol Abuse and Alcoholism
Bethesda, MD 20892

INTRODUCTION

In this chapter, we will attempt to review the literature with regard to the effects of alcohol exposure upon the levels of polyunsaturated fatty acids (PUFA) in various mammalian tissues. Our focus will be upon arachidonic (20:4n6) and docosahexaenoic acid (22:6n3), in particular, as these represent the two principal endpoints of n-6 and n-3 fatty acid metabolism, respectively, as well as the two principal long-chain polyunsaturates found in mammalian tissues. Data from our own studies of the effects of ethanol inhalation upon lipid composition will also be presented in an effort to define some of the variables that must be controlled if consistent results are to be obtained in this regard.

EFFECTS IN HUMANS

Tissue levels of polyunsaturated fatty acids are known to be altered in many pathological states (see review by Holman and Johnson, 1982) and several authors have reported alterations of 20:4n6 and 22:6n3 concentrations following long-term alcohol abuse (Table 1). Generally, 20:4n6 levels were found to be reduced in erythrocyte membranes of alcoholics (Alling, et al., 1984b; Alvaro, et al., 1982; Glen, et al., 1984) as well as in the serum (Johnson, et al., 1985; Marzo, et al., 1970) and plasma (Alling, et al., 1984b; Glen, et al., 1984) although at least one study has reported increases in the plasma concentrations (Szebeni, et al., 1986) and another found no difference in erythrocyte concentrations (Clemens, 1988). Reductions in 22:6n3 have also been reported in the erythrocyte membrane (Alling, et al., 1984b) and serum (Johnson, et al., 1985) while no change was seen in plasma levels (Alling, et al., 1984b). One study obtained brains from alcoholics at autopsy and reported decreases in the levels of 20:4n6 in the phosphatidylethanolamine (PE) fraction in cerebral gray matter and increases in 22:6n3 in the PE fraction in cerebral white matter, cerebellum, and medulla oblongata (Lesch, et al., 1973).

Since these changes were, in many cases, observed in patients with alcoholic liver disease, they may not be directly attributable to ethanol consumption, but may be secondary to liver damage caused by long-term alcohol abuse. For example, one study reported that decreases in both 20:4n6 and 22:6n3 in the serum were found only in alcoholics with cirrhosis of the liver: in long-term alcoholics without liver disease, the levels did not differ from those of controls (Johnson, et al., 1985). On the other hand, at least one study has reported reductions in both 20:4n6 and 22:6n3 in erythrocyte

Table 1. Polyunsaturated Fatty Acid Composition in Human Alcoholics.*

Author	Tissue	20:4n6	22:6n3	Comments
Alling, et al., 1984b	erythrocyte membranes	↓	↓	- measured only in PC and PE fraction
	plasma	↓	n.-s.	- measured only in PC fraction
Alvaro, et al., 1982	erythrocyte membranes	↓	n.r.	- total lipid fraction only
Benedetti, et al., 1987	erythrocyte membranes	↓	↓	- non cirrhotic alcoholics
Clemens, 1988	erythrocyte membranes	n.-s.	n.r.	- similar results seen in both cirrhotic and non-cirrhotic alcoholics
Glen, et al., 1984	erythrocyte membranes	↓	n.r.	- total phospholipids and PC and PE fractions
	plasma	↓	n.r.	-
Johnson, et al., 1985	serum	↓	↓	- for cirrhotic alcoholics only: n.-s. in non-cirrhotic alcoholics
Lesch, et al., 1973	cerebral gray matter	↓	n.-s.	- PE fraction only
	cerebral white matter	n.-s.	↑	- 22:6n3 measured only in PE fraction
	cerebellum	n.-s.	↑	- 22:6n3 measured only in PE fraction
	medulla oblongata	n.-s.	↑	- 22:6n3 measured only in PE fraction
Marzo, et al., 1970	serum	↓	n.r.	-
Szebeni, et al., 1986	plasma	↑	n.r.	- in alcoholics with or without liver disease - not found in previous abusers following 12 months abstinance

* ↓: decrease   PC: phosphatidylcholine
  ↑: increase   PE: phosphatidylethanolamine
  n.r.: not reported
  n.-s.: non-significant

membranes of alcoholics without liver disease (Benedetti, et al., 1987). Still, the subjects in these studies may have suffered from other alcohol-related health problems affecting, among other functions, intake, absorption and metabolism of nutrients. Even when alcoholics do not differ greatly from non-alcoholics in terms of gross nutrient intake, they may differ considerably in the way in which they absorb and utilize available nutrients, including dietary fatty acids (see Watson and Watzl, 1992). Given the practical and ethical considerations of obtaining data from populations of human alcoholics, it is not surprising that researchers have often relied on animal research to understand the relationship between ethanol consumption and tissue concentrations of long-chain PUFAs.

ANIMAL STUDIES

Conceptual and Methodological Considerations

Although animal studies can offer a degree of experimental control and precision not available to clinical investigators, they do present a number of procedural and interpretive limitations which must be considered when evaluating the results of animal research into ethanol effects on nutrient metabolism. For example, ethanol administration in animals can be problematic since few laboratory animals voluntarily consume quantities of ethanol sufficient to mimic human consumption (see review by Lieber, *et al.*, 1989). Therefore, when ethanol is administered in the drinking water, even if it is the sole source of fluid, the daily fluid intake and resulting blood alcohol concentrations (BAC's), are likely to be relatively low. One way in which researchers have attempted to deal with this problem has been to provide ethanol as a major ingredient in a liquid diet which is the sole source of nutrients for the animal. Although this technique produces higher BAC's, it still relies upon voluntary consumption and, if the amount of alcohol in the diet is too great, reductions in food intake, and subsequent undernutrition, are likely to result. Three techniques which can produce very high peak BAC's are intragastric intubation, intraperitoneal injection, and inhalation of ethanol vapor. However, although these techniques do not require voluntary consumption by the subjects, reductions in food intake may still result due to ethanol-induced loss of consciousness and/or appetite. Therefore, when comparing animal studies, contrasting results must be considered in light of potential differences in the concentrations of ethanol to which the tissue is exposed, and the pattern of that exposure in addition to the nutritional status of the animal. Unfortunately, such a consideration is made difficult by the fact that in a number of studies discussed in this review, authors have not published BAC's or information about relative nutrient intake among treatment groups. Furthermore, even when sufficient blood alcohol concentrations are obtained in animal studies, the long-term exposure typically found in the human alcoholic is usually not simulated in the animal model.

A related issue is the choice of sample size in animal research. In many cases, non-significant results have been reported in spite of a trend towards significance in the predicted direction. Since some of the studies described in the following sections used very small numbers of animals in each experimental group, it is possible that the lack of significance was due to the lack of statistical power of the analysis rather than to the reality of the null hypothesis. In other words, the non-significant findings reported in some studies may be due to type II statistical errors (Keppel, 1973).

Effects on the Liver

Research into the effect of ethanol on polyunsaturated fatty acid composition in the liver is summarized in Table 2. Generally, ethanol consumption for several weeks or longer in the rat has been reported to reduce levels of 20:4n6 in whole liver homogenates (Corbett, *et al.*, 1991, Salem, *et al.*, 1987) as well as in mitochondrial (Cunningham, *et al.*, 1982; French, *et al.*, 1970; 1971; Thompson and Reitz, 1976; 1978) and microsomal fractions (Ellingson, *et al.*, 1991; French, *et al.*, 1971; Reitz, *et al.*, 1981, Salem, *et al.*, 1987), while decreases in whole liver 20:4n6 concentrations (Salem, *et al.*, 1987) or the microsomal 20:4n6/18:2 ratio (Reitz, *et al.*, 1981) have been reported following a single day of ethanol exposure. On the other hand, some authors have reported no decrease in 20:4n6 concentrations following ethanol exposure (French, *et al.*, 1970; 1971; Mendenhall, *et al.*, 1969; Waring, *et al.*, 1981) although, in two of these studies (French, *et al.*, 1970; 1971), the ethanol was delivered in the drinking water so that BAC's were likely to be relatively low. It is interesting to note that golden hamsters, which do voluntarily consume significant amounts of ethanol in their drinking water, exhibit subsequent reductions in whole liver 20:4n6 concentrations as a result of this procedure (Cunnane, *et al.*, 1985; 1987). Studies of the mouse reported no effect of ethanol inhalation for ten days (Littleton, *et al.*, 1979) or I.P. injections for 7 days (Wing, *et al.*, 1982) on 20:4n6 levels in whole liver homogenates. In one of these studies, however (Littleton, *et al.*, 1979), there was a trend towards lower 20:4n6 levels in the liver, suggesting that the sample size used (5 in each group) may have been insufficient to detect significant differences.

Of the few studies published on the effect of ethanol on 22:6n3 in the liver in the rat, most have reported decreases in mitochondria (Cunningham, *et al.*, 1982; French, *et al.*, 1971; Thompson and Reitz, 1978), or microsomes (French, *et al.*, 1971), while others have reported no effect on whole liver (Corbett, *et al.*, 1991; Littleton, *et al.*, 1979) or

Table 2. Effects of Ethanol on Polyunsaturated Fatty Acid Composition in the Liver.*

Author	Species	Route of Etoh Administration	Fraction	20:4n6	22:6n3	Comments
Arai, et al., 1984	baboons	liquid diet (4 months to 9 years)	mitochondria	↓	↓	-
Corbett, et al., 1991	rats	liquid diet (21 days)	whole liver	↓	n.-s.	-
Cunnane, et al., 1985	hamsters	drinking water (8 weeks)	whole liver	↓	↓	- triacylglycerols and phospholipids
Cunnane, et al., 1987	hamsters	drinking water (54 weeks)	whole liver	↓	↓	- triacylglycerols only - n.s in phospholipids
Cunningham, et al., 1981	macaques	liquid diet (18 months)	microsomes	↑	n.r.	- diacyl PC and diacyl PE fraction
Cunningham, et al., 1982	rats	liquid diet (31 days)	microsomes	n.-s.	n.r.	- lysophosphatidyl-choline fraction
		"	mitochondria	↓	↓	- 22:4n6 in PC and PE only - 22:6n3 in PI only
Cunningham, et al., 1983	macaques	liquid diet (10 months)	whole liver	↓	n.-s.	- microsomes and mitochondria
Ellingson, et al., 1991	rats	liquid diet (35 days)	microsomes	↓	↑	- 20:4n6 in PI and PS only - 22:6n3 in PS only
French, et al., 1970	rats	drinking water	mitochondria	↓	n.r.	
		(14 weeks)	"	n.r.	n.r.	-
		(22 weeks)	microsomes	n.-s.	n.r.	-
French, et al., 1971	rats	drinking water (14 weeks)	mitochondria	n.-s.	n.r.	-
		(44 weeks)	mitochondria	↓	↓	-
		"	microsomes	↓	↓	-
Littleton, et al., 1979	Swiss mice	inhalation (10 days)	whole liver	n.-s.	n.-s.	-
Mendenhall, et al., 1969	rats	liquid diet (46 days)	whole liver	↑	n.r.	- PC fraction in corn oil group: n.-s. in - in coconut group
Reitz, et al., 1981	rats	liquid diet (1 day to 11 weeks)	microsomes	↓	n.r.	- results shown as 20:4/18:2ratio
Thompson & Reitz, 1976	rats	liquid diet (48 days)	mitochondria	↓	n.-s.	-
Thompson & Reitz, 1978	rats	liquid diet (1 and 3 months)	mitochondria	↓	↓	- found in females of high-fat diet group only
Salem, et al., 1987	rats	inhalation (1 or 14 days)	whole liver	↓	n.-s.	-
		liquid diet (35 days)	microsomes	↓	n.-s.	-

(Continued on next page)

| Waring, *et al.*, 1981 | rats | liquid diet (30-35 days) | mitochondria | n.-s. | n.r. | - |
| Wing, *et al.*, 1982 | mice | I.P. injection twice daily for 7 days | whole liver | n.-s. | n.-s. | - |

* ↓: decrease	PC: phosphatidylcholine
↑: increase	PE: phosphatidylethanolamine
n.r.: not reported	PI: phosphatidylinositol
n.-s.: non-significant	PS: phosphatidylserine

mitochondria (Thompson and Reitz, 1976), or an increase in microsomal 22:6n3 (Ellingson, *et al.*, 1991).

Some researchers have studied ethanol effects on liver composition in primates fed ethanol in liquid diets for periods ranging from 10 months to several years and, again, the results have been inconsistent. Although one study found reductions in both 20:4n6 and 22:6n3 in baboons fed ethanol for up to 9 years (Arai, *et al.*, 1984), another study reported increases in 20:4n6 in microsomes of macaques following consumption for only 18 months (Cunningham, *et al.*, 1981). This same group did, however, find decreases in 20:4n6 in whole liver preparations in this species following a shorter period of consumption (10 months). Therefore, the inconsistencies in the results may have reflected different methodologies, or some other undefined variable.

Effects on the Blood

As indicated in Table 3, the results of research into effects of ethanol on PUFA in the blood are also inconsistent. Although several studies have reported ethanol-induced reductions in 20:4n6 concentrations in both erythrocyte membranes (Corbett, *et al.*, 1991; LeDroitte, *et al.*, 1984a; Rao, *et al.*, 1981, Salem, *et al.*, 1987) and serum (Alling, *et al.*, 1982) in rats, others have reported no effect (Alling, *et al.*, 1984a) or an increase in 20:4n6 (LaDroitte, *et al.*, 1984b). Two studies in the mouse have failed to find effects of ethanol in erythrocyte membranes (LaDroitte, *et al.*, 1984b; Wing, *et al.*, 1982), although neither study exposed the animals for longer than 7 days. One study in the rat (LaDroitte, *et al.*, 1984a) found effects after 21 days of ethanol inhalation but not after 7 days, and this same lab, in a separate study, found increases in 20:4n6 after 7 and 14 days, but not after 21 days (LaDroitte, *et al.*, 1984b), raising the possibility that the duration of exposure in the mouse studies may not have been of sufficient duration to lead to significant reductions. Interestingly, the only report of reduced 22:6n3 concentrations in the blood following ethanol exposure is in mice following only 7 days of treatment (Wing, *et al.*, 1982).

The PUFA status of the animal prior to, and during, ethanol treatment may have an effect on the fatty acid composition of the blood. When rats were fed diets low in essential fatty acids for more than one generation, ethanol administration led to reduced serum levels of 22:6n3 even though ethanol-treated rats exhibited increased concentrations when fed diets containing adequate amounts of EFA (Alling, *et al.*, 1984a). Even if the overall EFA level is sufficient, however, the levels of specific fatty acids may still be important. For example, rats fed diets containing no 18:3n6, an intermediate between 18:2n6 and 20:4n6, exhibited ethanol-induced reductions in platelet 20:4n6 levels while those fed diets containing both 18:2n6 and 18:3n6 did not (Engler, *et al.*, 1991).

Effects on the Brain

Table 4 summarizes research into ethanol-induced changes in PUFA composition in the brain in animals. Again, the results are inconsistent, with a few studies reporting reductions in 20:4n6 in microsomes (Aloia, *et al.*, 1985) synaptosomes (Alling, *et al.*, 1982; Corbett, *et al.*, 1992; Littleton and John, 1977; Littleton, *et al.*, 1979) and synaptic plasma

Table 3. Effects of Ethanol on Polyunsaturated Fatty Acid Composition in the Blood.*

Author	Species	Route of Etoh Administration	Fraction	20:4n6	22:6n3	Comments
Alling, *et al.*, 1982	rats	drinking water (135 days)	serum	↓	n.-s.	- PC only
Alling, *et al.*, 1984a	rats	daily I.P. injections (23 days)	serum	n.-s.	↑	- PC only - 22:6n3 reduced in low EFA dietary condition
Corbett, *et al.*, 1991	rats	liquid diet (21 days)	erythrocytes	↓	n.-s.	-
Cunnane, *et al.*, 1987	hamsters	drinking water (54 weeks)	plasma	n.-s.	n.-s.	-
			erythrocytes	n.-s.	n.-s.	-
Engler, *et al.*, 1991	rats	inhalation (6 days)	platelets	↓	n.-s.	- found only in sesame oil- and linseed oil-enriched groups
French, *et al.*, 1970	rats	drinking water (25 weeks)	erythrocytes	↓	n.-s.	-
LaDroitte, *et al.*, 1984a	rats	inhalation (7 weeks)	erythrocytes	n.-s.	n.-s.	-
		inhalation (21 days)	"	↓	n.-s.	-
		withdrawal (10 hrs.)	"	n.-s.	n.-s.	-
LaDroitte, *et al.*, 1984b	rats	inhalation (7 days)	erythrocytes	↑	n.-s.	-
		(14 days)	"	↑	n.-s.	-
		(21 days)	"	n.-s.	n.-s.	-
		withdrawal (6 days)	"	↑	n.-s.	-
		withdrawal (10 days)	"	↑	n.-s.	-
	Swiss, DBA, & C57 mice	I.P. injection twice daily for 7 days	"	n.-s.	n.-s.	-
Rao, *et al.*, 1981	rats	liquid diet (4 weeks)	erythrocytes	↓	n.r.	- PC fraction
Salem, *et al.*, 1987	rats	inhalation (1 day)	erythrocytes	↓	n-.s.	-
		(14 days)	"	n.-s.	n.-s.	-
Wing, *et al.*, 1982	mice	I.P. injection twicedaily for 7 days	erythrocytes	n.-s.	↓	

* ↓: decrease   n.r.: not reported
  ↑: increase   n.-s.: non-significant
  PC: phosphatidylcholine

Table 4. Effects of Ethanol on Polyunsaturated Fatty Acid Composition of the Brain.*

Author	Species	Route of Etoh Administration	Fraction	20:4n6	22:6n3	Comments
Aloia, *et al.*, 1985	rats	liquid diet (35 days)	microsomes	↓	↓	- 20:4n6 in PI & PS only - 22:6n3 in PS & PE only
Alling, *et al.*, 1982	rats	drinking water (20 weeks)	mitochondria	n.-s.	n.-s.	-
			myelin	n.-s.	n.-s.	-
			synaptosomes	↓	n.-s.	- PE only
Alling, *et al.*, 1984b	rats	daily I.P. injections (23 days)	whole brain	n.-s.	n.s	-
Corbett, *et al.*, 1992	rats	liquid diet (21 days)	synaptosomes	↓	n.r.	-
Crews, *et al.*, 1983	rats	single intubation	synaptosomes	n.-s.	n.-s.	-
		repeated intubations (4 days)	"	n.-s.	n.-s.	-
		withdrawal	"	n.-s.	n.-s.	-
Gustavsson & Alling, 1989	rats	liquid diet (3 wks)	synaptosomes	n.-s.	n.-s.	-
			myelin	n.-s.	n.-s.	-
			mitochondria	n.-s.	↑	-
Hansson, *et al.*, 1987	newborn rats	medium (3-8 days)	astroglial primary culture	n.-s.	n.-s.	-
Harris, *et al.*, 1984	DBA mice	liquid diet (7 days)	synaptic plasma membrane	—	↓	- PS only
LaDroitte, *et al.*, 1984a	rats	inhalation (21 days)	synaptosomes	n.-s.	n.-s.	-
Littleton & John, 1977	Swiss mice	inhalation (2 hours)	synaptosomes	↓	↓	-
		(10 days)	"	↓	n.s	-
Littleton, *et al.*, 1979	Swiss mice	inhalation (2 or 10 hours)	synaptosomes	n.-s.	↓	-
		(10 days)	"	↓	n.-s.	-
Smith & Gerhart, 1982	C57 mice	liquid diet (8 days)	synaptosomes	n.-s.	n.-s.	-
			myelin	n.-s.	n.-s.	-
Sun & Sun, 1979	guinea pigs	liquid diet (3 weeks)	synaptic plasma membrane	n.-s.	↑	- PE only
Wing, *et al.*, 1982	mice	inhalation (14 days)	myelin	n.-s.	↓	-
			synaptic	n.-s.	n.-s.	-

(Continued on next page)

Zérouga, et al,. 1991	rats	intubation (21 days) membrane	synaptic plasma diet only	↓	↓	- differences found in low n-3

* ↓: decrease   PI: phosphatidylinositol
  ↑: increase   PE: phosphatidylethanolamine
  n.r.: not reported   PS: phosphatidylserine
  n.-s.: non-significant

membranes (Zérouga, *et al.*, 1991), while most studies have reported no effect. Concentrations of 22:6n3 appear to be less sensitive to ethanol effects, with most studies reporting no effect of ethanol exposure, even following several weeks of I.P. injections (Alling, *et al.*, 1984b) or inhalation (LaDroitte, *et al.*, 1984a). On the other hand, one study has reported decreases in 22:6n3 in brain microsome PS and PE in rats fed ethanol in liquid diets for 35 days (Aloia, *et al.*, 1985), while another study has reported decreases in synaptic plasma membrane concentrations following 21 days of intubation in rats fed a diet low in n-3 fatty acids, but not in rats fed diets with high n-3 concentrations (Zérouga, *et al,*. 1991). Therefore, as discussed above regarding fatty acid levels in the blood, dietary PUFA status of the animal may be an important factor in determining the effects of ethanol on brain fatty acid composition.

Interestingly, mice appear to be more susceptible than rats to the effects of ethanol on PUFA metabolism. Mice have been reported to exhibit decreases in both 20:4n6 and 22:6n3 in synaptosomes following ethanol exposure for periods as short as 2 hours (Littleton and John, 1977) although other studies using longer periods of exposure have found decreases only in levels of 22:6n3 (Harris, *et al.*, 1984; Littleton, *et al.*, 1979; Wing, *et al.*, 1982). Again, however, these studies differed in their methodologies, making direct comparisons difficult.

The fact that mice have been reported to exhibit decreases in 22:6n3 following 2 hours of exposure to ethanol vapor is interesting in light of the fact that one study measured fatty acid composition in synaptosomal plasma membranes in rats following a single intubation of a large dose (5 g/kg) of ethanol (Crews, *et al.*, 1983). Although this study reported no significant difference in 22:6n3 levels, there was a distinct trend towards decreases in the treated group. Again, given the small sample size in this study (5-6 rats per group) it is possible that an effect of ethanol would not have been statistically detectable.

Effects of Ethanol Exposure on Development

The past two decades have been a time of increased interest in the effects of ethanol consumption on development. The identification of Fetal Alcohol Syndrome and its devastating neurological and behavioral consequences (Abel, 1982) has led to a few investigations into the effects of ethanol on fatty acid composition in the developing offspring. The fact that the developing brain is a major site of PUFA deposition coupled with the fact that brain development is particularly affected by early ethanol exposure has led a number of researchers to focus specifically on effects on brain composition; these studies are summarized in Table 5. The results of these studies thus far have been inconsistent, perhaps due to their methodological differences. For example, in mice fed moderate amounts of ethanol throughout the last two weeks of pregnancy, The levels of 22:6n3 in individual phospholipids in the offspring were increased both on day 22 and day 32 post-conception (approximately 3 and 13 days after birth, respectively), while levels of 20:4n6 were not affected by ethanol at either age (Wainwright, *et al.*, 1989). Furthermore, there was an interaction between ethanol and the amount of long-chain n-3 PUFA in the maternal diet on 22:6n3 concentrations in the PE fraction on day 22 in that the greatest increase was seen in the group in which the pregnant dams received part of their dietary fat as fish oil, which contains long-chain n-3's. In a separate study using lower concentrations of long-chain n-3 PUFA in the diet, mice given higher ethanol concentrations for the same duration gave birth to offspring with no difference in phospholipid 22:6n3 levels on day 22, but reduced levels on day 32 (Wainwright, *et al.*, 1990). Once again, however, in the PE fraction there was an interaction between the effects of ethanol and those of dietary fat, with the decrease occurring only in the pups of dams supplemented with long-chain n-3 PUFA: in the pups of dams fed safflower oil, which contains no long-chain n-3's, and very little 18:3n3, 22:6n3 levels did not change. In this study, 20:4n6 concentrations in the PC fraction of the brain were increased on day 22 in the offspring prenatally exposed to ethanol regardless of the dietary fat condition, but decreased on day 32. However, the decrease occurred only in the group receiving safflower oil. These two studies were designed specifically to measure interactions between ethanol exposure and dietary fat content, and their results confirm that a number of factors, including the fat composition of the diet, must be taken into account when evaluating the role of ethanol in PUFA metabolism.

In both of the previous studies, ethanol administration ended just prior to parturition, with the brain tissue being collected several days later. Therefore, it cannot be determined whether the changes observed were due directly to ethanol *per se* or were secondary to a developmental consequence of the prior ethanol treatment. In a recent study

Table 5. Effects of Prenatal Ethanol Administration on Polyunsaturated Fatty Acid Composition in Offspring.*

Author	Species	Method of Etoh Administration	Tissue	20:4n6	22:6n3	Comments
Duffy, *et al.*, 1991	rats	liquid diet (3 weeks before mating until 20 days postnatal)	whole brain (10 days postnatal)	n.-s.	n.-s.	-
			(20 days postnatal)	↑	n.-s.	- 20:4n6 higher in PS fraction only
Wainwright, *et al.*, 1989	mice	liquid diet (20% EDC) (gestation day 7-17)	whole brain (day 22 post-conception)	n.-s.	↑	- PC and PE fractions - in PE fraction, greater increase in pups of fish oil-fed than of safflower oil-fed group
			(day 32 post-conception)	n.-s.	↑	- 22:6n3 higher in PC fraction only
Wainwright, *et al.*, 1990	mice	liquid diet (25% EDC) (gestation day 7-17)	whole brain (day 22 post-conception)	↑	n.-s.	- PC fraction only
			(day 32 post-conception)	↓	↓	- 20:4n6 lower only in PC fraction in pups of group not fed long-chain n-3 PUFAonly - 22:6n3 lower in both PC and PE: in PE, lower only in pups of group fed long-chain n-3 PUFA

* ↓: decrease   PC: phosphatidylcholine
  ↑: increase   PE: phosphatidylethanolamine
  n.-s.: non-significant   PS: phosphatidylserine
  EDC: ethanol-derived calories

in rats, the dams were maintained on an ethanol diet throughout lactation, thereby exposing the pups to ethanol, via the mother's milk, up until the time that the tissue was collected (Duffy, *et al.*, 1991). In this study, there was no change in concentrations of 20:4n6 or 22:6n3 in the brain at 10 postnatal days of age while, at 20 days of age, there was an increase in 20:4n6 in the phosphatidylserine fraction. Further research will be needed to elaborate some of the critical factors which must be considered when determining what effect, if any, ethanol exposure during development has on PUFA concentrations in offspring.

INHALATION STUDIES

From the above discussion, it should be apparent that there are many variables that were often uncontrolled that may lead to differences in the measured effects of alcohol on lipid composition in general and on fatty acyl composition, in particular. These include the amount, duration and route of administration of alcohol; the age, animal species, tissue, subcellular fraction and lipid class studied; and a very important but often ignored variable, the diet composition. In relation to the latter, the fatty acyl distribution and fat content as well as the antioxidant vitamin and mineral status are two critical components.

In our laboratory, an animal model of alcoholism has been used with inhalation as the route of administration (Karanian *et al.*, 1986). This method has been shown to produce investigator controllable BAC's with typical levels in the 100-200 mg % range in Sprague Dawley rats. Dependence is achieved within 7 days and withdrawal effects are pronounced after 14 days of continuous exposure (Karanian *et al.*, 1986).

Table 6. Effects of 1 or 14 Days of Ethanol Inhalation on the Fatty Acyl Composition of Rat Liver and Erythrocytes

Fatty Acid	Control	Liver 24 hrs Etoh	14 d Etoh	Control	Erythrocytes 24 hrs Etoh	14 d Etoh
16:0	18.7	20.9	18.6	28.7	29.0	29.5
18:0	14.1	10.0	12.0	12.0	12.9	13.2
18:1	11.0	14.8	14.4	9.3	9.6	9.3
18:2n6	23.0	19.6	28.1	12.1	12.0	11.5
20:4n6	20.7	16.5	14.5	18.1	17.1	18.5
22:6n3	5.2	5.8	5.2	2.0	1.8	2.0

* Data are expressed as weight %. Animals sacrificed after 1 or 14 days of ethanol inhalation had mean BAC's of 173 and 182 mg /dL, respectively. These animals were chow fed and weighed about 200 g at the beginning of the experiment (from Salem, *et al.*, 1987).

Initial observations of the fatty acyl composition of total lipid extracts have indicated that the liver is very responsive to alcohol exposure but that erythrocytes are relatively resistant (Table 6). Alcohol exposure for various durations between 1 and 14 days led to significant differences in acyl composition. In rat liver total lipid extracts, there were pronounced differences in the composition of both essential and non-essential fatty acids with respect to time. For example, after only 24 hours of inhalation 18:0 had decreased and 16:0 and 18:1n9 had increased. The n-6 essential fatty acids, 18:2n6 and 20:4n6 had both declined from control values after only 24 hours of ethanol inhalation. For 20:4n6, 14 days of exposure led to a greater decline, whereas the 18:2n6 level rebounded above control levels. Very little change in the erythrocyte total lipid extracts were observed in this experiment. Of course, more detailed analysis of individual organelles and lipid fractions may uncover compositional effects of ethanol. In any case, it is clear that the liver responds to a much greater degree than erythrocytes to an alcohol challenge and this should be considered when designing or interpreting data from human studies.

A second experiment illustrates both the critical effect of the fatty acyl composition of the diet as well as the varying extent of alterations induced in different lipid classes. In this experiment, animals were fed a semi-synthetic diet containing 10% fat composed of 1% corn oil and 9% of the experimental oil listed in the table (eg., olive or borage oil). Young rats (70 g at onset) were given this diet for three weeks and in the last week were exposed to ethanol inhalation to obtain a mean BAC of 180 mg% at sacrifice. In the total lipid extract derived from animals fed the olive oil-based diet with lower essential fatty acids, ethanol exposure led to a 44% decline in total 20:4n6. In contrast, in a diet (borage oil) containing the 20:4n6 metabolic precursor, 18:3n6, the ethanol-induced decline in 20:4n6 was reduced to 26%. It thus appeared that this higher level of the 20:4n6 precursor was able to partially prevent the 20:4n6 decline caused by alcohol. Examination of the total phospholipid and cholesterol ester data in Table 7 indicates a qualitatively similar situation with the protective effects of the borage oil particularly impressive in the latter lipid class. The triglyceride fraction acyl composition showed little sensitivity to ethanol. In the liver total lipid and cholesterol ester fractions, the borage oil diet also appeared to have a protective action on the ethanol-induced 22:6n3 loss; however, it should be noted that the control level of this fatty acid was very low in the borage oil diet and a further decrease becomes difficult to observe due to a "floor effect". Still, in some cases the 22:6n3 appears to be increased by ethanol in the borage oil group. There were also significant increases in 20:3n6 in the borage oil group induced by ethanol.

Table 7. Modulation by Diet of the Effects of Ethanol Inhalation on Rat Liver Fatty Acyl Composition Within Various Lipid Classes.

	Borage Oil Diet		Olive Oil Diet	
	Control	Ethanol	Control	Ethanol
Total Lipid Extract				
20:4n6	23.6	17.5	12.4	7.0
22:6n3	1.3	2.2	2.7	2.0
Total Phospholipid				
20:4n6	30.6	28.6	26.9	19.5
22:6n3	2.4	5.2	6.3	6.0
Cholesterol Ester				
20:4n6	18.8	14.6	13.6	3.2
22:6n3	0.8	1.0	2.7	0.6

From these studies, it should be apparent that a variety of variables must be carefully controlled if reliable, precise, reproducible and comparable data are to be obtained by investigators in different laboratories. Of course, much research will be required to define the differences that are to be expected in lipid and fatty acyl composition in different animal models of alcoholism, the mechanisms involved and the pathological consequences for various organ systems.

REFERENCES

Abel, E. A., 1982, Consumption of alcohol during pregnancy: A review of effects on growth and development of offspring. *Human Biol.* 54,421.

Alling, A., and Gustavsson, L., 1986, Effects of ethanol on concentration and acyl group composition of acidic phospholipids, *in:* "Phospholipid Research and the Central Nervous System: Biochemical and Molecular Pharmacology," L. A. Horrocks, L. Freysz, and G. Toffano, eds., Liviana Press, Padova.

Alling, A., Becker, W., Jones, A. W., and Änggärd, E., 1984a, Effects of chronic ethanol treatment on lipid composition and prostaglandins in rats fed essential fatty acid deficient diets. *Alcohol. Clin. Exp. Res.* 8:238.

Alling, C., Gustavsson, L., Kristensson-Aas, A., and Wallerstedt, S., 1984b, Changes in fatty acid composition of major glycerophospholipids in erythrocyte membranes from chronic alcoholics during withdrawal. *Scand. J. Clin. Lab. Invest.* 44:283.

Alling, C., Liljequist, S., and Engel, J., 1982, The effect of chronic ethanol administration on lipids and fatty acids in subcellular fractions of rat brain. *Med. Biol.* 60:149.

Aloia, R. C., Paxton, J., Daviau, J. S., Van Gelb, O., Mlekusch, W., Truppe, W., Meyer, J. A., and Braver, F. S., 1985, Effect of chronic alcohol consumption on rat brain microsome lipid composition, membrane fluidity and Na^+-K^+-ATPase activity. *Life Sci.* 36:1003.

Alvaro, D., Angelico, M., Attili, A. F., De Santis, A., Piéche, U., and Capocaccia, L., 1982, Abnormalities in erythrocyte membrane phospholipids in patients with liver cirrhosis. *Biochem Med.* 28:157.

Arai, M., Gordon, E. R., and Lieber, C. S., 1984, Decreased cytochrome oxidase activity in hepatic mitochondria after chronic ethanol consumption and the possible role of decreased cytochrome aa_3 content and changes in phospholipids. *Biochim. Biophys. Acta.* 797:320.

Benedetti, A., Birarelli, A. M., Brunelli, E., Curatola, G., Ferretti, G., Del Prete, U., Jezequel, A. M., and Rolandi, F., 1987, Modification of lipid composition of erythrocyte membranes in chronic alcoholism. *Pharmacol. Res. Commun.* 19:651.

Chin, J. H., and Goldstein, D. B., 1985, Effects of alcohols on membrane fluidity and lipid composition, *in:* "Membrane Fluidity in Biology," R. C. Aloia and J. M. Boggs, eds., Academic Press, New York.

Clemens, M. R., 1988, The relationship between lipid composition and lipid peroxidation of erythrocytes from alcoholics. *Adv. Biosci.* 71:101.

Corbett, R., Berthou, F., Leonard, B. E., and Ménez, J.-F., 1992, The effects of chronic administration of ethanol on synaptosomal fatty acid composition: Modulation by oil enriched with gamma-linolenic acid. *Alcohol Alcohol.* 27:11.

Corbett, R., Floch, H. H., Ménez, J.-F., and Leonard, B. E., 1991, The effects of chronic ethanol administration on rat liver and erythrocyte lipid composition: Modulatory role of Evening Primrose Oil. *Alcohol Alcohol.* 26:459.

Crews, F. T., Majchrowicz, E., and Meeks, R., 1983, Changes in cortical synaptosomal plasma membrane fluidity and composition in ethanol-dependent rats. *Psychopharmacology.* 81:208.

Cunnane, S. C., Manku, M. S., and Horrobin, D. F., 1985, Effect of ethanol on liver triglycerides and fatty acid composition in the golden Syrian hamster. *Ann. Nutr. Metab.* 29:246.

Cunnane, S. C., McAdoo, K. R., and Horrobin, D. F., 1987, Long-term ethanol consumption in the hamster: Effects on tissue lipids, fatty acids and erythrocyte hemolysis. *Ann. Nutr. Metab.* 31:265.

Cunningham, C. C., Bottenus, R. E., Spach, P. I., and Rudel, L.L., 1983, Ethanol-related changes in liver microsomes and mitochondria from the monkey, *Macaca fascicularis. Alcohol. Clin. Exp. Res.* 7:424.

Cunningham, C. C., Filus, S., Bottenus, R. E., and Spach, P. I., 1982, Effect of ethanol consumption on the phospholipid composition of rat liver microsomes and mitochondria. *Biochim. Biophys. Acta.* 712:225.

Cunningham, C. C., Sinthusek, G., Spach, P. I., and Leathers, C., 1981, Effect of dietary ethanol and cholesterol on phospholipid composition of hepatic mitochondria and microsomes from the monkey, Macaca nemestrina. *Alcohol. Clin. Exp. Res.* 5:417.

Cunningham, C. C., and Spach, P. I., 1987, The effect of chronic ethanol consumption on the lipids in liver mitochondria. *Annals N.Y. Acad.Sci.* 492:181.

Curstedt, T., 1982, Biosynthesis of molecular species of hepatic glycerophosphatides during metabolism of $(1,1\text{-}^2H_2)$ ethanol in rats. *Biochim. Biophys. Acta.* 713:589.

Duffy, O., Menez, J-F., Floch, H. H., and Leonard, B. E., 1991, Changes in whole brain membranes of rats following pre- and post-natal exposure to ethanol. *Alcohol Alcohol.* 26:605.

Ellingson, J. S., Janes, N., Taraschi, T. F., and Rubin, E., 1991, The effect of chronic ethanol consumption on the fatty acid composition of phosphatidylinositol in rat liver microsomes as determined by gas chromatography and ^1H-NMR. *Biochim. Biophys. Acta.* 1062:199.

Engler, M. M., Karanian, J. W., and Salem, Jr., N., 1991, Ethanol inhalation and dietary n-6, n-3, and n-9 fatty acids in the rat: Effect on platelet and aortic fatty acid composition. *Alcohol. Clin. Exp. Res.* 15:483.

French, S. W., Ihrig, T. J., and Morin, R. J., 1970, Lipid composition of RBC ghosts, liver mitochondria and microsomes of ethanol-fed rats. *Quart. J. Stud. Alc.* 31:801.

French, S. W., Ihrig, T. J., Shaw, G. P., Tanaka, T. T., and Norum, M. L., 1971, The effect of ethanol on the fatty acid composition of hepatic microsomes and inner and outer mitochondrial membranes. *Res. Comm. Chem. Path. Pharmacol.* 2:567.

Glen, E., MacDonnell, L., Glen, I., and MacKenzie, J., 1984, Possible pharmacological approaches to the prevention and treatment of alcohol-related CNS impairment: Results of a double-blind trial of essential fatty acids, *in:* "Pharmacological Treatments for Alcoholism," G. Edwards, and J. Littleton, eds., Methuan, New York.

Gustavsson, L., and Alling, C., 1989, Effects of chronic ethanol exposure on fatty acids of rat brain glycerophospholipids. *Alcohol.* 6:139.

Hansson, E., Gustavsson, L., Jönsson, G., Alling, C., and Rönnbäck, L., 1987, Astroglial primary cultures: A model to study ethanol effects on the cell membrane lipid composition. *Alcohol Alcohol.* Suppl. 1:679.

Harris, R. A., Baxter, D. M., Mitchell, M. A., and Hitzemann, R. J., 1984, Physical properties and lipid composition of brain membranes from ethanol tolerant-dependent mice. *Molec. Pharmacol.* 25:401.

Holman, R. T., and Johnson, S., 1982, Changes in essential fatty acid profile of serum phospholipids in human disease. *Prog. Lipid Res.* 20:67.

John, G. R., Littleton, J. M., and Jones, P. A., 1980, Membrane lipids and ethanol tolerance in the mouse. The influence of dietary fatty acid. *Life Sci.* 27:545.

Johnson, S. B., Gordon, E., McClain, C., Low, G., and Holman, R. T., 1985, Abnormal polyunsaturated fatty acid patterns of serum lipids in alcoholism and cirrhosis: Arachidonic acid deficiency in cirrhosis. *Proc Nat. Acad. Sci. USA.* 82:1815.

Karanian, J.W., Yergey, J., Lister, R., D'Souza, N., Linnoila, M., and Salem, Jr., N., 1986, Characterization of an automated apparatus for control of inhalation chamber ethanol vapor and blood ethanol concentrations. *Alcohol. Clin. Exp. Res.* 10:443.

Keppel, G., 1973, "Design and Analysis: A Researcher's Handbook," Prentice-Hall, Inc., Englewood Cliffs.

Koblin, D. D., and Deady, J. E., 1981, Sensitivity to alcohol in mice with an altered brain fatty acid composition. *Life Sci.* 28:1889.

La Droitte, P., Lamboeuf, Y., and De Saint-Blanquat, G., 1984a, Lipid composition of the synaptosome and erythrocyte membranes during chronic ethanol-treatment and withdrawal in the rat. *Biochem. Pharmacol.* 33:615.

La Droitte, P., Lamboeuf, Y., and Saint Blaquat, G., 1984b, Membrane fatty acid changes and ethanol tolerance in rat and mouse. *Life Sci.* 35:1221.

Lesch, P., Schmidt, E., and Schmidt, F. W., 1973, Effects of chronic alcohol abuse on the fatty acid composition of major lipids in the human brain. *Z. Klin. Chem. Klin. Biochem.* 11:159.

Lieber, C. S., DeCarli, L. M., and Sorrell, M. F., 1989, Experimental models of ethanol administration. *Hepatology.* 10:501.

Littleton, J. M., and John, G. R., 1977, Synaptosomal membrane lipids of mice during continuous exposure to ethanol. *J. Pharm. Pharmnacol.* 29:579.

Littleton, J. M., John, G. R., and Grieve, S. J., 1979, Alterations in phospholipid composition in ethanol tolerance and dependence. *Alcohol. Clin. Exp. Res.* 3:50.

Marzo, A., Ghirardi, P., Sardini, D., Prandini, B. D., and Albertini, A., 1970, Serum lipids and total fatty acids in chronic alcoholic liver disease at different stages of cell damage. *Klin. Wschr.* 48:949.

Mendenhall, C. L., Bradford, R. H., and Furman, R. H., 1969, Effect of ethanol on fatty acid composition of hepatic phosphatidylcholine and phosphatidylethanolamine and on microsomal fatty acyl-CoA: Lysophosphatide transferase activities in rats fed corn oil or coconut oil. *Biochim. Biophys. Acta.* 187:510.

Miceli, J. N., and Ferrell, W. J., 1973, Effects of ethanol on membrane lipids III. Quantitative changes in lipid and fatty acid composition of nonpolar and polar lipids of mouse total liver, mitochondria and microsomes following ethanol feeding. *Lipids.* 8:722.

Pirozhkov, S. V., Eskelson, C. D., Watson, R. R., Hunter, G. C., Piotrowski, J. J., and Bernhard, V., 1992, Effect of chronic consumption of ethanol and vitamin E on fatty acid composition and lipid peroxidation in rat heart tissue. *Alcohol.* 9:329.

Rao, G. A., Goheen, S. C., and Larkin, E. C., 1981, Changes in relative levels of linoleate to arachidonate in erythrocyte phosphatidylcholine in rats fed ethanol and arachidonate. *Toxicol. Letters.* 7:469.

Reitz, R. C., 1975, A possible mechanism for the peroxidation of lipids due to chronic ethanol ingestion. *Biochim. Biophys. Acta.* 380:145.

Reitz, R. C., 1984, Relationship of the acyl-CoA desaturases to certain membrane fatty acid changes induced by ethanol consumption. *Proc. West. Pharmacol. Soc.* 27:247.

Reitz, R. C., Wang, L., Schilling, R. J.,S tarich, G. H., Bergstrom, J. D., and Thompson, J. A., 1981, Effects of ethanol ingestion on the unsaturated fatty acids from various tissues. *Prog. Lipid Res.* 20:209.

Salem, Jr., N., Yoffe, A., Kim, H-Y., Karanian, J. W., and Taraschi, T. F., 1987, Effects of fish oils and alcohol on polyunsaturated lipids in membranes. *in:* "Proceedings of the AOCS Short Course on Polyunsaturated Fatty Acids and Eicosanoids," W.E.M. Lands, ed., American Oil Chemists' Society, Champaign, Illinois.

Smith, T. L., and Gerhard, M. J., 1982, Alterations in brain lipid composition of mice made physically dependent to ethanol. *Life Sci.* 31:1419.

Sun, G. Y., and Sun, A. Y., 1979, Effect of chronic ethanol administration on phospholipid acyl groups of synaptic plasma membrane fraction isolated from guinea pig brain. *Res. Comm. Chem. Path. Pharmacol.* 24:405.

Sun, G. Y., and Sun, A. Y., 1985, Ethanol and membrane lipids. *Alcohol. Clin. Exp. Res.* 9:164.

Szebeni, J., Eskelson,, C., Sampliner, R., Hartmann, B., Griffin, J., Dormandy, T., Watson, R. R., 1986, Plasma fatty acid pattern including diene-conjugated linoleic acid in ethanol users and patients with ethanol-related liver disease. *Alcohol. Clin. Exp. Res.* 10:647.

Thompson, J. A., and Reitz, R. C., 1976, Studies on the acute and chronic effects of ethanol ingestion on choline oxidation. *Annal. N. Y. Acad. Sci.* 273:194.

Thompson, J. A., and Reitz, R. C., 1978, Effects of ethanol ingestion and dietary fat levels on mitochondrial lipids in male and female rats. *Lipids.* 13:540.

Wainwright, P. E., Huang, Y. S., Mills, D. E., Ward, G. R., Ward, R. P., and McCutcheon, D., 1989, Interactive effects of prenatal ethanol exposure and n-3 fatty acid supplementation on brain development in mice. *Lipids.* 24:989.

Wainwright, P. E., Huang, Y-S., Simmons, V., Mills, D. E., Ward, R. P., Ward, G. R., Winfield, D., and McCutcheon, D., 1990, Effects of prenatal ethanol and long-chain n-3 fatty acid supplementation on development in mice. 2. Fatty acid composition of brain membrane phospholipids. *Alcohol. Clin. Exp. Res.* 14:413.

Wang, D. L., and Reitz, R. C., 1983, Ethanol ingestion and polyunsaturated fatty acids: Effects on the acyl-CoA desaturases. *Alcohol. Clin. Exp. Res.* 7:220.

Waring, A. J., Rottenberg, H., Ohnishi, T., and Rubin, E., 1981, Membranes and phospholipids of liver mitochondria from chronic alcoholic rats are resistant to membrane disordering by alcohol. *Proc. Nat. Acad. Sci. USA.* 78:2582.

Watson, R. R., and Watzl, B, eds., 1992, "Nutrition and Alcohol," CRC Press, Boca Raton.

Wing, D. R., Harvey, D. J., Hughes, J., Dunbar, P. G., McPherson, K. A., and Paton, W. D. M., 1982, Effects of chronic ethanol administration on the composition of membrane lipids in the mouse. *Biochem. Pharmacol.* 31: 3431.

Zérouga, M., Beaugé, F., Niel, E., Durand, G., and Bourre, J. M., 1991, Interactive effects of dietary (n-3) polyunsaturated fatty acids and chronic ethanol intoxication on synaptic membrane lipid composition and fluidity in rats. *Biochim. Biophys. Acta.* 1086:295.

REVELANT EFFECTS OF DIETARY POLYUNSATURATED FATTY ACIDS ON SYNAPTIC MEMBRANE RESPONSES TO ETHANOL AND CHRONIC ALCOHOL INTOXICATION

Françoise J. Beaugé

Centre de Recherche Pernod-Ricard
94015 Créteil Cedex
France

INTRODUCTION

Well-defined alterations in synaptic membrane fluidity were shown to be actual clues of tolerance to and dependence on ethanol (Beaugé et al., 1990). Dietary polyunsaturated fatty acids (PUFA) influence the brain membrane fatty acid pattern and can disturb membrane associated events (Bourre et al., 1984, Sun and Sun, 1985).

The effects of various dietary oils such as : soya, sunflower, cod liver, safflower and evening primrose oils were studied on synaptic membrane responses to acute and chronic alcohol intoxication. The oils differed mainly by their amount in alpha-linolenate (Corbett et al., 1991; Zérouga et al., 1991). They were given either separately (pharmacological study) during a short term period or in the diet (nutritional study) over two rat generations.

ANIMAL TREATMENT

PHARMACOLOGICAL STUDY

The safflower and evening primrose oils were used in this study. Male adult Srague-Dawley received a milk solution as a component of the dietary regimen. With respect to alcohol groups, ethanol was added to this solution. Ad libitum access to a standard laboratory chow was maintained. The control rats were pair-fed with isocaloric milk diet and the milk solution was the only fluid provided in the case of all the groups. The oil enrichment was administered p.o. as appropriate at a dose of 1 ml oil/kg, once daily. Reference groups (control and ethanol-treated) received 1ml/kg saline p.o. (Corbett and Leonard, 1984). The animals received oils or saline over two weeks before the initiation of milk or milk alcohol regimen. Milk solutions were supplemented with a vitamin B complex enriched mixture such that rats would obtain minimum daily requirement (e.g: of thiamine HCl). The safflower and evening primrose oils contain about 75% of linoleate and are totally devoid of alpha-linolenate as sunflower oil but evening primrose oil contains 9.8% of gamma-linolenate (Corbett et al., 1991).

This dietary regimen was maintained for a period of 20 weeks. Body weights and dietary intakes were monitored regularly. Housing environnement was maintained at 21°C±1°C. The light cycle was from 08.00 to 20.00 hrs. At the end of this period, the rats were killed by decapitation in the morning and brains excised immediately.

NUTRITIONAL STUDY

Soya, sunflower and cod liver oils were used in this second study. Male adult Wistar from the 2rd generation totally fed with controlled semi-synthetic diets, were used throughout. The diets were called: Soya (S), Sunflower (SF), Cod liver (CL) obtained from INRA, Jouy en Josas and a "Standard" (usual lab chow from Extra-Labo, Ets Pietrement, France). S contains soybean oil (1.9%/100g, i.e.: 955 mg of n-6 FA and 130 mg n-3 FA per 100g diet); CLO contains sunflower oil (2v/3v) and cod liver oil (1v/3v) (2.0%/100g, i.e.: 925 mg of n-6 FA and 150 mg n-3 FA per 100g diet); SF contains sunflower oil (1.6%/100g, i.e.: 1035 mg of n-6 FA and 3 mg n-3 FA per 100g diet). These two latter diets are deficient in alpha linolenate but the CL diet is supplemented in long chain n-3 FA (EPA, DHA). The three diets supplied equivalent amounts of n-6 FA. The "Standard" diet contains various animal and vegetal oils (5.0%/100g, i.e.: 1433 mg of n-6 FA and 103 mg n-3 FA per 100g diet).

Ethanol was administered daily to alcohol-treated animals for 21 days by gavage in increasing dosages (3 to 6 g ethanol/kg bwt) in order to counteract metabolic tolerance and to achieve comparable maximum blood alcohol levels over the study. Paired control diet animals received a calorical equivalent to ethanol as a starch solution.

After 3 weeks of ethanol treatment, the animal functional tolerance was determined by the importance of the hypothermic response (maximal fall in body temperature: ΔTmax °C) after a challenge dose of ethanol in both control and alcohol-treated animals (Beaugé et al., 1984).

The rats were sacrified by decapitation 20-24 hrs later when the blood alcohol level was lower than 5mM (Beaugé, 1991).

METHODS

The synaptic membranes from brain cortices were prepared by differential ultracentrifugation as previously described (Beaugé et al., 1988; Zérouga et al., 1991; Beaugé et al., 1992).

Membrane fluidity or microorganization was assessed on aliquots of membrane suspensions after protein determination, by steady-state fluorescence polarization of DPH and its polar derivatives: TMA-DPH and PROP-DPH (Beaugé et al., 1988; Beaugé et al., 1992). The degree of fluorescence polarization P was determined with T-format Fluofluidimetre (SEFAM, Nancy, France) at 37°C for the pharmacological study and 25°C for the nutritional study. P is related to the order parameter of the membrane lipids. High P values represent high structural order or low membrane fluidity and vice versa. Furthermore, the membrane sensitivity (or tolerance) to ethanol was determined by the linear variation of P after acute in vitro addition of ethanol (0.1-1 M). The slope of the regression line: ΔP thus obtained was taken as a measure of membrane tolerance to ethanol (Beaugé et al, 1988; Beaugé, 1991; Beaugé et al.,1992).

STATISTICAL ANALYSES

Membranes from naive and ethanol-treated animals fed the same way, were prepared on the same day, analysed at the same time and paired. Data were expressed as means and the standard deviation were calculated. The resulting means were compared by Student's t test.

RESULTS

PHARMACOLOGICAL STUDY

No significant differences could be seen between groups in body weight profiles and calorific intakes. The percentage of alcohol intake expressed as percent of total calorific intake was similar in all alcohol groups over the course of the study. As in previous works (Corbett et al., 1986; Duffy et al., 1992) carried out with the same animal treatment, the consumption of alcohol among the animals ranged between 10 and 15g/kg bwt. All the 3 alcohol groups: reference, safflower and evening primrose seemed to be equally intoxicated with a blood alcohol level at 20 weeks of treatment of 1.4 to 1.8 g/l.

The fluidity parameters are stated in Table 1.

The results from the reference groups were very similar to those obtained previously after various alcohol treatments (Beaugé et al., 1988; 1992; Zérouga et al., 1991; Zérouga and Beaugé, 1992). Tolerance to the ethanol acute fluidizing effect appeared at the DPH and TMA-DPH levels.

Table 1. Steady-state fluorescence polarization of DPH probes in synaptic membranes. Pharmacological study.

DIETS	Reference		Safflower		Evening Primrose	
Treatment	Control	Alcohol	Control	Alcohol	Control	Alcohol
P(DPH)	.271	.279	.272	.284*	.294°	.287
ΔP(DPH)	.0257	.0195*	.0272	.0225	.0208	.0147*
P(TMA)	.311	.306	.306	.314	.311	.313
ΔP(TMA)	.0062	.0035*	.0064	.0034	.0050	.0043
P(PROP)	.348	.356*	.342	.346	.344	.349
ΔP(PROP)	.0138	.0102	.0192	.0245	.0165	.0158

Results are means of 6 P determinations at 37°C for each of the 5 different rat membrane preparations in each diet group.
The standard deviations were below 0.005 for P values and 0.0025 for ΔP.
Statistics according to the Student 's t test :
  *: p < 0.01, comparison alcohol vs control
  °: p < 0.01, comparison evening primrose control vs reference control.

The control evening primrose treated animals showed synaptic membranes more rigid (+8%) and less sensitive to an acute dose of ethanol in the apolar part of the membrane bilayer than the reference and safflower treated rats. During the chronic ethanol intoxication duration, evening primrose rats developed a clear-cut membrane tolerance at the DPH level without any other changes while the safflower rats exhibit an increase in membrane rigidity without statistically significant membrane tolerance to ethanol and disturbances at the polar level, similar to what was found with other alpha-linolenate deficient diets (Zérouga et al., 1991; Beaugé et al., 1992).

It has been said that evening primrose oil administration could conteract the reduction of polyunsaturated fatty acids caused by ethanol in brain membranes (Corbett et al., 1986). Furthermore, administration of this oil seems to stabilize the membrane without interfering with the adaptive response to ethanol.

NUTRITIONAL STUDY

Food consumption, calorific intake and weight gain were similar in the three dietary groups and comparable with those elicited by the standard lab chow.

The results concerning the fluidity parameters after the treatment are shown in Table 2.

Table 2. Steady-state fluorescence polarization of DPH probes in synaptic membranes. Nutritional study.

DIETS	Standard		Soya		Sunflower		CL	
Treatment	C	A	C	A	C	A	C	A
P(DPH)	.331	.345*	.336	.337	.327	.339*	.325	.340*
ΔP(DPH)	.0240	.0150*	.0214	.0152*	.0233	.0198	.0209	.0217
P(TMA)	.359	.351*	.350	.351	.357	.364	.352	.351
ΔP(TMA)	.0112	.0062*	.0048	.0029*	.0062	.0051	.0118	.0083
P(PROP)	.368	.374	.372	.373	.372	.377	.369	.376
ΔP(PROP)	.0111	.0071	.0087	.0065	.0062	.0071	.0119	.0088

C: control diet (5-7 animals); A: alcohol-treated (5-7 animals).
Fluorescence polarization determinations at 25°C.
Results as in Table 1.

The fine tuning of the brain membrane fatty acid profiles according to the different lipid intake (Greenwood et al., 1989) went well with an apparent maintenance of the overall fluidity at each probed level in the synaptic membrane. Slight differences (Beaugé et al., 1988) existed between rats fed soya and sunflower diets at the DPH and TMA-DPH levels. The alpha-linolenate deficient diet gave more fluid and more ethanol sensitive membranes. It has to be noted that the region probed by PROP-DPH is definitely different from that probed by TMA-DPH, i.e.: more rigid and less sensitive to ethanol.

Despite of the limited diet-related alterations in the synaptic membranes, clear-cut differences were noted in response to chronic ethanol intoxication. They were more marked with this kind of alcohol treatment than in the first study. P(DPH) increased by about 4% after alcoholization but with the concomitant soya diet. We were able to show (Zérouga et al., 1991) that this increase could be put together with slight but significant changes in the membrane double bound index and total percentage in polyunsaturated fatty acids which both decreased after alcoholization in the rats fed these three diets.

The membrane intrinsic rigidification has been previously related to a behavioral aspect of dependence (Le Bourhis et al., 1986). We found in separate experiments that a high percentage (more than 70%) of the animals fed the sunflower and the cod liver diets developed dependence versus 50% of the animal fed the soya diet.

The most striking differences in the fluidity responses were observed in the changes in membrane sensitivity or tolerance to alcohol. The standard and the soya diets allowed the development of the membrane tolerance to ethanol at the DPH and TMA-DPH and eventually PROP-DPH levels but neither the sunflower nor the so called CL diets. It is noteworthy that the soya and standard diets let alcohol induce increases in membrane cholesterol content and maintained a very equilibriated ratio n-6/n-3 (about 1) (Beaugé et al., 1992).

The membrane resistance or tolerance has been previously related to functional tolerance to ethanol, i.e. to the importance of the hypothermic response to ethanol after alcoholization (Beaugé, 1991). The Fig. 1 shows that the animals fed the sunflower or the cod liver diets did not developed tolerance to ethanol as well functionally as at the synaptic membrane level, contrary to those fed standard or soya diets.

Concurrently to variations in membrane fluidity parameters, resulting at least partly from dietary lipid modulation, changes in behaviors related to tolerance and dependence towards alcohol could be evidenced.

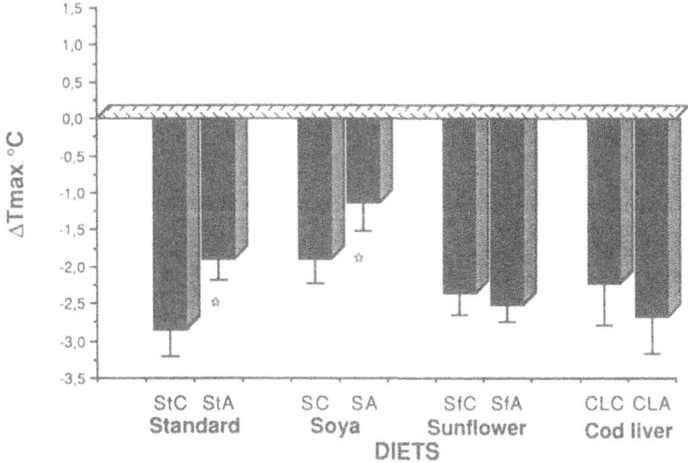

Figure 1. Maximal drop in body temperature following an acute i.p. administration of a challenge dose (3g/kg) of ethanol to chronic aly intoxicated animals. ΔT (°C) is the drop in temperature. Results are expressed as mean values and the bars represent the standard deviations. The animals are the same as in Table 2.

CONCLUSION

Taken together these two studies emphasize the importance of exogeneous lipid supply on synaptic membrane sensitivity and animal response to ethanol.

Whatever the protocol of administration, pharmacological or nutritional lipid modulations can alter brain membrane adaptation to ethanol and interfere with nervous tolerance development and probably with dependence acquisition.

A positive protective effect can apparently be obtained with dietary oils attenuating or preventing the alterations caused by alcohol intoxication in the polyunsaturated fatty acid composition and metabolism in the neuronal membranes. A structurally specific role of polyunsaturated fatty acids is also suggested by the negative effects of alpha-linolenate deficient diets partly counteracted by the presence of an isomer like gamma-linolenate (Ehringer et al., 1991). The importance of ethanol effects on brain membrane lipids is thus indirectly reappraised. More research is needed to explore the values of lipid nutrients along or not with other pharmalogical agents in preventing, alleviating or reversing ethanol addiction (Glen et al., 1987).

AKNOWLEDGMENTS

This work was supported by IREB (grant 88-09) and INSERM (U26). The author wishes to thank E. Meehan (University College, Galway, Ireland), M. Zérouga (U 26, Paris, France) and D. Choquart (CR PR, Créteil, France) for their contributions to the present work.

REFERENCES

Beaugé, F., Fleuret-Balter, C., Nordmann, J. and Nordmann, R., 1984, Brain membrane sensitivity to ethanol during development of functional tolerance to ethanol in rats, *Alcohol clin. exp. Res.* 8:67-171.

Beaugé, F., Zérouga, M., Niel, E., Durand, G. and Bourre, J.M., 1988, Effects of dietary linolenate/linoleate balance on neuronal membrane sensitivity to ethanol, in : "Biomedical and Social Aspects of Alcohol and Alcoholism," K. Kuriyama, A. Takeda and H. Ishui, eds, Elsevier Science Publishers, Amsterdam.

Beaugé, F., Aufrère, G., Niel, E., Zérouga, M. and Le Bourhis, B., 1990, Corrélats biophysiques membranaires de la tolérance et de la dépendance envers l'alcool, *Drug Alcohol Depend.* 25:57-65.

Beaugé, F., 1991, Physico-chemical membrane alterations during ethanol intoxication, *Alcohol Alcohol.* Suppl.1:233-239.

Beaugé, F., Zérouga, M., Durand, G. and Bourre, J.-M., 1992, Influence of dietary fatty acid balance on the development of tolerance during chronic ethanol intoxication in rats, *Alcohol Alcohol.* in press.

Bourre, J.-M., Pascal, G., Durand, G., Masson, M., Dumont, O. and Piciotti, M., 1984, Alterations in the fatty acid composition of brain cells (neurons, astrocytes and oligodendrocytes) and of subcellular fractions (myelin and synaptosomes) induced by a diet devoid of (n-3) fatty acids, *J. Neurochem.* 43:342-348.

Corbett, R. and Leonard B.E., 1984, Effects of carnitine on changes by chronic administration of alcohol, *Neuropharmacol.* 23:269-271.

Corbett, R., Ménez, J.F., Leonard, B.E. and Floch, H.H., 1986, Changes in neutral lipids in brain synaptosomes of rats after chronic administration of ethanol, *Neurosci. Lett.* 69:198-202.

Corbett, R., Ménez, J.-F., Floch, H. H. and Leonard, B. E., 1991, The effects of chronic ethanol administration on rat liver and erythrocyte lipid composition: modulatory role of 'evening primrose oil', *Alcohol Alcohol.* 26:459-461.

Duffy, O., Ménez, J.-F. and Leonard, B. E., 1992, Effects of an oil enriched in gamma linolenic acid on locomotor activity and behaviour in the Morris Maze, following in utero ethanol exposure in rats, *Drug Alcohol Depend.* 30:65-70.

Ehringer, W., Belcher, D., Wassall, S.R. and Stillwell, W., 1991, A comparison of the effects of α-linolenic and γ-linolenic acid in phospholipid bilayers, *Chem. Phys. Lipids* 57:87-96.

Glen, I., Skinner, F., Glen, E. and McDonnell, L., 1987, The role of essential fatty acids in alcohol dependence and tissue damage, *Alcohol. clin. exp. Res.* 11:37-41.

Greenwood, C.E., Mc Gee, C.D. and Dyers, J.R.,1989, Influence of dietary fat on brain membrane phospholipid fatty acid composition and neuronal function in mature rats, *Nutrition* 5:278-281.

Le Bourhis, B., Beaugé, F., Aufrère, G. and Nordmann, R., 1986, Membrane fluidity and alcohol dependence, *Alcohol. clin. exp. Res.* 10:337-342.

Sun, G. Y. and Sun, A. Y., 1985, Ethanol and membrane lipids, *Alcohol. clin. exp. Res.* 9:164-180.

Zérouga, M., Beaugé, F., Niel, E., Durand, G. and Bourre, J. M., 1991, Interactive effects of dietary (n-3) PUFA and chronic ethanol intoxication on membrane lipid composition and fluidity in rats, *Biochim. Biophys. Acta* 1086:295-304.

Zérouga, M. and Beaugé, F., 1992, Rat synaptic membrane fluidity and sensitivity to ethanol after intermittent exposures to ethanol in vivo, *Alcohol* 9:311-315.

CHRONIC ETHANOL TREATMENT INCREASES THE INCORPORATION OF [¹⁴C]-SERINE INTO PHOSPHATIDYLSERINE IN NEUROBLASTOMA X GLIOMA CELLS

F. David Rodríguez[1], Christer Alling[2] and Lena Gustavsson[2]

[1]Dept. of Biochemistry and Molecular Biology, University of
Salamanca, 37007 Salamanca, Spain
[2]Dept. of Psychiatry and Neurochemistry, Lund University,
P.O. Box 638, S-220 09 Lund, Sweden

INTRODUCTION

The phospholipid base-exchange reactions catalyze the calcium dependent, non-energy requiring incorporation of serine, ethanolamine and choline into phosphatidylserine (PS), phosphatidylethanolamine (PE) and phosphatidylcholine (PC), respectively (Porcellati et al., 1971; Kanfer, 1972). This may be the only pathway for the *de novo* synthesis of PS in mammalian cells, whereas different pathways exist for the synthesis of PC and PE. PS can be translocated to the mitochondria where it is decarboxylated to form PE (Butler and Morell, 1983). Moreover, progressive methylation of PE leads to PC synthesis at the endoplasmic reticulum level. Thus, base-exchange reactions, together with decarboxylation of PS and subsequent methylation of PE to form PC may play a key role in both the net synthesis of PS and the remodeling of pre-existing phospholipids (Bjerve, 1985; Voelker, 1990; Vance, 1991). Experimental evidence suggests that base-exchange reactions are regulated by phosphorylation-dephosphorylation processes (Kanfer et al., 1988), long chain bases such as oleoylamine and sphingosine (Singh et al., 1992), unsaturated fatty acids (Kanfer and McCartney, 1991) and phospholipids (Corazzi et al., 1991). The three base-exchange activities are present in rat brain synaptosomal plasma membranes (Holbrook and Wurtman, 1988) and plasma membranes isolated from rat liver (Siddiqui and Exton, 1992). Recently, it has been shown that phospholipid base-exchange activity can be regulated upon G-protein and receptor activation (Siddiqui and Exton, 1992). These findings indicate that base-exchange activities may play a role in signal transduction-related events.

Several reports have focused on the effects of ethanol on the phospholipid composition of neuronal cell membranes. Sun et al. (1983, 1984) showed that chronic ethanol treatment induces an augmentation of negatively charged phospholipids in synaptic plasma membranes of rodents. We have also found an increase in the amount of acidic phospholipids in different organs of ethanol-fed rats (Alling et al., 1984; Gustavsson and Alling, 1987). However, Magruder et al. (1985) reported a decrease in the amount of anionic phospholipids after chronic ethanol treatment. On the other hand, we have shown that binge but not continuous alcohol exposure induces a decrease of acidic phospholipids

Alcohol, Cell Membranes, and Signal Transduction in Brain
Edited by C. Alling *et al.*, Plenum Press, New York, 1993

in neuroblastoma x glioma cells maintained in culture (Alling et al., 1991). The discrepancies found in the literature may be due to the use of different experimental models as well as the pattern of exposure and ethanol concentration. By measuring the mass of phospholipids it may not be possible to detect a specific alteration induced by ethanol on a defined target, such as an enzyme implicated in a given phospholipid metabolic pathway. In this study we have investigated the effect of acute and chronic ethanol treatment on the incorporation of radioactive serine into PS in a hybrid neuroblastoma x glioma tumor cell line (NG 108 15).

MATERIALS AND METHODS

Cell Cultivation and Ethanol Exposure

NG 108 15 cells (passage 20-40) were cultivated in 35 mm diameter plastic dishes (NUNC S/A, Denmark) containing 1 ml of Dulbecco's modified Eagle's medium supplemented with 5 % fetal calf serum, 2 mM l-glutamine, 50 x HAT (final concentration: hypoxantine 0.1 mM, aminopterine 4μM and thymidine 16 μM), 100 μg/ml streptomycin and 100 IE/ml penicillin (Simonsson et al., 1989). Antibiotics were purchased from Astra Pharmaceuticals, Sweden; other cell culture chemicals were from Flow Laboratories (U.K.). Cells were kept at 37 ° C in an incubator under a humidified atmosphere containing 10 % CO_2 / 90 % air. Medium was changed every day. Ethanol (final concentration 100 mM) was added directly to the medium. The dishes were placed inside tightly caped plastic boxes together with an open dish containing an appropriate amount of ethanol. By using this method ethanol concentrations were maintained constant over time (Rodríguez et al., 1992)

Incorporation of [^{14}C]-Serine into Phosphatidylserine

Control and ethanol-treated cells were labelled with 0.5 μCi/ml of L-[^{14}C(U)]-serine (specific radioactivity 180.5 mCi/mmol) for different periods of time. A group of dishes were radiolabelled in the presence of 100 mM ethanol in order to investigate possible acute effects of ethanol on the incorporation of label into phospholipids. In some experiments, the cells were radiolabelled for 180 min. Afterwards, the label was withdrawn and the radioactivity present in PS determined after different periods of time. These experiments were also carried out in control and ethanol-treated cells. At the indicated time-points, cells were washed with Hank's balanced salt solution (without Mg^{2+} and Ca^{2+}) and harvested in ice-cold 50 mM TRIS-HCl buffer containing 2 mM EDTA (pH 7.4).

Lipid Extraction and Separation

The lipids were extracted according to Bligh and Dyer (1959). The lipid-containing chloroform phase was evaporated under gentle N_2 flux, redissolved in 1 ml chloroform:methanol (2:1, by volume) and stored at -20°C until analysis. Phospholipids were isolated using a double one-dimensional separation method (Kennerly, 1987). The lipid extracts were applied 8 cm from the bottom of 20x20 cm TLC plates (silica gel 60, Merk, Germany) previously kept at 100 °C for 15 minutes. The solvent system used in the first direction was chloroform:methanol:aqueous ammonia 28% (65:35:7.5, by volume). The outer lane containing phospholipid standards was removed and stained by exposure to iodine vapours to ascertain the location of the spot corresponding to PC. Afterwards, the plate was cut horizontally just below the extrapolated PC spots. The upper part contained PC and PE spots. The lower part of the plate was turned 180 °C and chromatographed in the solvent system: chloroform:acetone:methanol:acetic acid:water (50:15:10:10:5, by volume). By using this procedure acidic phospholipids such as PS, phosphatidylinositol (PI) and phosphatidic acid (PA) remained at the origin in the first solvent system but separated in the second solvent system. The silica gel areas corresponding to PC, PE, PI and PS were identified with authentic standards and scraped off after visualization with iodine vapours. The radioactivity in individual spots was analysed in a scintillation counter. An aliquot of the organic phase was taken to determine the radioactivity present in the lipid phase. The phosphorus content in phospholipids was measured by a colorimetric method (Bartlett, 1959). Protein determination was carried out by following the method of Bradford (1976).

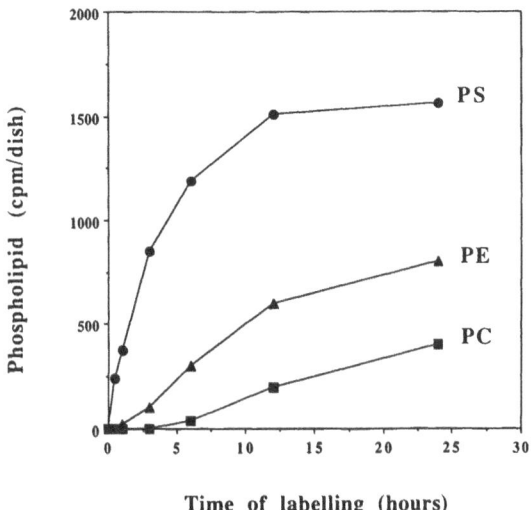

Figure 1. Uptake of [¹⁴C]-serine into NG 108 15 cell phospholipids. Cells were incubated in the presence of 0.5 μCi/ml label for the indicated times, and the radioactivity in phospholipids measured after separation on TLC plates as described in materials and methods. The results are the mean of three determinations in one representative experiment.

RESULTS

Incorporation of [¹⁴C]-Serine into Phospholipids

NG 108 15 cells were incubated together with radiolabelled serine for 30 minutes to 24 hours. The radioactivity present in phospholipids was measured at different time-points. Figure 1 shows the time-dependent uptake of [¹⁴C]-serine into PS, PE and PC. The radioactivity in PS increased up to approximately 6 hours, after which the uptake curve indicated that equilibrium had been reached. Radioactivity in PE was not detectable until three hours after addition of the label. This finding indicates that there is a time-gap between the synthesis of PS and subsequent decarboxylation of newly synthesized PS to form PE. This result may reflect that PS was translocated to the mitochondria.

Furthermore, after 12 hours of labelling a small amount of radioactivity was measured in PC, indicating the conversion of PE into PC through the methylation pathway. The radioactivity present in PS represented 80-100 % of the radioactivity in the phospholipids analyzed (PS + PI + PE + PC) up to 6 hours after addition of the label. After that the percentage decreased to 57 % at 24 hours (Figure 1).

Effect of Acute Ethanol Exposure on Serine Uptake

The effect of ethanol exposure on the incorporation of serine into PS was investigated in cells incubated with [^{14}C]-serine and in the presence or absence of 100 mM ethanol for different time-periods. Table 1 shows that ethanol did not affect the rate of uptake of label into PS. Moreover, ethanol did not alter the incorporation of label into PE or PC (data not shown).

Table 1. Effect of ethanol (100 mM) exposure on the incorporation of [^{14}C]-serine into PS.

Time of labelling and ethanol exposure (hours)	Radioactivity in phosphatidylserine (cpm/dish)	
	Control	Ethanol
0.5	241 ± 52	270 ± 102
1	375 ± 123	479 ± 177
3	848 ± 165	710 ± 190
6	1187 ± 175	1213 ± 282
12	1510 ± 90	1780 ± 300

All values are the mean ± S.E.M. of at least four experiments done in triplicate.

Effect of Long-Term Ethanol Exposure on Serine Uptake

The effect on serine base-exchange activity was investigated by studying the in vivo incorporation of radioactive serine into PS in control and ethanol-exposed cells (100 mM for two days). After 30 or 60 minutes of labelling, no significant differences in the incorporation of label were seen (Figure 2). However, after 6 hours, a 15 % increased radioactivity was observed in ethanol treated cells. This augmentation was already significant after 3 hours of incubation in the presence of label and at a concentration of 25 mM ethanol (data not shown).

In another set of experiments, the degradation of PS was studied. Control and ethanol-exposed cells were labelled with radioactive serine. After 3 hours, the label-containing medium was washed out and replaced by fresh medium containing no label. At different time-points after washing out the label, radioactivity was determined in PS. We found that ethanol did not affect the rate of initial decay of label from PS (Figure 3).

We also investigated the effect of ethanol on the mass of PS by measuring the phosphorus content. An approximately 20 % increase in PS mass was observed after treating the cells with 100 mM ethanol for 2 days (Figure 4). This result correlated with the studies where serine base-exchange activity was assessed by analysing the uptake of [^{14}C]serine into PS.

DISCUSSION

Our results show that NG 108 15 cells utilized exogenously added serine for PS biosynthesis. Furthermore, 2 days of ethanol exposure potentiated both the incorporation of label into this phospholipid as well as increased the mass of PS present in the cells. The increase in PS mass is apparently not due to a decrease in the catabolism of the phospholipid, since the initial decay of incorporated radioactive serine into PS was not altered by ethanol, either acutely or after chronic treatment. Moreover, there were no significant changes in the mass of PC and PE in NG 108 15 cells after long-term ethanol exposure (data not shown) indicating that the effect was specific for PS.

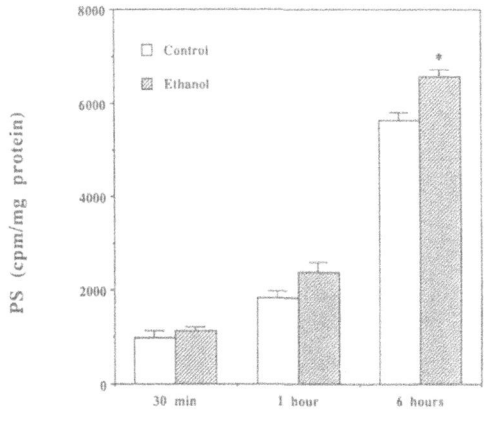

Time of labelling

Figure 2. Incorporation of [¹⁴C]serine into PS after different periods of time in control (blank bars) and ethanol-treated cells (2 days, 100 mM) (hatched bars). Values are the mean $\pm$ S.E.M. of three separate experiments done in triplicate. * $p < 0.05$ (Student's t-test).

It has previously been shown (see Alling et al., 1984; Gustavsson and Alling, 1987; Alling et al., 1991; Sun and Sun, 1983; Sun et al. , 1984; Magruder et al., 1985) that ethanol alters the composition of membrane phospholipids in different systems. These changes mainly include negatively charged phospholipids. Also, it has been reported that specific changes in anionic phospholipids of ethanol-treated rats conferred tolerance to ethanol in liposomes made from control rats (Taraschi et al., 1986; Taraschi et al., 1991). However, a clear elucidation of the alterations induced by ethanol on the membrane phospholipids has not yet been reached. The study of the effect of ethanol on the distribution and metabolism of specific lipid species may help to unravel the molecular mechanisms of damage induced by ethanol on the membrane phospholipid composition.

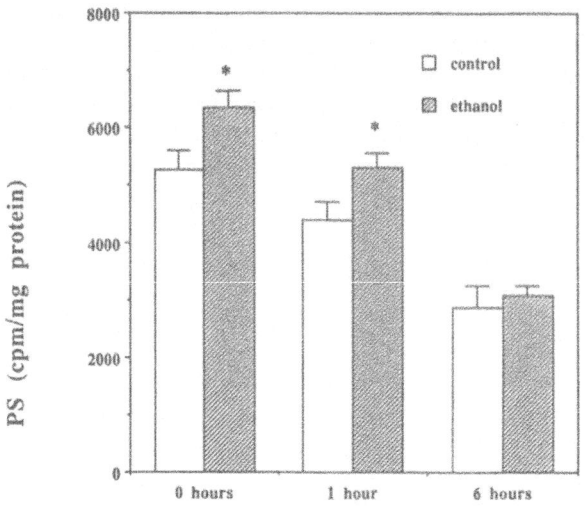

Time after withdrawal of label

Figure 3. <u>Decay of radioactivity in PS.</u> Control cells (blank bars) and ethanol-exposed cells (2 days, 100 mM) (hatched bars) were incubated in the presence of [^{14}C]serine for 3 hours. Label was withdrawn and fresh medium containing no label was added. Radioactivity in PS was measured after different periods of time. Values are the mean $\pm$ S.E.M. of three experiments done in triplicate. * $p < 0.05$ (Student's t-test).

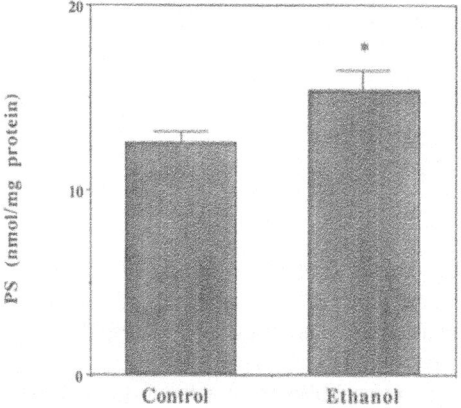

Figure 4. <u>PS phosphorus content in control and ethanol-tretated cells</u> (2 days, 100 mM). Values are the mean $\pm$ S.E.M. of three experiments done in triplicate. * $p < 0.05$ (Student's t-test).

The activity of several membrane-bound enzymes is in many cases dependent on or regulated by membrane lipids. Although the molecular basis of these interactions is not well understood, it is clear that in some situations phospholipids play an important role in the activation of some enzymes. It is known for instance that PS is indispensable for the activation of protein kinase C (Nishizuka, 1984). In this study we have observed that the levels of PS increased after chronic ethanol exposure but not after acute treatment. This increase in PS levels is apparently due to an effect of ethanol on the incorporation of serine into the glycerolipid moiety. The augmentation in PS levels seen after ethanol treatment may have consequences of great interest in the control of protein kinase activation or in the regulation of receptor function (Baudry et al., 1991).

In conclusion, this study indicates that ethanol enhances serine base-exhange activity. The consequences of this finding need to be further investigated.

ACKNOWLEDGEMENTS

This study was supported by the Medical Faculty, University of Lund, The Swedish Alcohol Research Fund, the Albert Påhlson Foundation and the Swedish Medical Reserch Council (project 05249). The technical assistance of Berit Färjh is gratefully acknowledged. F.D.R.is a posdoctoral fellow of the joint program University of Lund (Sweden)-University of Alcalá de Henares (Spain).

REFERENCES

Alling. C., Becker, W., Jones, A.W. and Änggård, E., 1984, Effects of chronic ethanol treatment on lipid composition and prostaglandins in rats fed essential fatty acid deficient diets, *Alcohol Clin.Exp.Res.*8:238.

Alling, C, Rodríguez, F.D., Gustavsson, L. and Simonsson, P., 1991, Continuous and intermittent exposure to ethanol: Effect on NG 108 15 cell membrane phospholipids, *Alcohol & Alcoholism* Suppl.1:227.

Bartlett, G.R., 1959, Phosphorus assay in column chromatography, *J.Biol.Chem.*234:466.

Baudry, M, Massicotte, G and Hauge, S., 1991, Phosphatidylserine increases the affinity of the AMPA/Quisqualate receptor in rat brain membranes, *Behav. and Neural Biol.*55:137.

Bjerve, K., 1985, The biosynthesis of phosphatidylserine and phosphatidylethanolamine from L-[3¹⁴C]serine in isolated rat hepatocytes, *Biochim.Biophys.Acta.*833:396.

Bligh, E.G. and Dyer, W., 1959, A rapid method of total lipid extraction and purification, *Can.J.Biochem.Physiol.*37:911.

Bradford, M.M., 1976, A rapid and sensitive method for the quantitation of microgram quantities of protein utilizing the principle of protein-dye binding,*Anal.Biochem.*72:248.

Butler, M. and Morell, P., 1983, The role of phosphatidylserine decarboxylase in brain phospholipid metabolism, *J.Neurochem.*41:1445.

Corazzi, L., Pistolesi, R. and Arienti, G, 1991, The fusion of liposomes to rat brain microsomal membranes regulates phosphatidylserine synthesis, *J.Neurochem.*56:207.

Gustavsson, L. and Alling, C., 1987, Increase in synaptosomal acidic phospholipids after intermittent but not continuous ethanol exposure, *Alcohol & Alcoholism* 24:196.

Holbrook, P.G. and Wurtman, R.J., 1988, Presence of base-exchange activity in rat brain nerve endings:dependence on soluble substrate concentrations and effect of cations, *J.Neurochem.*50:156.

Kanfer, J.N., 1972, Base exchange reactions of the phospholipids in brain particles, *J.Lipid Res.*13:468.

Kanfer, J.N., 1980, The base exchange enzymes and phospholipase D in mammalian tissues, *Can.J.Biochem.*58:1370.

Kanfer, J.N., McCartney, D and Hattori, H., 1988, Regulation of the choline, ethanolamine and serine base exchange enzyme activities of rat brain microsomes by phosphorylation and dephosphorylation, *FEBS Lett.*240:101.

Kanfer, J.N. and McCartney, D., 1991, Sphingosine and unsaturated fatty acids modulate the base exchange activities of rat brain membranes,*FEBS Lett.*291:63.

Kennerly, D.A., 1987, Diacylglycerol metabolism in mast cells, *J. Biol. Chem.*262:16313.

Magruder, J.D., Waid-Jones, M. and Reitz, 1985, Ethanol-induced alterations in rat synaptosomal plasma membrane phospholipids, *Mol. Pharmacol.*27:256.

Nishizuka, Y., 1984, Turnover of inositol phospholipids and signal transduction. *Science* 225:1365.

Porcellati, G., Arienti, M, Pirotta and Giorgini, D., 1971, Base-exchange reactions for the synthesis of phospholipids in nervous tissue: the incorporation of serine and ethanolamine into phospholipids of isolated brain microsomes, *J. Neurochem.*18:1395.

Rodríguez, F.D., Simonsson, P. and Alling, C., 1992, A method for maintaining constant ethanol concentrations in cell culture media, *Alcohol & alcoholism* 27:309.

Siddiqui, R.A. and Exton, J.H., 1992, Phospholipid base exchange activity in rat liver plasma membranes, *J. Biol. Chem.*267:5755.

Simonsson, P., Hansson, E. and Alling, C, 1989, Ethanol potentiates serotonin stimulated inositol lipid metabolism in primary astroglial cell cultures, *Biochem. Pharmacol.*38:2801.

Singh, I.N., Sorrentino, G., Massarelli, R. and Kanfer, J.N., 1992, Oleoylamine and sphingosine stimulation of phosphatidylserine synthesis by LA-N-2 cells is protein kinase C independent, *FEBS Lett.*296:166.

Sun, G.Y. and Sun, A., 1983, Chronic ethanol administration induced an increase in phosphatidylserine in guinea pig synaptic plasma membranes, *Biochem. Biophys. Res. Com.*113:262.

Sun, G.Y., Huang, H-M., Lee, D.Z. and Sun, A., 1984, Increased acidic phospholipids in rat brain membranes after chronic ethanol administration, *Life Sci.*25:2127.

Taraschi, T.F., Ellingson, J.S., Wu, A., Zimmerman, R. and Rubin, E., 1986, Phosphatidylinositol from ethanol-fed rats confers membrane tolerance to ethanol, *Proc. Natl. Acad. Sci. USA* 83:9398.

Taraschi, T.F., Ellingson, J.S., Janes, N. and Rubin, E., 1991, The role of anionic phospholipids in membrane adaptation to ethanol, *Alcohol & Alcoholism* Suppl.1:241.

Vance, J.E., 1991, Newly made phosphatidylserine and phosphatidylethanolamine are preferentially translocated between rat liver mitochondria and endoplasmic reticulum, *J. Biol. Chem.*266:89.

Voelker, D.R., 1990, Characterization of phosphatidylserine synthesis and translocation in permeabilized reticulum., *J. Biol. Chem.*266:89.

ETHANOL-INDUCED CHANGES IN SIGNAL TRANSDUCTION VIA FORMATION OF PHOSPHATIDYLETHANOL

Lena Gustavsson[1], Christofer Lundqvist[1], Elisabeth Hansson[2], F. David Rodríguez[3], Per Simonsson[1] and Christer Alling[1]

[1]Department of Psychiatry and Neurochemistry
University of Lund, P.O. Box 638, S-220 09 Lund
[2]Department of Cell Biology, Faculty of Health Sciences
S-581 85 Linköping, Sweden
[3]Department of Biochemistry and Molecular Biology
University of Salamanca, 37007 Salamanca, Spain

INTRODUCTION

In 1984 we demonstrated that phosphatidylethanol (Peth) was formed in different organs of ethanol-treated rats (Alling et al., 1984). Peth is a unique, anionic phospholipid formed in cell membranes only in the presence of ethanol. The highest concentration of Peth was found in membrane-rich organs such as brain and kidney but it could also be detected in liver, skeletal muscle and heart (Alling et al., 1983; Alling et al., 1984; Benthin et al., 1985). In addition to rats intoxicated with ethanol, Peth was also formed in cultured cells exposed to ethanol (Lundqvist et al., 1993).

The formation of Peth in mammalian cells is catalysed by phospholipase D (Gustavsson and Alling, 1987; Kobayashi and Kanfer, 1987). Formation by base-exchange enzymes or via de novo synthesis from diacylglycerol have been excluded as possible mechanisms (Gustavsson and Alling, 1987; Kobayashi and Kanfer, 1987; Pai et al., 1988). Normally phospholipase D carries out a hydrolysis of phospholipids leading to the formation of phosphatidic acid (PA) (Kanfer, 1980). The enzyme also utilizes short-chain primary alcohols as substrate leading to the formation of the corresponding phosphatidylalcohol via a transphosphatidylation reaction. This phenomenon was first observed in plants (Dawson, 1967, Yang et al., 1967) but has also been demonstrated in mammalian tissues (Gustavsson and Alling, 1987, Kobayashi and Kanfer, 1987). Beside ethanol, phospholipase D also incorporates methanol, 1-propanol and 1-butanol into phospholipids. The ability of phospholipase D to use ethanol as a substrate leading to the formation of Peth makes it possible that this enzyme activity is one of the mechanisms by which ethanol interferes with cell function.

During recent years phospholipase D has emerged as a component of signal transduction cascades. Activation of several different receptor complexes as well as stimulation of protein kinase C lead to an increased phospholipase D activity in a variety of cell types, and thereby to Peth formation when ethanol is present (for review see Exton, 1990; Billah and Anthes, 1990; Shukla and Halenda, 1991). In some systems the receptor-mediated stimulation of phospholipase D activity is regulated by a, so far, unidentified guanine nucleotide binding protein (Exton, 1990; Billah and Anthes, 1990; Shukla and Halenda, 1991). Protein kinase C and the intracellular concentration of Ca^{2+} are other important mediators of the enzyme activity. Thus, phospholipase D is involved in the signal transduction processes which include formation of lipid-derived second messengers (Fig. 1). Because of the rapid interconversion between PA, formed by phospholipase D, and 1,2-diacylglycerol (1,2-DAG), formed by phospholipase C, there is a close relationship between these different signal transduction pathways (Fig. 1).

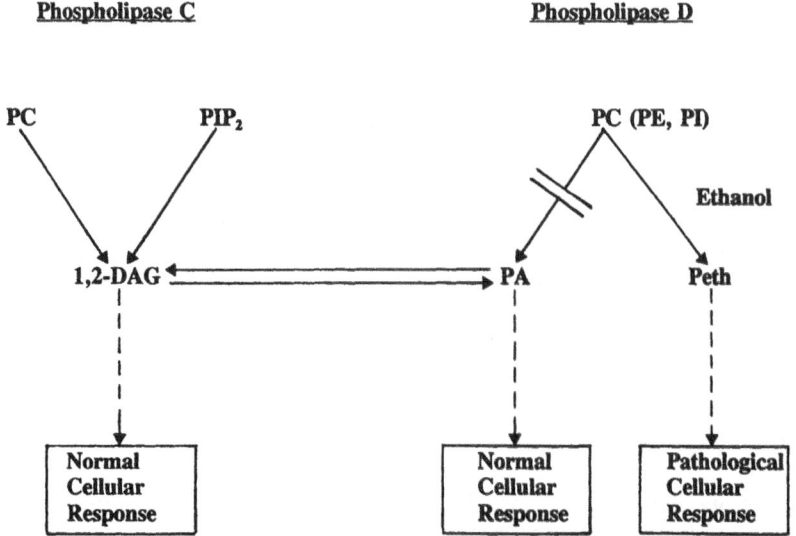

Figure 1. Scheme of receptor-mediated degradation of phospholipids by phospholipases C and D. Phospholipases C acting on phosphatidylcholine (PC) and phosphatidylinositolbisphosphate (PIP$_2$) and phospholipase D, which mainly degrades PC, are activated by a variety of receptor agonists. There is a rapid interconversion of the products formed via these two enzyme systems. 1,2-diacylglycerol (1,2-DAG) is metabolized to PA via diacylglycerolkinase and the opposite reaction is catalysed by phosphatidate phosphohydrolase. When ethanol is present, it is used by phospholipase D for the synthesis of Peth. Because of substrate competition the hydrolysis, leading to PA formation, is inhibited simultaneously. (PE=phosphatidylethanolamine, PI=phosphatidylinositol)

There are several possible sites at which ethanol may interact with signal transduction through phospholipase D and Peth formation. First, ethanol inhibits the normal receptor-mediated formation of PA at the same time as Peth is formed. Secondly, Peth in itself may accumulate in the cell membrane and affect other membrane-associated processes. Thus, ethanol acts via phospholipase D through at least two mechanisms.

EFFECTS OF ETHANOL ON PHOSPHOLIPASE D-MEDIATED SIGNAL TRANSDUCTION

Although most reports on receptor-mediated activation of phospholipase D are from studies on peripheral cells, there is now considerable evidence that this signal transduction pathway is also present in the nervous system. Table 1 summarizes the receptors reported to be coupled to phospholipase D activity in cells and tissues originating from brain.

α₁-Adrenergic Receptors
    -Rat brain cortical slices (Llahi & Fain, 1992)

Bradykinin Receptors
    -PC12 cells (Horwitz, 1991)

P₂-Purinergic Receptors
    -Astroglial cells in primary culture (Gustavsson et al., 1993)

Muscarinic Acetylcholine Receptors
    -Synaptosomes from canine brain (Qian & Drewes, 1989)
    -Astrocytoma cells 1321N1 (Martinsson et al., 1990)
    -LA-N-2 human neuroblastoma cells (Sandmann & Wurtman, 1991)
    -Astroglial cells in primary culture (Gustavsson et al., 1993)

Ethanol and Protein Kinase C-Mediated Phospholipase D Activity in Astroglial Cells

In a series of experiments on primary cultures of astroglial cells, we have studied the effect of ethanol on phospholipase D-mediated product formation induced by stimulation of protein kinase C. The cultures were prepared from cerebral cortex of new-born rats and contained mostly astrocytes (Hansson, 1988). Activation of protein kinase C by phorbol ester and diacylglycerol induced phospholipase D activity in this cell type (Gustavsson and Hansson, 1990). Both the hydrolysis and the transphosphatidylation reactions were stimulated and there was a linear increase in the amount of PA and Peth during the first 20 minutes of stimulation with TPA (12-O-tetradecanoyl 13-phorbol acetate), whereafter the time-courses reached a steady-state. The formation of Peth after stimulation with TPA was dependent on the concentration of ethanol (Fig. 2). Maximal formation of Peth was observed with 150 mM ethanol in the cell culture medium, but a significant amount was formed with ethanol concentrations as low as 25 mM. Parallel with the increasing Peth formation, there was a concentration-dependent inhibition by ethanol of the PA formation and both curves reached maximum and minimum respectively at the same ethanol concentration (Fig. 2). Thus, Peth was formed at the expense of PA, indicating a redirection of the reaction towards the use of ethanol as a substrate. The decrease in PA formation with increasing ethanol concentrations was smaller compared to the Peth increase. This may be due to rapid metabolism of PA to 1,2-DAG or other metabolites. Peth on the other hand has been demonstrated to be metabolically more stable (see next section). However, at 150 mM ethanol the TPA-induced increase in the concentration of PA was almost completely inhibited (data not shown).

Ethanol and Receptor-Mediated Phospholipase D Activity in Astroglial Cells

Primary astroglial cell cultures also displayed receptors coupled to phospholipase D (Table 1) (Gustavsson et al., 1993). Among different receptor agonists tested, carbachol and ATP stimulated Peth formation to the greatest extent. Both agonists induced a rapid and transient increase in the amount of PA, and when ethanol was present, formation of Peth. The responses were linear during the first 30 seconds of stimulation and reached a maximum after one minute. As was observed after TPA stimulation, the ATP and carbachol-induced increases in PA concentration were inhibited by ethanol in parallel with the production of Peth.

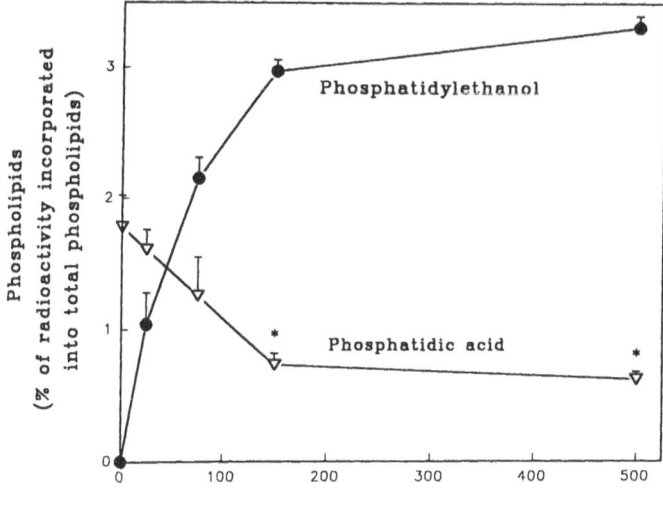

Ethanol concentration

Figure 2. Ethanol concentration dependence of TPA-induced formation of Peth and PA. Primary cultures of astroglial cells were prelabelled with [³H]glycerol (5 μCi/dish) for 20 hours. Ethanol at different concentrations was added 5 minutes prior to addition of TPA (100 nM). The reaction was stopped with ice-cold buffer 1 hour after the TPA addition and the lipids extracted according to Bligh and Dyer (1959). The lipids were separated on HPTLC-plates in the solvent system chloroform:acetone:methanol:concentrated acetic acid:water 50:15:10:10:5, stained with iodine and scraped into scintillation fluid for counting of radioactivity (Gustavsson and Hansson, 1990). The data are presented as means $\pm$ S.E.M. Statistical significances of differences: * $p < 0.05$ vs PA increase in the absence of ethanol (n=3).

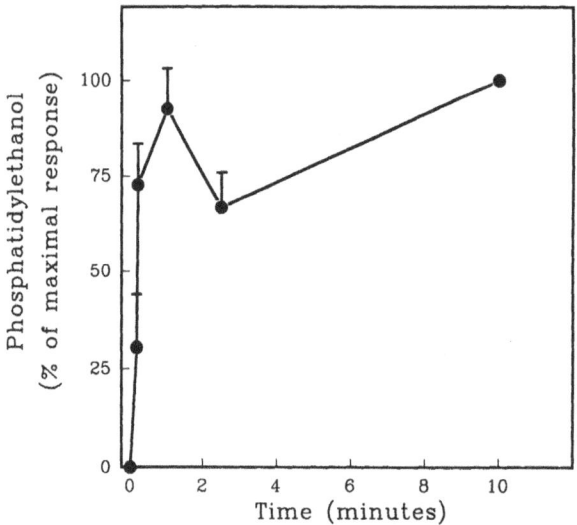

Figure 3. Time-course of carbachol-induced Peth formation in primary astroglial cell cultures. Cells were labelled with [³H]oleic acid (5 μCi/ml medium) for 6 hours and stimulated with 1 mM carbachol. Ethanol (150 mM) was added 5 minutes before the agonist. The reaction was stopped with ice-cold isopropanol and extracted according to Bligh and Dyer (1959). The lipids were separated on HPTLC-plates in the solvent system ethylacetate:iso-octane:acetic acid 90:50:20, stained with iodine and scraped into scintillation fluid for counting of radioactivity (Gustavsson et al., 1993). Data are presented as means$\pm$ S.E.M. (n=4 in each point).

Figure 3 shows the time-course of carbachol-induced Peth formation in rat astroglial cells. The carbachol-induced increase in phospholipase D activity was mainly mediated via muscarinic acetylcholine receptors, which was demonstrated by the use of receptor agonists and antagonists. Atropine, a muscarinic receptor antagonist, completely inhibited the carbachol-induced response whereas mecamylamine, an antagonist acting on the nicotinic receptor, did not have any significant effect. Phospholipase D therefore seems to be coupled to muscarinic acetylcholine receptors in astroglial cells. The carbachol-induced formation of Peth was concentration-dependent with a half maximal effective concentration of 5×10^{-5} M.

Table 2 demonstrates that ethanol significantly inhibited the initial rate of carbachol-induced PA formation in astroglial cells. In the absence of ethanol there was a 76% increase in the amount of PA after 30 seconds of stimulation. In the presence of ethanol, this increase was reduced to 13%. The significant reduction in PA levels remained during the first 2.5 minutes of stimulation.

Table 2. Effect of ethanol on carbachol-induced formation of PA. Astroglial cells in primary culture prelabelled with [³H]oleic acid were stimulated with 1 mM carbachol with or without ethanol (150 mM) in the cell culture medium. The methods of analyses were the same as in figure 3. Data are presented as means $\pm$ S.E.M. (n=3 in each point). Statistical significances of differences: * p<0.05, ** P<0.01 vs the amount of PA in cells stimulated with carbachol for the same time without ethanol.

	PA (% of radioactivity in phospholipids)	
Time of stimulation	No ethanol	150 mM ethanol
0	0.67 ± 0.03	0.62 ± 0.03
30 s	1.18 ± 0.13	0.70 ± 0.04 **
1 minute	1.07 ± 0.11	0.71 ± 0.02 *
2.5 minutes	1.21 ± 0.18	0.55 ± 0.03 **
10 minutes	0.87 ± 0.11	0.55 ± 0.03

Interaction With Other Signalling Systems

Thus, ethanol inhibits receptor- and protein kinase C-mediated formation of PA, the normal product of phospholipase D activity, and this may consequently induce changes in cell function. However, the extent to which the PA levels are decreased also depends on the activation of other phospholipases which may be induced by the same receptor stimuli. PA is also formed through phosphorylation of the phospholipase C product, 1,2-DAG, by diacylglycerolkinase (Fig. 1). Since phospholipases C and D often are activated by the same receptor agonists (for review see Shukla and Halenda, 1991), the ethanol-induced decrease in PA production depends on the relative contribution of phospholipases C and D to the total amount of lipid-derived second messengers formed. In muscarinic stimulation of primary astroglial cultures, there was an almost complete inhibition of the PA production indicating that phospholipase D was the dominating phospholipase activated.

However, in the human neuroblastoma cell line, SH-5YSY, phospholipase D seems to contribute to only a minor part of the total amount of 1,2-DAG and PA formed after muscarinic stimulation (Boyano-Adánez and Gustavsson, 1993). Therefore ethanol had almost no effect on the total amount of these lipid metabolites formed in SH-SY5Y cells. Similarly, the PA formation induced by bradykinin in PC12 cells was not completely inhibited even in the presence of high ethanol concentrations (1%) (Horwitz, 1991). In order to find the receptor systems at which ethanol exerts effects on cell function via phospholipase D, it is of crucial importance to identify the sites where phospholipase D is the dominating signalling pathway. Stimulation of human neutrophils with chemotactic peptide (fMLP) induces activation of both phospholipase D and polyphosphoinositide-specific phospholipase C as well as functional responses coupled to these two signalling pathways (Pai et al., 1988). In this cell type, ethanol inhibited PA formation after receptor stimulation but had no effect on the production of inositol 1,4,5-trisphosphate. Interestingly, it was demonstrated that the fMLP-mediated superoxide generation and degranulation was inhibited by ethanol in a concentration-dependent manner and it was concluded that the ethanol-induced effects was at least partly mediated via phospholipase D (Bonser et al., 1989, Kanaho et al., 1991, Nilsson et al., 1992). Ethanol effects exerted via phospholipase D-mediated signal transduction and its functional correlates in cells from the nervous system require further investigation.

EFFECTS OF PHOSPHATIDYLETHANOL

Accumulation of Peth in Cell Membranes

Peth is formed and accumulated in the cell membrane. It is possible that a normally not occuring phospholipid affects cell function and it may therefore be responsible for some of the effects of ethanol on the cell. Crucial for the possible effects of Peth is the amount formed. In kidneys from rats injected intraperitoneally with ethanol (3 g/kg body weight) the concentration of Peth reached 20 nmol/g wet weight (Benthin et al., 1985). Table 3 presents the amount of Peth formed in different cells. Significant amounts of Peth was found in neuroblastoma x glioma cells (NG 108-15) after two days of ethanol exposure and also in neutrophilic granulocytes isolated from chronic alcoholics. In the presence of a stimulus for phospholipase D, a considerably higher concentration of Peth was found in NG 108-15 cells, reaching up to 3% of the total phospholipid pool in the case of TPA stimulation. Thus, there is a high potential for Peth formation, especially at membrane sites where signal transduction via phospholipase D occurs. If the formation of Peth is mainly located in the plasma membrane, as may be expected for receptor agonist-induced responses, this would lead to a substantial change in the cell membrane composition.

Although a significant amount of Peth is formed in different systems in the presence of ethanol, the turnover rate is of major importance for the amount accumulating. In cell types and tissues so far studied, the metabolism of Peth is relatively slow. Figure 4 demonstrates the disappearance of Peth from NG 108-15 cells in which exogenously added Peth had been incorporated into the membrane (Lundqvist et al., 1993). After treatment with Peth, the cells were washed and incubated in cell culture medium without Peth. Initially, there was a rapid decrease which may be due to redistribution of Peth into the medium. After this rapid decline the disappearance was slower and Peth was still detectable 20 hours after removal of PEth from the medium. The average half-life of exogenous Peth was 5 hours in NG 108-15 cells. Endogenously synthesized Peth has also been demonstrated to be degraded slowly after ethanol has been removed (Metz and Dunlop, 1990; Gustavsson et al., 1991). In rat hepatocytes, the Peth formed after receptor stimulation had a half-life of more than one hour (Gustavsson et al., 1991).

Table 3. <u>Amount of Peth formed under different conditions.</u> The amount of Peth was analysed after lipid extraction of cells and separation on HPTLC-plates in the solvent system ethylacetate:iso-octane:acetic acid 90:50:20. The lipids were stained with Coomaissie Brilliant Blue according to Nakamura and Handa (1984) and analysed by densitometry.

NG108-15 cells, 2 days of ethanol exposure	40 pmol/mg protein
NG108-15 cells, stimulation with TPA in the presence of ethanol	$\leq$ 3 nmol/mg protein
Neutrophilic granulocytes, from chronic alcoholics	range: 0-80 pmol/10^6 cells

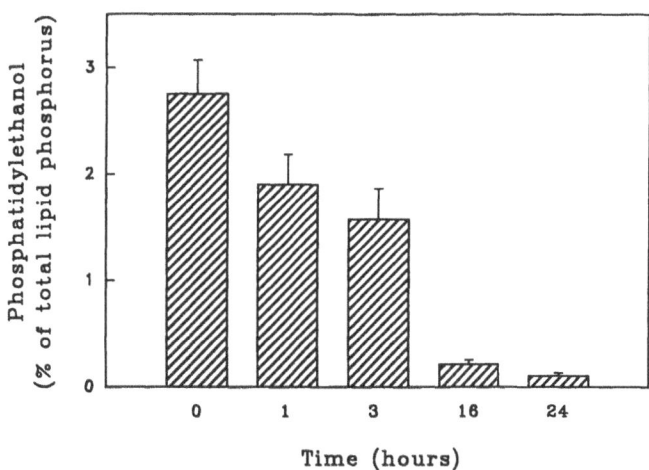

Time (hours)

Figure 4. <u>Degradation of Peth in NG 108-15 cells.</u> Exogenously added Peth was incorporated into cell membranes of NG 108-15 cells according to Lundqvist et al (1993). The cells were incubated with 30 μM Peth in the cell culture medium for 5 hours, washed with buffer and incubated in Peth-free medium for different periods of time. The amount of Peth was analysed by densitometry as described in table 3.

Effect of Peth on the Intracellular Concentration of Inositol 1,4,5-trisphosphate

It is possible that Peth accumulating in the cell membrane induces disturbances in the cell function. Membrane processes are possible targets for toxic effects of Peth and therefore different signal transduction mechanisms are possible sites of action. The incorporation of exogenous Peth into NG 108-15 cells (Lundqvist et al., 1993) made it possible to study the effects of this lipid itself in the absence of ethanol. After incubation of NG 108-15 cells with 30 μM Peth for 5 hours, 3 % of the phospholipids consisted of Peth. This is comparable to the amount of Peth that can be formed endogenously after stimulation of the cells with TPA in the presence of ethanol.

In order to study the effect of Peth on signalling processes, the concentration of inositol 1,4,5-trisphosphate was analysed in NG 108-15 cells containing exogenous Peth.

Peth did not affect receptor-mediated formation of inositol-1,4,5-trisphosphate in these cells, when studied with bradykinin as receptor agonist. On the other hand, there was a significant increase in the basal levels of this second messenger (Fig. 5A.). The increase in basal inositol 1,4,5-trisphosphate was dependent on the time of incubation with Peth (Lundqvist et al., 1993). A significant change in the inositol 1,4,5-trisphosphate levels was observed already 15 minutes after the addition of Peth. Furthermore, the effect of Peth was dependent on its concentration.

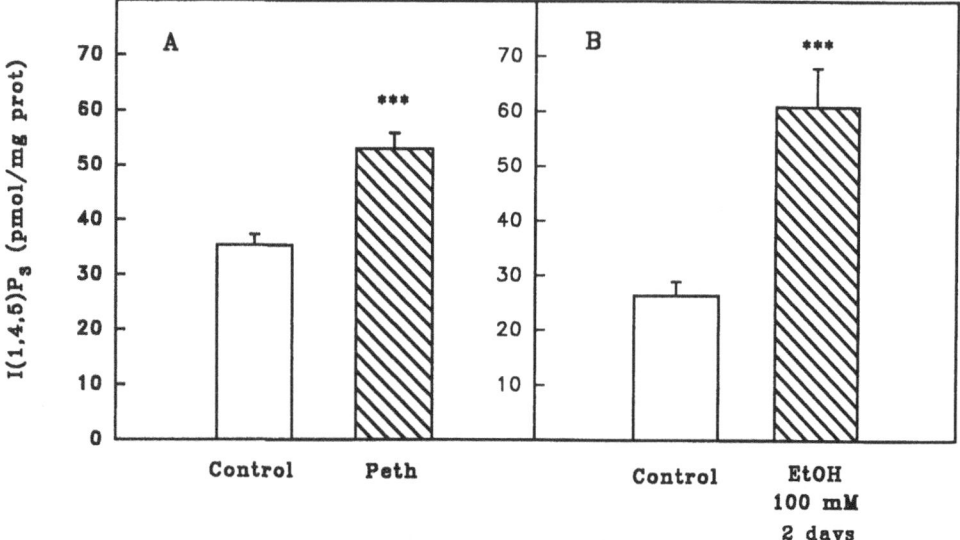

Figure 5. Effects of Peth (A) and ethanol exposure (B) on the basal levels of inositol 1,4,5-trisphosphate in NG108-15 cells. A: Exogenously added Peth was incorporated into NG 108-15 cells by incubation in medium containing 30 μM Peth for 3 hours (n=10 in each group). B: NG 108-15 cells were exposed to ethanol (100 mM) for two days (n=4 in each group). A and B: Incubations were stopped with TCA and the inositol 1,4,5-trisphosphate (I(1,4,5)P$_3$) concentration assayed by means of a radio receptor displacement assay using cerebellar membranes essentially according to Bredt et al (1989). Statistical significance of differences: *** $p < 0.001$ vs controls.

To examine whether the increased levels of inositol 1,4,5-trisphosphate could be obtained also with endogenously formed Peth, NG 108-15 cells were treated with ethanol (100 mM) for two days, which resulted in the formation of Peth as was demonstrated in table 3. Similarly, to what was observed with exogenously added Peth, the basal levels of inositol 1,4,5-trisphosphate were significantly enhanced also after ethanol treatment (Fig. 5B). A combination of both ethanol and Peth did not give any further increase compared to Peth or ethanol alone (Lundqvist et al., 1993). The lack of additivity may indicate that the cause of an increased inositol 1,4,5-trisphosphate concentration was due to Peth under both conditions. The possible consequences of the Peth-induced increase in inositol 1,4,5-trisphosphate levels including changes in the intracellular calcium homeostasis or the state of the inositol 1,4,5-trisphosphate receptors remain to be studied.

Other Effects of Peth

Other effects of Peth have also been reported as summarized in table 4. Peth affects membrane-associated enzymes as well as biophysical characteristics of the membrane. Of special interest is that Peth was reported to increase the membrane tolerance to an acute dose of ethanol (Omodeo-Salé et al., 1991; Rottenberg et al., 1992), an effect that may indicate a role for Peth in some of the adaptation phenomena developing during exposure to the drug.

Table 4. Effects of Peth.

-Changes in membrane fusion properties (Bondeson & Sundler, 1987)

-Activation of protein kinase C (Asaoka et al., 1988)

-Increased membrane fluidity and tolerance to ethanol-induced
fluidization (Omodeo-Salé et al., 1991)

-Reduction of ethanol-induced increase in Na^+,K^+-ATPase activity
(Omodeo-Salé., 1991)

-Increased resistance to ethanol-induced membrane disordering (Rottenberg et al., 1992)

-Increased basal level of inositol 1,4,5-trisphosphate (Lundqvist et al., 1993)

CONCLUSION

Phospholipase D is an enzyme which accepts ethanol as substrate leading to the formation of Peth. This may be one of the means through which ethanol exerts its adverse effects on the function of cells in the nervous system. In the presence of ethanol, signal transduction via phospholipase D deteriorates by a change in the relative proportions of the lipid second messengers formed. Maximal effects were obtained with an ethanol concentration of 150 mM, but significant changes have been observed with concentrations as low as 25 mM. The decrease in the amount of PA formed may lead to functional disturbances in receptor systems where phospholipase D is the dominating pathway for cell signalling. Furthermore, the formation and accumulation of Peth, a normally not occuring phospholipid, in the cell membrane induces changes in membrane-associated processes. Taken together, these results indicate that phospholipase D may be the mediator of some of the ethanol-induced disturbances on receptor-regulated changes in brain cell function.

ACKNOWLEDGEMENTS

The studies were supported by grants from the Swedish Medical Research Council (proj. nos. 21X-05249 and 03P-08895), the Albert Påhlsson Foundation, the Swedish Alcohol Research Fund, the Crafoord Foundation and the Medical Faculty, University of Lund.

REFERENCES

Alling, C., Gustavsson, L., and Änggård E., 1983, An abnormal phospholipid in rat organs after ethanol treatment. *FEBS Lett.* 152:24.

Alling, C., Gustavsson, L., Månsson, J-E., Benthin, G., and Änggård, E., 1984, Phosphatidylethanol formation in rat organs after ethanol treatment. *Biochim.Biophys.Acta* 793:119.

Asaoka, Y., Kikkawa, U., Sekiguchi, K., Shearman, M.S., Kosaka, Y., Nakano, Y., Satoh, T., and Nishizuka, Y., 1988, Activation of a brain-specific protein kinase C subspecies in the presence of phosphatidylethanol. *FEBS Lett.* 231:221.

Benthin, G., Änggård, E., Gustavsson, L., and Alling, C., 1985, Formation of phosphatidylethanol in frozen kidneys from ethanol-treated rats. *Biochim.Biophys.Acta* 835:385.

Billah, M.M., and Anthes, J.C., 1990, The regulation and cellular functions of phosphatidylcholine hydrolysis. *Biochem.J.* 269:281.

Bligh, E.G., and Dyer, W.J., 1959, A rapid method for total lipid extraction and purification. *Can.J.Biochem.Physiol.* 37:911.

Bondeson, J, and Sundler, R., 1987, Phosphatidylethanol counteracts calcium-induced membrane fusion but promotes proton-induced fusion. *Biochim.Biophys.Acta* 899:258.

Bonser, R.W., Thompson,N.T., Randall, R.W., and Garland, L.G., 1989, Phospholipase D activation is functionally linked to superoxide generation in the human neutrophil. *Biochem.J.* 264:617.

Boyano-Adánez, M.C., and Gustavsson, L., 1993, Effects of ethanol on receptor-mediated phospholipase D activity in human neuroblastoma cells. This volume.

Bredt, D.S., Mourey, R.J., and Snyder, S.H., 1989, A simple, sensitive and specific radioreceptor assay for inositol 1,4,5-trisphosphate in biological tissues. *Biochem.Biophys.Res.Commun.* 159:976.

Dawson, R.M., 1967, The formation of phosphatidylglycerol and other phospholipids by the transferase activity of phospholipase D. *Biochem.J.* 102:205.

Exton J.H., 1990, Signaling through phosphatidylcholine breakdown. *J.Biol.Chem.* 265:1.

Gustavsson, L., and Alling C., 1987, Formation of phosphatidylethanol by phospholipase D in rat brain by phospholipase D. *Biochem.Biophys.Res.Commun.* 142:958.

Gustavsson, L., and Hansson, E., 1990, Stimulation of phospholipase D activity by phorbol esters in cultured astrocytes. *J.Neurochem.* 54:737.

Gustavsson, L., Moehren, G., and Hoek, J.B., 1991, Phosphatidylethanol formation in rat hepatocytes. *Ann.N.Y.Acad.Sci.* 625:438.

Gustavsson, L., Lundqvist, C., and Hansson, E., 1993, Receptor-mediated phospholipase D activity in primary astroglial cultures. *Glia*, in press.

Hansson, E., 1988, Astroglia from defined brain regions as studied with primary cultures. *Prog.Neurobiol.* 30:369.

Horwitz, J., 1991, Bradykinin activates a phospholipase D that hydrolyzes phosphatidylcholine in PC12 cells. *J.Neurochem.* 56:509.

Kanaho, Y., Kanoh, H., Saitoh, K., and Nozawa, Y., 1991, Phospholipase D activation by platelet-activating factor, leukotriene B_4, and formyl-methionyl-leucyl-phenylalanine in rabbit neutrophils. *J.Immunol.* 146:3536.

Kanfer, J.N., 1980, The base-exchange enzymes and phospholipase D of mammalian tissues. *Can.J.Biochem.* 58:1370.

Kobayashi, M., and Kanfer, J.N., 1987, Phosphatidylethanol formation via transphosphatidylation by rat brain synaptosomal phospholipase D. *J.Neurochem.* 48:1597.

Llahi, S., and Fain, J.N., 1992, α_1-adrenergic receptor-mediated activation of phospholipase D in rat cerebral cortex. *J.Biol.Chem.* 267:3679.

Lundqvist, C., Rodríguez F.D., Simonsson P., Alling, C., and Gustavsson L., 1993, Phosphatidylethanol affects inositol 1,4,5-trisphosphate levels in NG108-15 neuroblastoma x glioma hybrid cells. *J.Neurochem.* 60:738-744.

Martinsson, E.A., Trilivas, I., and Brown, J.H., 1990, Rapid protein kinase C-dependent activation of phospholipase D leads to delayed 1,2-diglyceride accumulation. *J.Biol.Chem.* 265:22282.

Metz, S.A., and Dunlop, M., 1990, Production of phosphatidylethanol by phospholipase D phosphatidyl transferase in intact or dispersed pancreatic islets: Evidence for the in situ metabolism of phosphatidylethanol. *Arch.Biochem.Biophys.* 283:417.

Nakamura, K. and Handa, S., 1984, Coomaissie brilliant blue staining of lipids on thin-layer plates. *Anal.Biochem.* 142:406.

Nilsson, E., Andersson, T., Fällman, M., Rosendahl, K., and Palmblad, J., 1992, Effects of ethanol on the chemotactic peptide-induced second messenger generation and superoxide production in polymorphonuclear leukocytes. *J.Infect.Dis.* 166:854.

Omodeo-Salé, F., Lindi, C., Palestini, P., and Masserini, M., 1991, Role of phosphatidylethanol in membranes. Effects on membrane fluidity, tolerance to ethanol, and activity of membrane-bound enzymes. *Biochemistry* 30:2477.

Pai, J-K., Siegel, M.I., Egan, R.W., and Billah, M.M., 1988, Phospholipase D catalyzes phospholipid metabolism in chemotactic peptide-stimulated HL-60 granulocytes. *J.Biol.Chem.* 263:12472.

Qian, Z., and Drewes, L.R., 1989, Muscarinic acetylcholine receptor regulates phosphatidylcholine phospholipase D in canine brain. *J.Biol.Chem.* 264:21720.

Rottenberg, H., Bittman, R., and Li, H-L., 1992, Resistance to ethanol disordering of membranes from ethanol-fed rats is conferred by all phospholipid classes. *Biochim.Biophys.Acta* 1123:282.

Sandmann, J., and Wurtmann, R.J., 1991, Stimulation of phospholipase D activity in human neuroblastoma (LA-N-2) cells by activation of muscarinic acetylcholine receptors or by phorbol esters: Relationship to phosphoinositide hydrolysis. *J.Neurochem.* 56:1312.

Shukla, S.D., and Halenda, S.P., 1991, Phospholipase D in cell signalling and its relationship to phospholipase C. *Life Sci.* 48:851.

Yang, S.F., Freer, S., and Benson, A.A., 1967, Transphosphatidylation by phospholipase D. *J.Biol.Chem.* 242:477.

PHOSPHATIDYLETHANOL EFFECTS ON
INOSITOL 1,4,5-TRISPHOSPHATE LEVELS
AND PROTEIN KINASE C ACTIVITY IN
NG108-15 CELLS

Christofer Lundqvist, Christer Larsson, Christer Alling
and Lena Gustavsson

Department of Psychiatry and Neurochemistry
University of Lund
PO Box 638
S-220 09 Lund
Sweden

INTRODUCTION

In animal tissues exposed to ethanol, the formation of an abnormal phospholipid, phosphatidylethanol (PEth), has been demonstrated (Alling et al., 1983; Alling et al., 1984). This phospholipid is formed through a transphosphatidylation reaction mediated by phospholipase D (Gustavsson and Alling, 1987; Kobayashi and Kanfer, 1987). The ability to activate phospholipase D and thus stimulate phosphatidylethanol formation if ethanol is present has been ascribed to several receptor agonists and protein kinase C activators (Exton, 1990; Billah and Anthes, 1990; Shukla and Halenda, 1991). Thus considerable quantities of phosphatidylethanol may be formed in ethanol-exposed tissues, including brain. Some recent studies have, in addition, suggested that PEth is degraded only slowly (Metz and Dunlop, 1991; Gustavsson et al, 1991). Therefore significant amounts of PEth may accumulate during a period of ethanol exposure.

PEth has by some authors been suggested to be a candidate cytotoxic substance which may be responsible for some of the adverse effects exerted on cells by alcohol (Urbano-Marquez et al., 1989; Diamond, 1989). Possible targets for such a toxic effect of PEth include membrane-associated enzymes and processes. PEth has been shown to affect Na^+/K^+-ATPase, 5'-nucleotidase, membrane fluidity and fusion of membrane vesicles (Omodeo-Salé et al., 1991; Bondeson and Sundler, 1987). We have recently reported that PEth affects the intracellular levels of inositol 1,4,5-trisphosphate $(I(1,4,5)P_3)$ (Lundqvist et al 1992). $I(1,4,5)P_3$ is formed through phospholipase C-mediated hydrolysis of phosphatidylinositol 4,5-bisphosphate and is of central importance as a mediator of calcium release from intracellular stores (Berridge 1984, Berridge and Irvine 1984).

Protein kinase C (PKC) is a key enzyme in signal transduction which has been shown to be activated by 1,2-diacylglycerol formed by phospholipase C. PKC itself has also been shown to mediate a regulatory influence on phospholipase C. It is therefore possible that the described increase in $I(1,4,5)P_3$ (Lundqvist et al, 1992) could be related to an effect of PEth on PKC. Asaoka et al (1988) have indeed demonstrated that PEth can act as an activator of PKC in an in-vitro, cell-free system. An effect of PEth on PKC-activity would be interesting especially since PKC has been suggested to be a common target for a large variety of cytotoxins including ethanol (Costa, 1990; Higashi and Hoek, 1991; Messing et al., 1991).

The aim of the present study was therefore to examine whether PKC down-regulation or inhibition would abolish the $I(1,4,5)P_3$ response to exogenous PEth and whether exogenous PEth influences PKC phosphorylation activity.

MATERIALS AND METHODS

Materials

Materials were obtained from the following manufacturers: tissue culture dishes, Costar (U.S.A.); Dulbecco's modified Eagle's medium and cell medium supplements, GIBCO (Scotland); [^{3}H]-inositol 1,4,5-trisphosphate (specific activity, 15-20 Ci/mmol), New England Nuclear (U.S.A.); inositol 1,4,5-trisphosphate, phosphatidylcholine from egg yolk, phospholipase D from peanut, 12-O-tetradecanoyl phorbol 13-acetate (TPA), myelin basic protein (MBP), H7 (1-(5-isoquinolinylsulfonyl)-2-methyl-piperazine), staurosporine, phenylmethylsulphonylfluoride (PMSF), Sigma chemical co. (U.S.A.); Beckman Ready Safe, Beckman Instruments (U.S.A.). [γ-^{32}P]-ATP was from the department of Medical and Physiological Chemistry, University of Lund, Sweden. MBP_{4-14} was kindly donated by Dr. T.Saermark, The Panum Institute, Copenhagen, Denmark. PEth was synthesized from egg yolk phosphatidylcholine by using peanut phospholipase D according to the method by Eibl and Kovatchev (1981). The purity was checked by HPTLC before addition to the cells.

Cell cultivation

NG108-15 cells (passages 20-32) were cultured in 35 mm diameter plastic dishes containing 1 ml of Dulbecco's modified Eagle's medium supplemented with 5% fetal calf serum, 2 mM L-Glutamate, 2% 50 x HAT (final concentrations, 0.1 mM hypoxanthine, 4 μM aminopterin, 16 μM thymidine), 100 μg/ml of streptomycin and 100 IU/ml of penicillin. The culture dishes were maintained for 5 days after passage at 37°C in an incubator under a humidified atmosphere with 10% CO_2. Medium was changed daily. At the time of the experiments, the cells had reached confluence.

Incubation with exogenous PEth

PEth in chloroform : methanol (2:1 by volume) was evaporated under nitrogen gas.

1 ml of the cell culture medium without fetal calf serum was added and the suspension was mixed and sonicated in a water bath. The concentration was adjusted to the required PEth concentration using medium without fetal calf serum. The cells were then incubated with this PEth-containing medium for various times with or without the PKC inhibitors staurosporine or H7. Control cells were incubated for the same times with the same medium but without PEth. After the incubation, cells were harvested as described and processed further for lipid analyses or $I(1,4,5)P_3$ assays.

Inositol (1,4,5) trisphosphate assays

A receptor binding assay developed by Bredt et al was used with slight modifications (Bredt et al., 1989). Harvested cells in trichloroacetic acid (TCA, 0.5 M) were centrifuged at 2000 x g for 15 min at 4°C. The pellet was mixed, resuspended in 250 μL 1 M NaOH and stored for protein analysis (Bradford, 1976). The supernatant was transferred to glasstubes and extracted four times with 1.25 ml water-saturated diethylether. The extract was neutralized by adding 20 μL 500 mM Tris buffer, pH 8.4. Rat or bovine cerebella were homogenized in cold buffer A (50 mM Tris, 1 mM EDTA, 1 mM 2-mercaptoethanol, pH 7.7) and washed by repeated centrifugations at 15 000 x g for 15 min. Cerebellar membranes were stored in aliquots at -70°C. 0.2 pmol [^{3}H]-$I(1,4,5)P_3$ (20 cpm/fmol) was added to Eppendorf Microfuge tubes together with aliquots of the cerebellar membranes diluted with buffer B (same as buffer A but with pH 8.6) to yield 60 μg protein/tube. Standard solutions of $I(1,4,5)P_3$ or neutralized cell extracts were added to a final volume of 275 μL/tube. After mixing, tubes were left on ice for 10 min. The displacement of [^{3}H]-$I(1,4,5)P_3$ bound to cerebellar membranes was stopped by centrifugation at 12 000 x g in a Beckman Microfuge. The supernatant was carefully aspirated and the pellet was resuspended in 500 μL water and taken to scintillation vials for scintillation counting. A standard curve of 1-100 nM $I(1,4,5)P_3$ was used. 50% displacement of bound [^{3}H]-$I(1,4,5)P_3$ was obtained at 10-15 nM $I(1,4,5)P_3$. The displacement curves from rat and bovine cerebella had similar profiles. Identical concentrations of $I(1,4,5)P_3$ were detected using membranes from the two sources. Coefficient of variation was 6.6% (within assay variation) and 6.8% (between assay variation).

Phorbol ester pretreatment

In order to down-regulate the PKC activity in NG108-15 cells, the cells were incubated with 100 nM TPA for 16 hours in the cell culture medium. The TPA-containing medium was removed, cells were washed with fresh medium and the PEth-containing medium was added as described above. After varying time periods, incubations were terminated through removal of the medium and addition of ice-cold TCA (0.5 M) after which the cells were harvested and $I(1,4,5)P_3$ was assayed.

PKC activity

PKC activity was assayed in isolated membranes from NG108-15 cells, as described by Chakravarthy et al (1991) using the specific PKC substrate MBP_{4-14} (Yasuda et al, 1990) or the non-specific substrate MBP and [^{32}P]-ATP. Cells which had been incubated for 5 h in the presence or absence of PEth, were scraped in ice cold lysation buffer (50 mM Tris-HCl, pH 7.5, 5 mM $MgCl_2$, 5 mM EGTA, 100 μM PMSF) and homogenized with a Polytron^R homogenizer. The suspension was centrifuged at 48400 x g for 25 mins to obtain a crude membrane pellet. Membranes were subsequently incubated in 50 mM Tris-HCl (pH 7.5), 10 mM $MgCl_2$, 2 μM $CaCl_2$, 200 μM sodium pyrophosphate, 2 mM NaF, 200 μM PMSF, 40 μM MBP_{4-14} and 100 μM [^{32}P]-ATP (100-200 cpm/pmol). After 10 minutes at 20°C, stimulations were interrupted by addition of 40 μL 1 M TCA and 80 μL of the solution was spotted on Whatman P81 filters. These were washed in 75 mM phosphoric acid and taken for scintillation counting.

Statistical analyses

Statistical analyses were performed using Student's two-tailed t-test. Differences were considered significant if $p < 0.05$. Data are presented as means $\pm$ S.E.M.

RESULTS

Phosphatidylethanol effects on cellular $I(1,4,5)P_3$

PEth was incorporated into membranes of NG108-15 cells. The incorporation caused an increase in the basal cellular levels of $I(1,4,5)P_3$. This increase was time- and dose-dependent and similar to the increase caused by exposure of the cells to ethanol with a concomittant endogenous PEth synthesis (Lundqvist et al, 1992). The effect was significant and was not due to a general effect of exogenous phospholipid since phosphatidylcholine at equimolar concentrations did not elicit the same change (Fig 1).

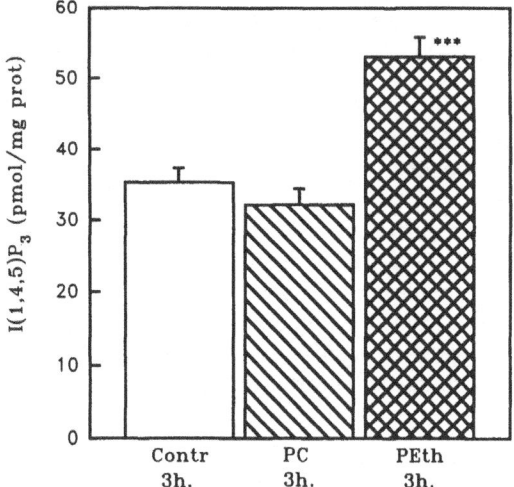

Fig 1. Effects of PEth and phosphatidylcholine on $I(1,4,5)P_3$ levels in NG108-15 cells. NG108-15 cells cultivated as described were incubated for 3 hours in the presence of PEth or phosphatidyl-choline (30 μM). Cells were harvested and cellular $I(1,4,5)P_3$ was assayed by means of a radio-receptor assay as described. n=10, error bars represent S.E.M.

Effects of PKC inhibition and down-regulation

When cells were preincubated with the PKC inhibitors staurosporine or H7, the same PEth-induced increase in I(1,4,5)P$_3$ was observed relative controls as when the inhibitors were not present (Fig 2). Thus the inhibition of PKC activity did not affect the ability of PEth to induce changes in cellular inositol phosphate metabolism.

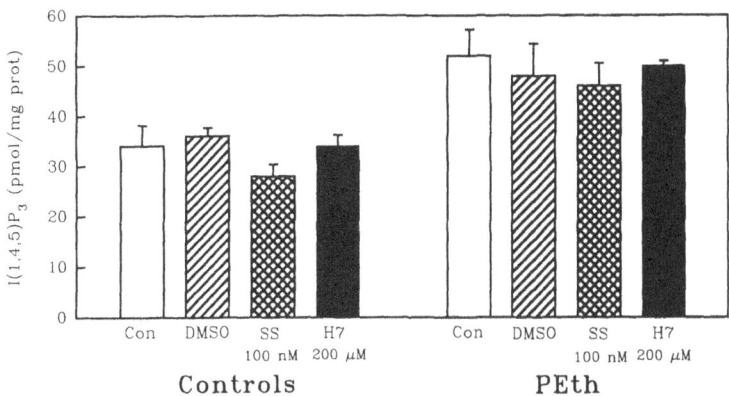

Fig 2. PKC inhibition and PEth-induced I(1,4,5)P$_3$ increase. NG108-15 cells were incubated for 1 hour in 30 μM PEth in the presence or absence of staurosporine or H7 as described under Methods. Cellular I(1,4,5)P$_3$ was assayed by means of a radio-receptor assay as described. Con=control (water), DMSO=dimethylsulphoxide solvent only, SS=staurosporine in DMSO, H7=H7 in water. n=5, error bars represent S.E.M.

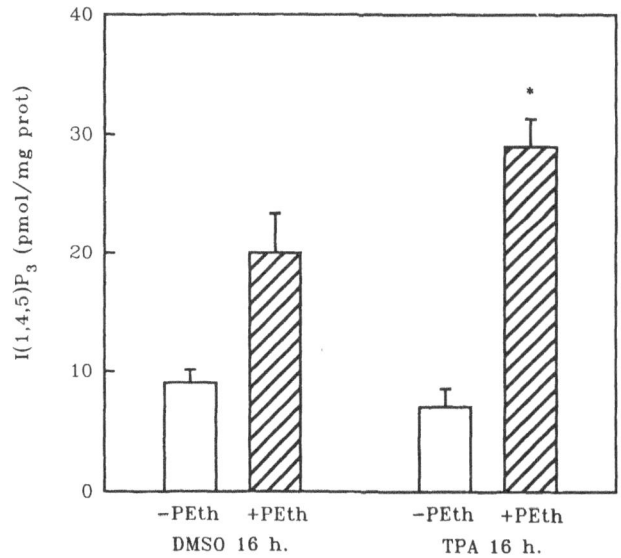

Fig 3. PKC down-regulation and PEth-induced I(1,4,5)P$_3$ increase. NG108-15 cells were incubated with TPA over night (16 h, 100 nM) before being exposed to 30 μM PEth for 3 hours. Cellular I(1,4,5)P$_3$ was assayed by means of a radio-receptor assay as described. n=6, error bars represent S.E.M. *; p<0.05 relative DMSO incubated, PEth-incorporated cells.

Down-regulation of PKC activity through phorbol ester treatment over night did not abolish the PEth response either (Fig 3). On the contrary, there was a slight enhancement of the PEth-induced increase in I(1,4,5)P$_3$ through this treatment.

Phosphatidylethanol effects on PKC activity

PEth incorporation into membranes from NG108-15 cells had no effect on the phosphorylation of the PKC-specific substrate MBP$_{4-14}$ thus indicating that PEth does not affect PKC activity in this system (Fig 4a). When the non-specific phosphorylation substrate MBP was used instead of the PKC-specific peptide, a (non-significant) tendency towards an increase in phosphorylation activity was seen in PEth-treated membranes compared to controls (Fig 4b).

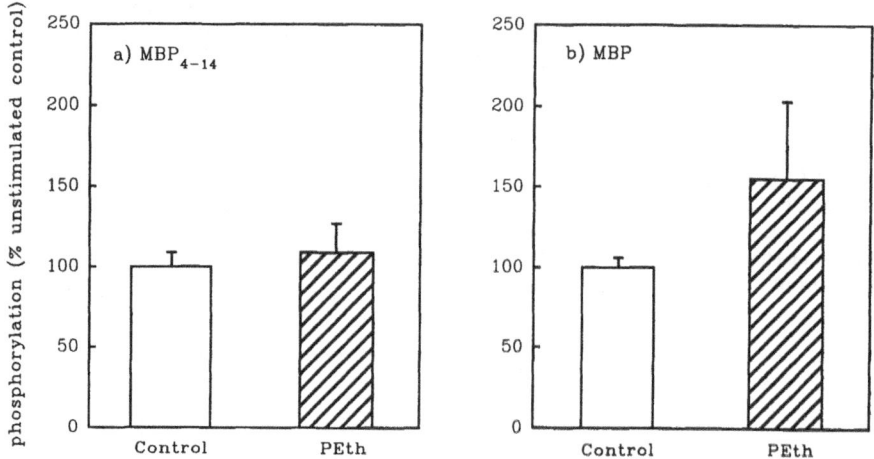

Fig 4.a) <u>PEth effect on PKC activity.</u> The phosphorylation by PEth-treated membranes or controls from NG108-15 cells of the PKC-specific substrate MBP$_{4-14}$ in the presence of [^{32}P]-phosphate was assayed as described under Methods. Results are from two separate assays each performed in triplicate. Error bars represent S.E.M. **b)** <u>PEth effect on MBP phosphorylation.</u> The phosphorylation by PEth-treated membranes or controls from NG108-15 cells of the non-specific phosphorylation substrate MBP in the presence of [^{32}P]-phosphate was assayed as described under Methods. Results are from three separate assays each performed in triplicate. Error bars represent S.E.M.

DISCUSSION

In the present paper we further characterize the previously described stimulatory effect of PEth on levels of inositol 1,4,5-trisphosphate in NG108-15 cells (Lundqvist et al 1992). We demonstrate that protein kinase C does not seem to be involved in the observed PEth-specific changes since no change in PKC activity was observed in cells treated with PEth. Moreover, neither PKC inhibition nor down-regulation abolished the effects of exogenous PEth.

The described negative results of the effects of PEth on PKC are in contrast to those of Asaoka et al (1988) who demonstrated that PEth was able to partially mimic the effects of phosphatidylserine in a cell-free system. PEth was thus by these authors shown to have a stimulatory effect on PKC in this system. NG108-15 cells normally contain phosphatidylserine at levels equivalent to approximately 5-10% of total phospholipids in their cell membranes and it seems reasonable to suppose that the endogenous levels of this activator is in great excess of that which is required to activate the PKC present in the cells. This would then conceal the effect which PEth might have on its own if present in membranes without or with a small amount of phosphatidylserine. In addition, Asaoka has also demonstrated a PKC subclass specificity in the activation by PEth, with PKCγ being the isozyme most activated at physiological Ca^{2+} concentrations similar in size to the concentrations used here (2-3 μM) (Asaoka, 1989). In the present study, we have not specifically examined the effects of PEth on the different PKC isozymes and theoretically therefore, a small isozyme-specific effect might remain undetected in our assay of total PKC phosphorylation activity, especially since the total PKC activity in NG108-15 cells is rather small compared to many other cell lines. In any case, since non-specific PKC inhibition and down-regulation did not remove the PEth effect on $I(1,4,5)P_3$ levels it seems unlikely that PKC activity changes play a major role in mediating this efffect.

An interesting side-effect of PEth became evident when we used the non-specific phosphorylation substrate MBP. In these assays it seemed that PEth stimulated the phosphorylation or alternatively inhibited the dephosphorylation of this substrate. The specificity and functional role of such an effect remain to be elucidated.

There are several possibilities for other mechanisms through which the PEth effect on $I(1,4,5)P_3$ may be mediated. PEth has in other systems been shown to have specific effects on certain membrane-bound enzymes including Na^+/K^+-ATPase and 5'-nucleotidase (Omodeo-Salé et al 1991). Therefore a direct effect on the enzymes involved in the metabolism of $I(1,4,5)P_3$ may also be a possibility but other more indirect effects can of course not yet be excluded. PEth has, for example, also been shown to have various effects on the physical properties of membranes and an indirect effect on membrane-associated enzymes through changes in membrane parameters is also a possibility which merits further study (Omodeo-Salé et al, 1991; Bondeson and Sundler, 1987).

In conclusion, we have presented evidence suggesting that PEth increases $I(1,4,5)P_3$ levels in NG108-15 cells through a mechanism which does not involve a change in the activity of PKC. The exact mechanism behind the increase in basal $I(1,4,5)P_3$ levels remains to be elucidated.

ACKNOWLEDGEMENTS

The authors would like to thank Monica Mihailescu and Berit Färjh for excellent technical assistance. Financial support was obtained from the Swedish Medical Research Council (Project nr 21X-05249 and 03P-08895), the Swedish Alcohol Research Fund, the Swedish Society of Medicine, the Albert Påhlsson Foundation and the Medical Faculty of Lund University which is hereby gratefully acknowledged.

REFERENCES

Alling C., Gustavsson L. and Änggård E. (1983) An abnormal phospholipid in rat organs after ethanol treatment. *FEBS Lett.* 152, 24-28.

Alling C., Gustavsson L., Månsson J-E., Benthin G. and Änggård E. (1984) Phosphatidylethanol formation in rat organs after ethanol treatment. *Biochim.Biophys.Acta* 793, 119-122.

Asaoka Y., Kikkawa U., Sekiguchi K., et al. (1988) Activation of a brain-specific protein kinase C subspecies in the presence of phosphatidylethanol. *FEBS Lett.* 231, 221-224.

Asaoka Y. (1989) Distinct effects of phosphatidylethanol on three types of rat brain protein kinase C. *Kobe J. Med. Sci.* 35, p. 229-237.

Berridge M.J. (1984) Inositol trisphosphate and diacylglycerol as second messengers. *Biochem.J.* 220, 345-360.

Berridge M.J. and Irvine R.F. (1984) Inositol trisphosphate, a novel second messenger in cellular signal transduction. *Nature* 312, 315-321.

Billah M.M. and Anthes J.C. (1990) The regulation and cellular functions of phosphatidylcholine hydrolysis. *Biochem.J.* 269, 281-291.

Bondeson J. and Sundler R. (1987) Phosphatidylethanol counteracts calcium-induced membrane fusion but promotes proton-induced fusion. *Biochim.Biophys Acta* 899, 258-264.

Bradford M.M. (1976) A rapid and sensitive method for the quantitation of microgram quantities of protein utilizing the principle of protein dye binding. *Anal.Biochem.* 72, 248-254.

Bredt D.S., Mourey R.J. and Snyder S.H. (1989) A simple, sensitive, and specific radioreceptor assay for inositol 1,4,5-trisphosphate in biological tissues. *Biochem.Biophys.Res.Commun.* 159, 976-982.

Chakravarthy B.R., Bussey A., Whitfield J.F., Sikorska M., Williams R.E. and Durkin J.P. (1991) The direct measurement of protein kinase C (PKC) activity in isolated membranes using a selective peptide substrate. *Anl. Biochem.* 196, p. 144-150.

Costa L.G. (1990) The phosphoinositide protein kinase C system as a potential target for neurotoxicity. *Pharm.Res.* 22, 393-408.

Diamond I. (1989) Alcoholic myopathy and cardiomyopathy. *N.Engl.J.Med.* 320, 458-459.

Eibl H. and Kovatchev S. (1981) Preparation of phospholipids and their analogs by phospholipase D. *Methods Enzymol.* 72, 632-639.

Exton J.H. (1990) Hormonal regulation of phosphatidylcholine breakdown. *Biology and Medicine of Signal Transduction* 24, 152-157.

Gustavsson L. and Alling C. (1987) Formation of phosphatidylethanol in rat brain by phospholipase D. *Biochem.Biophys.Res.Commun.* 142, 958-963.

Gustavsson L., Moehren G. and Hoek J.B. (1991) Phosphatidylethanol formation in rat hepatocytes. *Ann.N.Y.Acad.Sciences* 625, 438-440.

Higashi K. and Hoek J.B. (1991) Ethanol causes desensitization of receptor-mediated phospholipase C activation in isolated hepatocytes. *J.Biol.Chem.* 266, 2178-2190.

Kobayashi M. and Kanfer J.N. (1987) Phosphatidylethanol formation via transphosphatidylation by rat brain syaptosomal phospholipase D. *J.Neurochem.* 48, 1597-1603.

Lundqvist C., Rodriguez F.D., Simonsson P., Alling C. and Gustavsson L. (1993) Phosphatidylethanol affects inositol 1,4,5-trisphosphate levels in NG108-15 neuroblastoma x glioma hybrid cells. *J.Neurochem.* 60, 738-744.

Messing R.O., Peterson P.J., and Henrich C.J. (1991) Chronic ethanol exposure increases levels of protein kinase-C-delta and kinase-C-epsilon and protein kinase-C-mediated phosphorylation in cultured neural cells. *J.Biol.Chem.* 266, 23428-23432.

Metz S.A. and Dunlop M. (1991) Inhibition of the metabolism of phosphatidylethanol and phosphatidic acid, and stimulation of insulin release, by propranolol in intact pancreatic islets. *Biochem.Pharmacol.* 41, R1-R4.

Omodeo-Salé F., Lindi C., Palestini P. and Masserini M. (1991) Role of phosphatidylethanol in membranes - effects on membrane fluidity, tolerance to ethanol, and activity of membrane-bound enzymes. *Biochemistry.* 30, 2477-2482.

Shukla S.D. and Halenda S.P. (1991) Phospholipase D in cell signalling and its relationship to phospholipase C. *Life Sci.* 48, 851-866.

Urbano-Marquez A., Estruch R., Navarro-Lopez F., Grau J.M., Mont L. and Rubin E. (1989) The effects of alcoholism on skeletal and cardiac muscle. *N.Engl.J.Med.* 320, 409-415.

Yasuda I., Kishimoto A., Tanaka S-I., Tominaga M., Sakurai A., Nishizuka Y. (1990) A synthetic peptide substrate for selective assay of protein kinase C. *Biochem.Biophys.Res.Comm.* 166, 1220-1227.

MULTIPLE ACTIONS OF ETHANOL ON ACETYLCHOLINE RECEPTORS

Keith W. Miller

Department of Anesthesia
Massachusetts General Hospital
and
Department of Biological Chemistry and Molecular Pharmacology
Harvard Medical School
Boston, MA 02114

INTRODUCTION

The molecular mechanisms underlying the many acute actions of ethanol remain unknown (I have deliberately used the plural here because it seems unlikely that a single mechanism could underlie all the diverse actions of ethanol). There are a number of reasons why our understanding is so poor. First, the pharmacology of ethanol is complex. It has access to all parts of the body and exerts its action at very high concentrations (typically 10-100 mM) and in a nonspecific way. Furthermore, it lacks pharmacological antagonists. Thus, most of the tools of pharmacology are rendered impotent. Second, and not unrelated, the physiological sites of ethanol's actions remain ill-defined, so that detailed studies of the type required to elucidate molecular mechanisms have been discouraged. Who would want to risk the years of research required to elucidate such mechanisms when the system chosen for study might turn out to be irrelevant? Thirdly, ethanol most likely acts on membrane-embedded proteins, which makes the most powerful structural techniques of biochemistry (x-ray crystallography, two dimensional NMR) difficult to apply for technical reasons. In the face of these uncertainties work has tended to concentrate on rather general models such as the lipid bilayer or purified proteins such as luciferase (see Chapter by Rubin in this volume for an excellent review).

REQUIREMENTS FOR ELUCIDATING THE MOLECULAR MECHANISMS OF ETHANOL'S ACTION

What are the requirements for elucidating ethanol's action at a molecular level? Experience with other proteins suggests that we need a target protein that can be obtained in a highly purified, preferably homogeneous, form. Such a protein's function should be studied in great detail because most proteins exist in several conformational states that differ little in free energy (G). The free energy of each state is defined by thousands of interactions arising from the attractive and repulsive forces experienced by each atom as well as by the protein's interaction with it environment, including solvation by water and lipids and contact with neighboring lipids. In this delicately balanced system, the addition of drugs may perturb any of these interactions, altering the relationships between the conformations and changing the net behavior of the system. Therefore, each of the individual kinetic steps defining the pathways connecting these various conformational states must be experimentally resolved; that is by determining which rate constants are affected by ethanol, the locus (or loci) of its actions will be precisely defined.

Alcohol, Cell Membranes, and Signal Transduction in Brain
Edited by C. Alling *et al.*, Plenum Press, New York, 1993

A priori, ethanol might act in two fundamentally different ways. First, it might stabilize one conformational state of the protein (for example, state C in Figure 1) relative to the others, slowing the rate constants of the conformational change from C to B (k_{-2} in Figure 1) but not that from B to C (k_{+2}), with the result that a higher fraction of the protein would occupy this state in the presence rather than in the absence of ethanol. This situation is equivalent to that of the classical allosteric interaction; it might result either from ethanol binding to a site or pocket only available when the protein adopts the spatial configuration required for that conformational state or from a physico-chemical interaction that might alter, for example, the protein's solvation. Secondly, ethanol might change the activation energy between two states (states A and B in Figure 1) altering the forward and backward rate constants (k_{+1} and k_{-1}) without affecting the proportion of the protein occupying each of the ground states at equilibrium. Such an action could result from a number of causes, amongst them a decrease in lipid viscosity (equivalent to an increase in lipid fluidity). Whether the action would have any physiological consequences would then depend on whether the *rates* of the reaction or the proportions of the protein in conformation A and B were important to the net behavior of the system. Only if rates are important would there be any outcome for this action of ethanol.

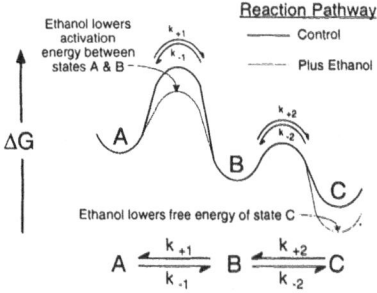

Figure 1. Illustration of possible modes of action of ethanol on a protein that can adopt three different conformational states, represented A, B and C, that are in equilibrium with each other as defined in the diagram. The highest proportion of the protein will be in the state with the lowest free energy, etc., such that [C]>[B]>[A], where [ ] denotes concentration. The forward and backward rate constants between the states are denoted by k_{+i} and k_{-i} respectively, where the subscript i defines which step is being referred to (i = 1 or 2). The equilibrium constants, K_i, are given by K_1 = [B]/[A] = k_{+1}/k_{-1} and K_2 = [C]/[B] = k_{+2}/k_{-2}, and the free energy difference between A and B is ΔG_1 = -RTlnK$_1$ and between B and C is ΔG_2 = -RTlnK$_2$. Two actions of ethanol are illustrated and discussed in the text above.

Once the kinetic sites of ethanol's action are established, hypotheses about the underlying mechanism may be formulated, enabling further studies to be designed to test them. For example, if one conformational state is found to be stabilized by ethanol (say state C in Figure 1), the mutually exclusive alternate hypotheses that (1) ethanol binds to a site only available in that state or that (2) ethanol changes the "fluidity" of the lipids surrounding the protein may be tested. In the first case the dependence of the fraction of receptor in state C upon ethanol's concentration would be hyperbolic and in the second case it would be linear.

The clearest conclusion that can be obtained from such kinetic experiments is that certain models are invalid. On the other hand, when the kinetic data are consistent with a model it always remains possible that some other unexamined model might do equally well. This possibility can only be taken care of by undertaking structural studies to establish whether or not the molecular events thought to underlie the model really happen. For example, in case (2) above, a linear dependence on ethanol concentration implies a class of possible physico-chemical mechanisms of action of which perturbation of the interactions between the protein and its surrounding lipid bilayer is but one. Structural studies, such as those involving spin-labeled lipids, carried out in a purified system would be required to test this hypothesis. In practice, such structural studies require much more protein than do kinetic studies. This is mainly because of the insensitivity of the methods available and partly because of the losses that occur during purification of proteins. Thus, while it is possible for an electrophysiologist to observe a single ion channel open and close, a spectroscopist doing an electron paramagnetic resonance experiment requires some 10^{15} molecules of spin labeled protein or lipid!

SELECTING TARGETS FOR MOLECULAR MECHANISM STUDIES

There is ample evidence that transmembrane ion channels are one important site of action for ethanol. Fortunately advances in molecular neurobiology in the last decade or so have led to the elucidation of the amino acid sequence of more and more of these channels, and it has become apparent on the basis of homology that these channels fall into several superfamilies distinguished by the number of putative transmembrane helices that are assigned to them by hydropathy analysis. In particular one superfamily of ligand-gated channels possessing four transmembrane helices per subunit has emerged as a target for ethanol (Stroud et al., 1990). The members of this family include the GABA, the glycine, the glutamate, the NMDA, the $5HT_3$ and the nicotinic acetylcholine (nAcChoR) receptors. Effects of ethanol on several of these receptors are described in this volume. It should be noted that ethanol's actions on the nAcChoR (this chapter) and the $5HT_3$ receptor (Lovinger, this volume) are quite similar.

While transitions between the open state and neighboring states of all these receptors can be studied by electrophysiological techniques, the nAcChoR has a unique advantage. One of its subtypes occurs in the electric organ of certain fish in great abundance. For example, per gram of tissue the nAcChoR is some 10,000 times more abundant in the electric tissue of *Torpedo* than is the GABA receptor in mammalian brain. This has enabled more detailed studies to be made of the nAcChoR than of any other member of the superfamily (Lester, 1992).

Extracellular

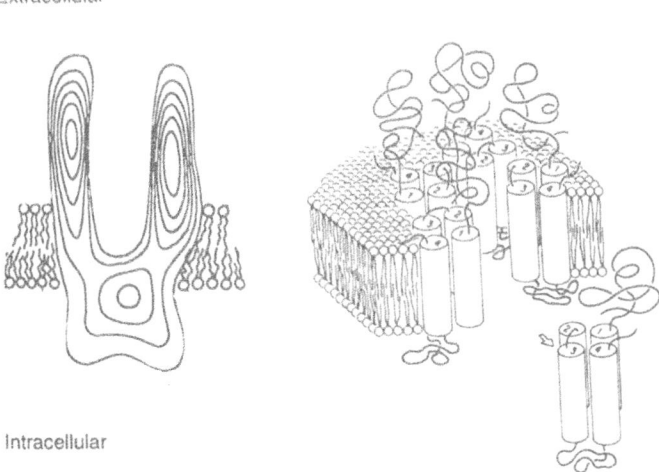

Intracellular

Figure 2. The structure of the nAcChoR. The cross-section through the receptor (left) shows contours of electron density and is adapted from Brisson and Unwin (1985). The phospholipid molecules have been superimposed to scale by Barrantes (1989); about 24 surround the receptor in each leaflet of the bilayer. The receptor, which is about 110 Å long and 85 Å in diameter, extends from the bilayer on the extracellular side forming a deep funnel-shaped structure presumably leading to a central channel or pore. On the right is shown a cartoon of the proposed oligomeric structure of the members of this superfamily of receptors (Olsen and Tobin, 1990). Left panel adapted from Barrantes (1989).

It has been established that the nAcChoR is a ~280kD protein composed of four different subunits of stoichiometry $\alpha_2\beta\gamma\delta$ (see for example Stroud et al., 1991; and Barrantes, 1989). The acetylcholine binding sites reside one each on the two alpha subunits. Photolabeling has established that the second trans-membrane helix (M2) of each subunit lines the channel in a centrosymmetric fashion and that the M1 and M4 helices are exposed to the lipid bilayer (Revah et al., 1990; Blanton and Cohen, 1992: reviewed in Rankin et al., 1993). The dimensions of the receptor have been established by imaging of electron micrographs and the number of lipids surrounding the receptor has been established by electron paramagnetic resonance spectroscopy to be around fifty (Reviewed by Stroud et al., 1990). Figure 2 summarizes the relatively high degree to which the topology of the receptor is known. The nAcChoR may be purified to homogeneity and reconstituted into defined lipids.

THE KNOWN CONFORMATIONAL STATES OF THE ACETYLCHOLINE RECEPTOR

Figure 3 summarizes what has been learnt from electrophysiological and biochemical studies about the conformational changes that the nAcChoR undergoes (Adams, 1981; Neubig, et al., 1982; Heidmann, et al., 1983) At peripheral synapses and in the absence of other drugs only the top line is functionally important because acetylcholine esterase limits the time over which the receptor is exposed to acetylcholine. However, it is quite possible that in the central nervous system exposure to acetylcholine may be more prolonged.

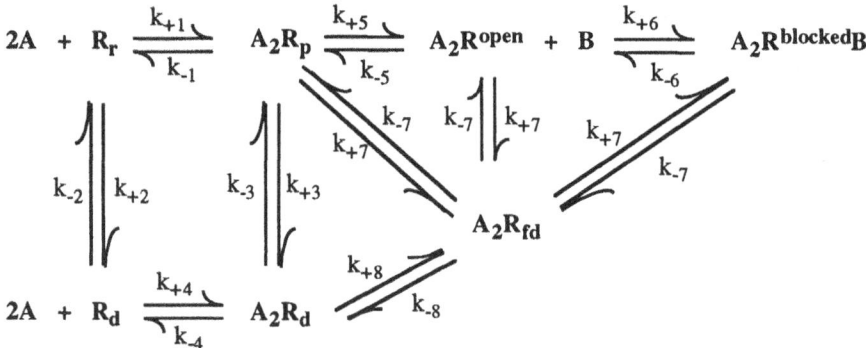

Figure 3. The scheme summarizes the known states of the AcChoR and the pathways between them. **A** denotes an agonist (for example, acetylcholine or nicotine) and **R** the receptor. Subscripts define the conformational state of the receptor, which are: the resting, R_r; desensitized, R_d; fast desensitized, R_{fd}; pre-open, R_p; open, R^{open}; and blocked, $R^{blocked}$, states. Rate constants between the states are denoted by k_i, where a + and − denote forward and backward rate constants. The associated dissociation constant is given by $K_i = k_{-i}/k_{+i}$. **B** denotes an agonist at high concentration or a noncompetitive inhibitor such as a local anesthetic.

In the absence of acetylcholine (**A**) the receptor may be thought of as existing in an equilibrium between a resting state (R_r) and a desensitized state (R_d; Figure 3). When acetylcholine is discharged from the presynaptic terminal, two acetylcholine molecules bind to the receptor. Binding to the desensitized state does not result in channel opening, whereas binding to the resting state leads to channel opening (A_2R^{open}) within a few hundred microseconds (Adams, 1981). The channel remains open for a short time (about a millisecond) before closing. It usually opens and closes several times before the agonist dissociates and it can conduct 10^4 ions per millisecond. If agonist persists in the vicinity of the receptor, other conformational changes occur. Receptors convert in some hundred milliseconds to the fast desensitized state (A_2R_{fd}) and within minutes to the slow desensitized state (A_2R_d); neither state conducts ions. The R_r state has the lowest, and the R_d state the highest, affinity for agonists (Neubig et al, 1982; Heidmann et al, 1983). Recently it has been established that high agonist concentrations (**B**) have an inhibitory action on ion flux ($A_2R^{blocked}B$; Sine & Steinbach 1984; Pasquale et al, 1983), termed self-inhibition, which occurs by a similar mechanism to local anesthetics action.

ACTIONS OF ETHANOL ON THE ACETYLCHOLINE RECEPTOR'S KINETIC PATHWAYS

Ethanol has actions on nearly all parts of the kinetic scheme in Figure 3. Although much remains to be learnt, the broad features of ethanol's action are established and some tentative working hypotheses can be proposed.

A word on effective ethanol concentrations is necessary. Much of the work described below is performed at 4°C, a physiological temperature for the *Torpedo*. Because ethanol becomes *less* potent as the temperature is lowered, the effective ethanol concentrations are not as high as they may seem. Indeed, the general anesthetic concentration of ethanol at 4°C is around 300-400 mM and the inebriating concentration is probably a tenth of this (reviewed in Forman et al., 1989).

Slow Desensitization

One general feature is that in the absence of agonist, when only the R_r and R_d conformations are available for ethanol to act upon, it has weak effects. Thus, Forman *et al.* (1989), working with nAcChoR from *Torpedo*, reported that preincubation with 1.0 M ethanol for 30 min. only caused sufficient desensitization to reduce the subsequent ion flux response of the system to agonist by 25%. Preincubation with agonist alone also causes a slow conversion from the R_r to R_d state because agonists have a higher affinity for the R_d than the R_r state. Ethanol has a much bigger effect on this rate of agonist-induced slow desensitization than in causing desensitization on its own. Agonist-induced slow desensitization is enhanced four-fold if 1.0 M ethanol is included with the agonist in the preincubation medium.

Overall, the important conclusion, that applies even more strongly to fast desensitization (see below), is that in the absence of agonist ethanol has rather modest effects on the nAcChoR. This suggests that even at concentrations above physiological the protein's structure is unaffected, a conclusion in agreement with some preliminary data we have obtained using spin–labels covalently attached to the nAcChoR (Dalton & Miller, unpublished work).

Fast Desensitization

Whereas slow desensitization occurs in minutes, fast desensitization occurs in a few hundred milliseconds and requires special techniques to be studied. In the nAcChoR there is only one report on fast desensitization but the results bear comparison with those for the $5HT_3$ receptor that are reported by Lovinger elsewhere in this volume.

Once again, preincubation with ethanol alone had little or no effect, but ethanol strongly modulated the agonist's ability to induce fast desensitization. The rate of agonist-induced fast desensitization increases sigmoidally with acetylcholine concentration (Figure 4). Addition of 0.5 M ethanol had two effects. First the maximum rate of fast desensitization was enhanced and second the sigmoid curve was shifted to the left, that is ethanol increases the apparent affinity of acetylcholine.

The result is that at low agonist concentrations the rate of fast desensitization can be enhanced nearly an order of magnitude whereas at saturating agonist concentrations the rate is enhanced one and a half times at 0.5 M ethanol. The effect of lower concentrations of ethanol have not been examined, but there are some indications that the threshold for ethanol's action on fast desensitization occurs at relatively low concentrations.

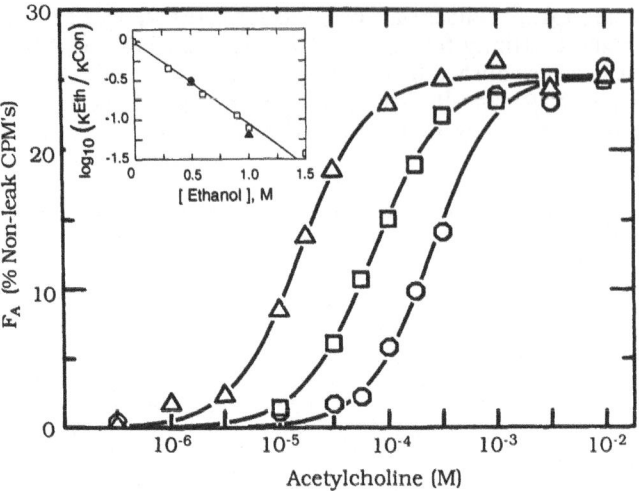

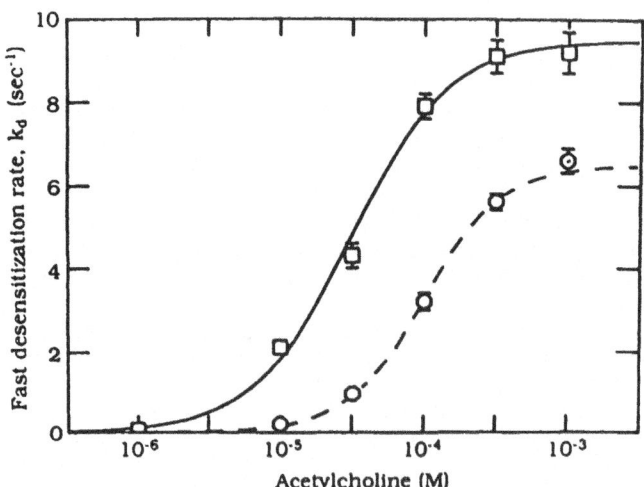

Figure 4. Ethanol shifts the acetylcholine concentration-response curve for both channel opening (upper panel) and fast desensitization (lower panel) to the left by equal amounts. **In the upper panel** cation flux was measured over as little as five milliseconds after adding acetylcholine and ethanol simultaneously to the vesicles. Control O, 0.5 M, $\square$, and 1.0 M ethanol, $\triangle$. **In the lower panel** fast desensitization was measured by pulsed quenched-flow.after preincubation for 100 to 500 milliseconds with the concentration of acetylcholine shown together with either no, O, or 0.5 M, $\square$, ethanol. The **inset** to the upper panel shows the relative shift in the apparent dissociation constant, K (obtained from the half maximum point of the curves), induced by different concentrations of ethanol. For fast desensitization a single point at 0.5 M ethanol is shown as O, and for channel opening two series of experiments are shown as $\square$ and $\blacktriangle$ (see Forman et al., 1989 for details).

Channel Activation

Early electrophysiological experiments indicated that ethanol had effects on the nAcChoR quite unlike other general anesthetics or longer chain alcohols. Instead of inhibition, enhancement is seen (Gage et al., 1975; Bradley et al., 1980).

The major effect of ethanol on acetylcholine-induced channel opening, as measured by cation flux from *Torpedo* vesicles, is to shift the agonist activation curve to the left without increasing maximum activation. This is an acute effect of ethanol because it is seen in experiments where acetylcholine and ethanol are added simultaneously to vesicles (mixing time ~10^2 µsec) and cation efflux out of the vesicles is measured over as little as the next five milliseconds (Figure 4). The enhancement of apparent agonist affinity for channel opening (or flux) is exactly the same as is that for fast desensitization (Figure 4 inset). However, ethanol's net effect on ion flux at any given acetylcholine concentration is a little smaller than that on fast desensitization because there is no increase in the maximum effect. Nonetheless, it has been calculated that at low agonist concentrations and 4°C, 50 mM ethanol (in the inebriating range, see above) causes a twenty percent enhancement of flux (Forman et al., 1989).

The ethanol-induced enhancement of acetylcholine's apparent affinity for channel opening can be attributed *a priori* to a number of possible mechanisms, including: (a) increasing agonist affinity for its binding site, K_1; (b) decreasing K_5 causing more occupied nAcChoR, R_p, to open, R^{open}, and (c) increasing single channel conductance.

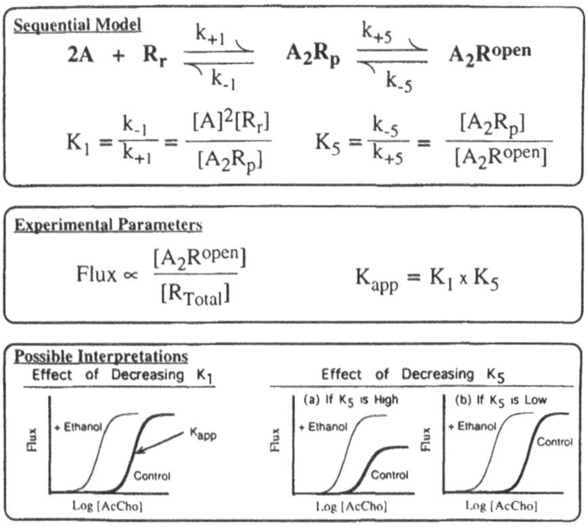

Figure 5. Models for the ethanol-induced increase in apparent affinity of acetylcholine for stimulating ion flux. *In the upper panel*, the top row presents the relevant part of Figure 3 which defines the sequential model; the simplification is achieved by carrying out the experiment on such a rapid time scale that the other reactions do not have time to take place. The second row in the upper panel defines the equilibrium dissociation constants, K_i. *In the middle panel*, the experimental parameters, Flux and K_{app}, are defined, the latter in terms of the sequential model assuming a steady state during the flux measurement. *The lower panel* shows the predicted behavior of the flux response when each of the parameters in the model is varied independently.

Of these putative mechanisms, the latter, c, is ruled out by single channel studies which show that ethanol does not change channel conductance (Nelson and Sachs, 1981: Aracava et al., 1991). However, distinguishing between the remaining two explanations is more difficult. The argument will be easier to follow with reference to figure 5, above.

The apparent dissociation constant for stimulation of flux is simply the product of the dissociation constants for the two steps (Figure 5, middle panel). If K_1 is decreased then acetylcholine binds more tightly to its site and the pre-open state becomes more highly populated at lower acetylcholine concentrations with the result that the flux concentration-response curve will shift to the left. The maximum flux will be unchanged because the ratio of pre-open to open receptors, K_5, is unaltered (Figure 5; lower panel, left-hand graph. Thus, explanation (a) above is quite consistent with the data. At first sight, explanation (b) now seems unlikely because decreasing K_5, while it does decrease K_{app}, would lead to an increase in the maximum observed flux since a higher fraction of the receptors will now be in the open state (Figure 5; lower panel, graph middle graph). Unfortunately however, single channel studies show that the ratio of microscopic channel opening rate, k_{+5} (Figure 3), to the closing rate, k_{-5}, is high (at least 32 for acetylcholine (Colquhoun and Ogden, 1988) meaning that $K_5 = 0.03$). Thus, at saturating acetylcholine concentrations, the probability of non-desensitized channels being open is ≥ 0.97 and displacing the equilibrium further towards the open state by decreasing K_5 would result in little additional *observable* flux (Figure 5; lower panel, graph (b) at right). Therefore, explanation (b) cannot be ruled out on the basis of flux studies with acetylcholine. Unfortunately single channel studies, which in principle would provide an unequivocal answer, are complicated by time dependent effects on kinetics because drugs (acetylcholine and ethanol in this case) must be in contact with the receptor for the long periods of time necessary to accumulate sufficient single channel events to provide a statistical sample. Consequently the preliminary data that have been published to date do not settle this question (Nelson and Sachs, 1981: Aracava et al., 1991).

To solve the problem of whether ethanol acts on the binding step or the opening step of the sequential model (that is on K_1 or K_5), we have recently sought to employ an agonist that is not so efficient at opening channels (ie in pharmacologists' terms a partial agonist having a higher K_5 and therefore a lower maximum flux). Unfortunately, most such agents have relatively high affinity for their noncompetitive inhibition site and are good self-inhibitors. For example a recent study with nicotine found that $K_{app} = (K_1 \times K_5)$ = 4 mM and $K_6 = 1$ mM (Tonner, et al., 1992), compared to acetylcholine for which K_{app} = 100 μM and $K_6 = 100$ mM. Thus, with acetylcholine the two effects are well separated and a bell-shaped concentration-response curve with a long plateau between activation and self-inhibition is observed, whereas with nicotine opened channels are immediately blocked and a very weak flux response is seen (Tonner, et al., 1992). Addition of ethanol to nicotine causes K_{app} to decrease and consequently the maximum flux is enhanced, reaching values close to those of acetylcholine, because channels can now open at lower nicotine concentrations without being blocked (there is an additional effect on K_7 which enhances this effect — see section following). These results with nicotine have some quite fascinating implications if they can be reproduced with neuronal nAcChoR! However, they do not resolve the issue of whether ethanol acts on K_1 or K_5.

Recently, we have found that with the partial agonist suberyldicholine the self-inhibition problem is not so severe (Tonner and Miller, unpublished data). It has a bell-shaped concentration-response curve with a long plateau like acetylcholine's but with a lower maximum flux. In this case our preliminary data show that ethanol increases the maximum flux response while lowering K_{app}, indicating that ethanol has a much stronger effect on K_5 than on K_1.

Whether this conclusion is unique to suberyldicholine or applies equally to acetylcholine is unclear. In their work on the other hand, Forman et al. (1989) had argued that in the case of acetylcholine it is more parsimonious to assume that a decrease in K_1 explains the leftward shift in the agonist concentration-flux curves because ethanol causes the same degree of leftward shift for fast desensitization as it does for flux (Figure 4). Since inspection of figure 3 shows that the apparent dissociation constant for fast desensitization is roughly equal to $(K_1 \times K_7)$ there is some merit to this argument. On the other hand the maximum fast desensitization rate increases so K_7 is not constant and must account for some of the leftward shift. Continuing studies will be needed to finally resolve some of these questions.

Self-inhibition

Recent studies of partial agonists have shown that their apparent affinity for the blocked state is decreased by ethanol (ie their dissociation constant, K_6, increases). Thus, Tonner *et al.* reported (1992) that K_7 for nicotine increased linearly with ethanol concentration from 0.9 mM in the control to 4.3 mM at 1.5 M ethanol. These data were obtained by stimulating at saturating acetylcholine concentrations so that essentially all the receptor is in the open state, and then determining the noncompetitive inhibition curve of the partial agonist nicotine. The simplest explanation for the ethanol-induced increase in K_6 is that, not only does ethanol stabilize the open state, but it also stabilizes the blocked state. The blocked state is stabilized less than the open state because K_{app} is more sensitive to ethanol than than K_6.   Similar results are now being obtained with suberyldicholine (Tonner & Miller, 1993).

General Conclusions from Kinetic Studies

Overall, considerable progress has been made in identifying the steps of the kinetic scheme (Fig. 3) defining the nAcChoR's behavior upon which ethanol acts.   This information has come largely from biochemical techniques.   These have begun to define the dependence of particular equilibrium constants on ethanol concentration and, consequently, the perturbation in free energy changes between conformations.   Further progress can be expected in studies on fast and slow desensitization, in particular the determination of rate constants.and ultimately of activation energies   Such advances may be aided by extending the methods used for following kinetics in biochemical studies to include fluorescent ligands that can report on conformations other than the open channel state.   However, the events associated with channel opening and closing will require techniques with even faster time resolution.

The biochemical techniques have been so useful because isolated membranes can be mixed more rapidly with agonists and ethanol than can physiological preparations.   Thus, while electrophysiological techniques are no doubt more elegant and powerful, their full force has been blunted by the effects of desensitization.   Fortunately, the development of concentration jump techniques in recent years is beginning to solve this problem (Brett et al., 1986).   However, it is still not clear whether these new techniques will enable single channel studies to define rate constants such as k_{-1}, k_{+5}, k_{-5} and k_{+7} because observations of single channel switching over prolonged periods of time are required to accumulate sufficient statistical data.   During such long observational periods desensitization will occur.

Overall, then we are still some way from achieving such a complete knowledge of the kinetics of the nAcChoR that unambiguous conclusions can be drawn.   However, although more studies are needed, some tentative tests of existing models may already be made.

SOME MECHANISTIC MODELS DERIVED FROM KINETIC STUDIES

Gage et al. (1975) suggested that short-chain alkanols, including ethanol, reduce the membrane dielectric constant near nAcChoR, thus slowing the rate-limiting reorientation of a protein dipole within the transmembrane electric field during channel closure.   Assuming that the change in membrane dielectric constant was a linear function of the weight fraction of membrane-phase alcohol, this model accounted for the exponential relationship between miniature endplate potential lifetime and ethanol concentration.   The hypothesis was indirectly supported by evidence that ethanol inhibits a crustacean neuromuscular junction with a voltage sensitivity opposite to that observed for vertebrates, which the model attributes to reversal of the gating dipole orientation (Adams et al., 1977).

The physical basis of the Gage mechanism requires that a transmembrane electric field exist for the action of ethanol to occur, yet in our work both flux and fast desensitization are enhanced by ethanol in the absence of an applied transmembrane voltage.

Bradley et al. (1984) proposed a "hydrophobic patch" site within the AcChoR ion channel.   In the Bradley model, ethanol binds to the open channel state without inhibiting cation flux, thus increasing the apparent agonist affinity by "uncompetitive" stabilization of the open channel state (i.e. increased channel lifetime, as in the Gage model).   They further assumed that long- and short-chain alkanols both interact with this channel site but only the bulkier long chain alcohols cause channel block.   This elegant model predicts that ethanol should compete with an inhibitory alcohol like octanol with the result that octanol's apparent inhibition constant should be increased in the presence of ethanol.   Our rapid kinetics studies show on the contrary that even in the presence of 1 M ethanol no such shift occurs (Wood et al., 1991).   Furthermore, the logarithm of the ethanol-induced increase in acetylcholine's apparent affinity, which is proportional to the free energy change, is linear with ethanol concentration suggesting that some sort of physical interaction is involved rather than binding to a saturable site.   An alternative explanation that is consistent with this observation is that ethanol might act by binding to one or more low affinity sites.   New approaches must be developed to answer these questions.

One recent new idea we have introduced is to take a molecular pharmacological approach. For example, what compounds enhance the affinity of agonists for opening the channel? Normal alkanols with chain lengths up to butanol do so, those with longer do not. We recently studied a series of cycloalkanemethanols and found that the first six members of this series also enhanced agonist affinity. By comparing molecular parameters, we could formulate the hypothesis that only molecules below a critical molecular length should enhance affinity (Wood, Hill & Miller, unpublished data). More alcohol analogues must be studied to thoroughly test this hypothesis, but this example illustrates the approach.

The ability of alcohols to stabilize the desensitized state, R_d, correlates with their ability to disorder the lipid bilayer in which the nAcChoR is embedded (Miller et al., 1987). Furthermore reconstituting the nAcChoR into different lipid mixtures also modulates the equilibrium between the R_r and R_d states (Fong and McNamee, 1986). Whether there is a causal relationship remains to be demonstrated. An alternative hypothesis might be based upon the observation of Revah *et al.* (1991) which demonstrates coupling between the channel's lining and the acetylcholine binding site. These workers showed that mutation of leucine 247 in the channel lining on the M2 helix to a more polar residue such as threonine, decreased the rate of desensitization and increased the apparent affinity for acetylcholine. Interestingly, a similar story is emerging in the case of the $5HT_3$ receptor (Maricq et al., 1991).

POSSIBLE PHYSIOLOGICAL SIGNIFICANCE OF NICOTINIC RECEPTORS IN ETHANOL'S ACTIONS

Although the major justification for studying the nAcChoR is the practical one of studying the molecular mechanisms by which ethanol acts on the only member of the four transmembrane helix family where this is technically feasible, it is nonetheless interesting to speculate on the possible direct involvement or nAcChoR in ethanol's physiological actions. There are several subtypes of nAcChoR and the study of those in the central nervous system is ongoing (Heinemann et al., 1990). The *Torpedo* nAcChoR belongs to the α-bungarotoxin-binding muscle subgroup which also occurs to a lesser extent in the brain.

The homologous neuronal receptors do not bind α-bungarotoxin and occur in the autonomic and central nervous systems.

In the periphery, the nAcChoR is implicated in the ethanol-induced elevation of circulating epinephrine, perhaps mediated by the adrenal gland, leading to the development of cardiac hypertrophy (Perman, 1958; Adams & Hirst, 1986). Cross tolerance between nicotine and ethanol has been reported for certain behavioral tests (Collins *et al.* 1988). In the CNS of chronically ethanol fed rats, the number of nicotine receptors increases in the thalamus and hypothalamus, but decreases in the hippocampus. In *in situ* perfused stellate ganglia Larrabee & Posternak (1952) showed that, at low ethanol concentrations and submaximal stimulation, ethanol increased the height of the post-synaptic action potential. Significantly, the enhancement lasted some ten minutes but was followed by net inhibition if stimulation was prolonged. Such biphasic effects of ethanol (for a review see Siggins *et al.* 1987) are consistent with the kinetic studies described above, since ethanol initially enhances the action of agonist in opening channels but after more prolonged exposure it enhances desensitization, attenuating the subsequent action of agonist. Interestingly, prior doses of nicotine protect against subsequent nicotine-induced convulsions in mice, an action thought to involve desensitization. Ethanol enhances nicotine's protective action while having none of its own (Collins et al., 1988), a result entirely consistent with our results for slow desensitization. Ethanol's anomalous action on nAcChoRs could have greater implications in the autonomic and central nervous systems where nAcChoRs are involved in the control of transmitter and hormone release (dopamine, catecholamines, enkephalins, serotonin, corticotropin releasing hormone and vasopressin). Significantly, dopamine has been implicated in the behavioral reinforcing effects of a number of drugs, including ethanol and nicotine (Chiara & Imperato, 1988). Nicotinic agonists also mediate longer term effects. Thus, nicotine not only stimulates release of enkephalin *but*, by a different intracellular pathway, affects its synthesis (Eiden *et al.* 1984). Furthermore, it induces transcription of the c-fos protooncogene (Greenberg *et al.* 1986).

LIPID THEORIES; THE NEXT STEP

Reconstituting the purified nAcChoR into Lipid Bilayers

One of the great advantages of having the nAcChoR in abundance at high purity is that it is relatively easy to purify it to homogeneity and insert it into lipid bilayers of defined composition. This enables one to approach one of the central questions of alcohol research. What is the relationship between adaptation of lipid composition and the function of excitable membrane proteins?

Techniques have been developed which allow one to simultaneously purify the nAcChoR and reconstitute it into bilayers of defined lipid composition (Jones et al., 1988). To achieve this, native membranes are dissolved in a detergent, sodium cholate, in the presence of excess lipid. This mixture is applied to an acetylcholine affinity column which retains the nAcChoR and allows all other proteins to elute. While the nAcChoR is attached to the column fresh detergent-lipid mixtures are allowed to equilibrate with it to ensure complete exchange of native for exogenous lipids. The nAcChoR cannot be totally stripped of lipids without permanently losing its native conformation, so this step is important. Finally, the nAcChoR is eluted by adding an agonist to the solution passing through the column. The resulting protein-lipid-detergent eluate is then extensively dialysed allowing formation of detergent-free vesicles.

Two conformational changes have been studied as a function of lipid composition (Fong and McNamee, 1986). These are slow desensitization (step 2 in Figure 3) and channel opening.(steps 1 and 5 in Figure 3). When the nAcChoR is reconstituted into dioleylphosphatidylcholine (DOPC), it undergoes no conformational changes, but the nAcChoR is still in a native conformation because its function can be recovered again by addition of suitable lipids to its bilayer. Addition of 50, but not 30, mole percent cholesterol to the DOPC allows the R_r to R_d transition to occur but not channel opening (A_2R_p to A_2R^{open}), whereas addition of phosphatidic acid enables both transitions to be observed as does addition of both these lipids together. If the second lipid is changed from phosphatidic acid to phosphatidylethanolamine, then the ability to open the channel is lost. The role of the acyl chains of the phosphatidylcholine were studied in lipid bilayers of composition XPC:PA:Chol 56:19:25 (mole ratio). When X was dioleoyl (18:1 *cis*) or dipalmitelaidoyl (16:1 *trans*), but not dilinoleyl (18:2 *cis cis*), full function was supported.

Thus, important roles in maintaining the ability to undergo transitions between native conformational states are indicated for negatively charged phospholipids, unsaturated acyl chains and cholesterol. All these are features that have been implicated in membrane adaptation to ethanol, so this model holds some promise, at least for one superfamily of membrane proteins, of contributing to our understanding the functional implications of such changes. Unfortunately, the studies to date have used rather rudimentary assays to examine the state of the reconstituted nAcChoR. To obtain the level of detail necessary for mechanistic studies, it will be necessary to develop more informative assays that will work with small amounts of reconstituted membranes. The flux studies already described were carried out in native membranes. They are difficult to carry out in reconstituted vesicles because these have varying ability to trap ions. As an alternative, we are beginning to explore the kinetics of interaction of various fluorescent ligands with the nAcChoR in a stopped flow fluorimeter, a technique that can provide a great deal of kinetic information for a small outlay of material (Heidmann and Changeux, 1979).

Another direction of research that purification and reconstitution facilitates is the application of structural techniques. While it has proved possible to study the structure of the lipid bilayer region of many native membranes, the structure of proteins cannot be studied because of the heterogeneity of the protein composition. To overcome this problem, we have recently begun studies of two aspects of nAcChoR structure using reconstituted membranes. These are the interactions between lipids and the nAcChoR using electron paramagnetic resonance (EPR, sometimes called electron spin resonance or ESR) of spin-labeled lipids and the structure of the whole complex using Fourier transform infra red spectroscopy (FTIR).

Our current EPR studies using reconstituted membranes are achieving very much better resolved spectra of the spin-labeled phosphatidylcholine interacting with the nAcChoR than we achieved previously using native membranes (Abadji, Raines Watts & Miller, in preparation). The spin-label phosphatidylcholine reports that some 55 phosphatidylcholine molecules interact with each nAcChoR, exchanging between the bilayer and the protein's surface at the rate of approximately 10^8 times per second. During the time that the phosphatidylcholine is "stuck" to the surface, the spin–label, which is on the fourteenth carbon down the stearic acid acyl group near the center of the bilayer, is rotating about 10^8 times per second. This is at least an order of magnitude slower than in the lipid bilayer. Ethanol has small effects on the lipid dynamics at the lipid–protein interface which are currently under investigation, but it does not alter the number of lipids in contact with the nAcChoR. Because ethanol probably spends most of its time in the lipid-aqueous interface, it would be of interest to examine spin–labels attached higher up the acyl chains. However, there is a technical problem which makes separating out the signal from spin–labels in the bilayer and at the lipid–protein interface very difficult when the spin–label is attached near the top of the chain. Some progress has been made with this problem, but it appears impossible to obtain information from labels much more than halfway up the chain (Raines & Miller, work in progress).

Our FTIR studies have been carried out in collaboration with Professor Rothschild of Boston University. The technique is particularly sensitive at detecting the difference in structure between two conformational states of a protein. We have demonstrated that the R_r to R_d transition can be detected and characterized (Baenziger et al., 1992a, 1992b). Agonist-induced desensitization turns out to involve quite a small change in protein structure that is very similar in both native and in nAcChoR reconstituted in PC:PA:Chol membranes. However, in DOPC membranes, where exposure to acetylcholine causes no conformation change, some structural changes are still identified by FTIR. These appear to be a subset of those observed in native membranes and they are probably associated with modest changes localized near the acetylcholine binding pocket without contributions from the accompanying propagated conformation change only observed in native and PC:PA:Chol membranes (Rankin, Miller & Rothschild, work in progress). Finally, the resonances arising from bound acetylcholine have been assigned by synthesizing isotopically substituted acetylcholine and identifying the consequent band shifts in the spectra. If this strategy can be extended to other drugs, including ethanol, then direct information on drug-protein interactions may be obtained. Similarly, it may be possible to assign resonances to lipid headgroup-protein interactions and thus to approach questions concerning ethanol's action in this region.

CONCLUSION

Studies on this abundant representative of the four transmembrane ligand-gated channel superfamily are dissecting the action of ethanol into several underlying components. It will be necessary to refine our techniques to focus on each of these individually before the underlying molecular mechanisms may be resolved. Considerable progress is being made and new tools are being developed, so that within the forseeable future aspects of the problem will be isolated and an attack made on ethanol's underlying mechanisms. One potential difficulty in providing a complete description of ethanol's action, is the incomplete knowledge that biochemists have of how lipid–protein interactions modulate membrane protein function.

Finally, with the development of new more efficient expression vectors and the enhancement of sensitivity in spectroscopic instrumentation, the time when the approach described herein can be applied to a wider range of excitable proteins is drawing nigh.

ACKNOWLEDGEMENTS

The support of the National Institute on Alcohol Abuse and Alcoholism (AA-07040) is gratefully acknowledged. None of this work would have been possible without the able and enthusiastic collaboration of my colleagues Stuart Forman, Susan Wood, John Baenziger, Adam Hill, Lauraine Dalton, Douglas Raines, Saffron Rankin, Peter Tonner, Birgitte Bugge, Tony Watts and Ken Rothschild.

REFERENCES

Adams D.J., Gage, P.W. and Hamill, O.P., 1977, Ethanol reduces the amplitude of excitatory postsynaptic currents at a crustacean neuromuscular junction, *Nature* 266:739.

Adams, M.A., and Hirst, M., 1986, The influence of adrenal medullectomy on the development of ethanol-induced cardiac hypertrophy, *Can. J. Physiol. Pharmacol.* 64:592.

Adams, P.R., 1981, Acetylcholine receptor kinetics, *J. Membrane Biol.* 58: 161.

Aracava, Y., Fróes-Ferrão, M.M., Pereira, E.F.R. and Alburquerque, E.X., 1991, Sensitivity of N-methyl-D-aspartate (NMDA) and nicotinic acetylcholine receptors to ethanol and pyrazole, *N. Y. Acad. Sci.* 625:451.

Baenziger, J.E., Miller, K.W. and Rothschild, K.J., 1992a, Incorporation of the nicotinic acetylcholine receptor into planar multilamellar films: characterization by fluorescence and Fourier transform infrared difference spectroscopy, *Biophys. J.* 61:983.

Baenziger, J.E., Miller, K.W. and Rothschild, K.J., 1992b, Probing conformational changes in the nicotinic acetylcholine receptor by Fourier transform infrared difference spectroscopy, *Biophys. J.* 62:64.

Barrantes, F. J. (1989). The lipid environment of the nicotinic acetylcholine receptor in native and reconstituted membranes. *Crit Rev Biochem Mol Biol*, 24: 437.

Blanton, M.P. and Cohen, J.B., 1992, Mapping the lipid-exposed regions in the *Torpedo californica* nicotinic acetylcholine receptor, *Biochemistry* 31:3738.

Bradley, R.J., Peper, K., and Sterz, R., 1980, Post synaptic effects of ethanol at the frog neuromuscular junction, *Nature* 284, 60.

Bradley, R.J., Sterz, R. and Peper, K., 1984, The effects of alcohols and diols at the nicotinic acetylcholine receptor of the neuromuscular junction, *Brain Res.* 295:101.

Brett, R.S., Dilger, J.P., Adams, P.R. and Lancaster, B., 1986, A method for the rapid exchange of solutions bathing excised membrane patches, *Biophys. J,* 50:987.

Brisson, A. and Unwin, P.N., 1985, Quaternary structure of the acetylcholine receptor, *Nature* 315:474.

Chiara, G. and Imperato, A., 1988, Drugs abused by humans preferentially increase synaptic dopamine concentrations in the mesolimbic of freely moving rats, *Proc. Natl. Acad Sci. US.* 85:5274.

Collins, A.C., Burch, J.M., F, C.M. and Marks, M.J., 1988, Tolerance to and cross tolerance between ethanol and nicotine, *Pharmacol. Biochem. Behav.* 29:365.

Colquhoun, D. and Ogden, D.C., 1988, Activation of ion channels in the frog end-plate by high concentrations of acetylcholine *J. Physiol. (London)* 395:131.

Eiden, L.E., Giraud, P., Dave, J.R., Hotchkiss, A.J. and Affolter, H.-U., 1984, Nicotinic receptor stimulation activates enkephalin release and biosynthesis in adrenal chromaffin cells, *Nature* 312:661.

Fong, T.M. and McNamee, M.G., 1986, Correlation between acetylcholine receptor function and structural properties of membranes, *Biochemistry* 25:830.

Forman, S.A., Righi, D.L. and Miller, K.W., 1989, Ethanol increases agonist affinity for nicotinic receptors from *Torpedo*, *Biochim. Biophys. Acta* 987:95.

Gage, P.W., McBurney, R.N., and Schneider, G.T., 1975, Effects of some aliphatic alcohols on the conduction change caused by a quantum of acetylcholine at the toad end-plate, *J. Physiol. (London)* 244:409.

Greenberg, M.E., Ziff, E.B. and Greene, L.A., 1986, Stimulation of neuronal acetylcholine receptors induces rapid gene transcription, *Science* 234:80.

Heidmann, T. and Changeux, J.-P., 1979, Fast kinetic studies on the allosteric interactions between acetylcholine receptor and local anesthetic binding sites. *Eur J Biochem* 94:281.

Heidmann, T., and Changeux, J.-P., 1979, Fast kinetic studies on the interaction of a fluorescent agonist with the membrane-bound acetylcholine receptor from *Torpedo marmorata Eur J Biochem* 94:255.

Heidmann, T., Bernhardt, J., Neumann, E., Changeux, J.-P., 1983, Rapid kinetics of agonist binding and permeability response analyzed in parallel on acetylcholine receptor rich membranes from *Torpedo marmorata*, *Biochemistry* 22:5452.

Heinemann, S., Boulter, J., Deneris, E., Conolly, J., Duvoisin, R., Papke, R., & Patrick, J. (1990). The brain nicotinic acetylcholine receptor gene family. *Prog Brain Res*, 86, 195-203.

Jones, O.T., Eubanks, J.H., Earnest, J.P. and McNamee, M.G., 1988, Reconstitution of the nicotinic acetylcholine receptor using a lipid substitution technique, *Biochim. Biophys. Acta* 944:359.

Larrabee, M.G. and Posternak, J.M., 1952, Selective action of anesthetics on synapses and axons in mammalian sympathetic ganglia, *J. Neurophysiol.* 15:91.

Lester, H.A., 1992, The permeation pathway of neurotransmitter-gated ion channels, *Ann. Rev. Biophys. Biomol. Structure* 21:267-92.

Maricq, A.V., Peterson, A.S., Brake, A.J., Myers, R.M. and Julius, D, 1991, Primary structure and functional expression of the 5HT3 receptor, a serotonin-gated ion channel., Science 254:432.

Miller, K.W., Firestone, L.L. and Forman S.A.,1987, General anesthetic and specific effects of ethanol on acetylcholine receptors, *Ann. N. Y. Acad. Sci.* 492:71.

Nelson, D.J. and Sachs, F., 1981, Ethanol decreases dissociation of agonist from nicotinic channels, *Biophys. J.* 33:121a.

Neubig, R.R., Boyd, N.D. and Cohen, J.B., 1982, Conformations of *Torpedo* acetylcholine receptor associated with ion transport and desensitization, *Biochemistry* 21: 3460.

Olsen, R.W. and Tobin, A.J., 1990, Molecular Biology of GABA$_A$ receptors, *FASEB J.* 4:1469.

Pasquale, E.B., Takeyasu, K., Udgaonkar, J.B., Cash, D.J., Severski, M.C. and Hess, G.P., 1983, Acetylcholine receptor: evidence for a regulatory binding site from investigations of suberyldicholine-induced transmembrane ion flux in *Electrophorus electricus* membranes vesicles, *Biochemistry* 22:5967.

Perman, E.S., 1958, The effect of ethyl alcohol on the secretion from the adrenal medulla in man, *Acta Physiol. Scand.* 44:304.

Rankin, S.E., Raines, D.E., Dalton, L.A. and Miller, K.W., 1993, Functional aspects of acetylcholine receptor-lipid interactions *in:* "Protein Lipid Actions", A. Watts ed., Elsevier, Amsterdam.(in press).

Revah, F. Bertrand, D., Galzi, J.L., Devillers, T.A., Mulle, C., Hussy, N., Bertrand, S., Ballivet, M. and Changeux, J.-P., 1991, Mutations in the channel domain alter desensitization of a neuronal nicotinic receptor, *Nature* 353:846.

Revah, F., Galzi, J.L., Giraudat, J., Haumont, P.Y., Lederer, F., Changeux, J.-P., 1990, The noncompetitive blocker [3][H]-chlorpromazine labels three amino acids of the acetylcholine receptor gamma subunit: implications for the alpha-helical organization of regions MII and for the structure of the ion channel, *Proc.Nat. Acad. Sci. U.S.A.,* 87: 4675.

Siggins, G.R., Bloom, F.E., French, E.D., Madamba, S.G. Mancillas, J., Pittman, Q.J. and Rogers, J., 1987, Electrophysiology of ethanol on central neurons, *N.Y. Acad.Sci.*492:351.

Sine, S.M. and Steinbach, J.H., 1984, Agonists block current through acetylcholine receptor channels, *Biophys. J.* 46:277.

Stroud, R.M., McCarthy, M.P. and Shuster, M., 1990, Nicotinic acetylcholine receptor superfamily of ligand-gated ion channels. *Biochemistry* 29:11009.

Tonner, P.H. and Miller, K.W., 1993, Ethanol enhances suberyldicholine's intrinsic efficacy at the nicotinic receptor of *Torpedo* electroplaques, *Br. J. Pharm.* (abstract, in press).

Tonner, P.H., Wood, S.C., Miller, K.W., 1992, Can nicotine self-inhibition account for its low efficacy at the nicotinic acetylcholine receptor from *Torpedo,* Mol. Pharm. (in press).

Wood, S.C., Forman, S.A. and Miller, KW., 1991, Short chain and long chain alkanols have different sites of action on nicotinic acetylcholine receptor channels from *Torpedo, Mol. Pharmacol.* 39:332.

SENSITIVITY OF NMDA RECEPTORS TO ACUTE AND IN UTERO ETHANOL EXPOSURE

Steven W. Leslie[1,2] and Melanie S. Weaver[2]

[1]Division of Pharmacology
College of Pharmacology
[2]Institute for Neuroscience
The University of Texas at Austin
Austin, Texas 78712

Sensitivity of NMDA receptors to ethanol

Electrophysiological and biochemical studies have shown that N-methyl-d-aspartate (NMDA) receptors are highly sensitive to inhibition by ethanol. However, some laboratories report that very low ethanol concentrations, in the range of 5 mM or less, inhibit NMDA receptor function, while others report that the confines of ethanol sensitivity may start at somewhat higher concentrations. The review that follows addresses the question of where the literature now stands on the issue of the sensitivity of the NMDA receptor to ethanol.

Lovinger et al. (1989) were the first to report that ethanol inhibits NMDA-activated ion currents in cultured embryonic hippocampal neurons with an IC_{50} of approximately 30 mM. Concentrations of ethanol as low as 5 mM were found to be inhibitory. Kainate- and quisqualate-induced ion currents in hippocampal neurons were much less sensitive to ethanol. Similar findings were reported using cultured cerebellar granule cells from 8-day old rats (Hoffman et al., 1989) and neonatal dissociated neurons isolated from whole brain (Dildy-Mayfield and Leslie, 1991; Dildy-Mayfield et al., 1991). Hoffman et al. (1989) showed that NMDA-stimulated $^{45}Ca^{2+}$ uptake was reduced by ethanol with an IC_{50} of approximately 40 mM. In agreement with Lovinger et al. (1989), Hoffman et al. (1989) found that low concentrations of ethanol (10 mM) were capable of reducing $^{45}Ca^{2+}$ uptake. Dildy and Leslie (1989) reported that ethanol, 25 to 100 mM, decreased NMDA-stimulated calcium entry into fura-2 loaded, acutely dissociated neurons isolated from whole brains of 1 day old rat pups. In this study, 50 and 100 mM ethanol significantly reduced NMDA-stimulated calcium entry. Ethanol, 25 mM, appeared to reduce NMDA-stimulated calcium entry but this apparent inhibition was not statistically significant. Thus, ethanol produced a concentration-dependent reduction of NMDA-stimulated calcium uptake into dissociated neurons but the potency was not as great as that observed by Lovinger et al. (1989) and Hoffman et al. (1989). Lima-Landman and Albuquerque (1989) reported two effects of ethanol using an outside-out patch clamp technique in cultured hippocampal cells isolated from rat embryos at 16 to 18 days of gestation. Low concentrations of ethanol (1.74 to 8.65 mM) increased the probability of NMDA channel opening. At concentrations of 8.65 to 17.4 mM ethanol, two of seven patches showed decreased frequency of channel opening while the remainder exhibited increased opening. High ethanol concentrations (86.5 to 174 mM) consistently decreased the probability of opening as well as the mean open time. The observation that low concentrations of ethanol may increase the probability of NMDA channel opening is consistent with the findings of Lovinger et al. (1989) who reported that concentrations of ethanol below 5 mM appeared to increase NMDA-activated currents in cultured hippocampal neurons.

The Lima-Landman and Albuquerque (1989) finding that 8.65 to 17.4 mM ethanol increased channel opening frequency in the majority of patches differs from the findings of Lovinger et al. (1989) and Hoffman et al.(1989) who reported a consistent inhibition of NMDA-stimulated responses with 5 and 10 mM ethanol, respectively. Lima-Landman and Albuquerque (1989) proposed that different subtypes of NMDA receptors may exist, some of which may be more sensitive to inhibition by ethanol than others. Monyer et al. (1992), using molecular cloning techniques, showed evidence for three NMDA subtypes in rat brain. Sugihara et al. (1992) reported that alternative splicing of cDNA clones isolated from a rat forebrain cDNA library generated seven isoforms of NMDA receptors. Thus, it may prove to be the case, as suggested by Lima-Landman and Albuquerque (1989), that NMDA receptor subunits may have different sensitivities to ethanol. The existence of NMDA receptor subtypes may explain why NMDA-stimulated calcium entry into acutely dissociated neurons isolated from whole brain was not inhibited by concentrations of ethanol at or below 25 mM (Dildy and Leslie, 1989). The dissociated neuron preparation is composed of heterogeneous populations of neurons on which various NMDA receptor subtypes may reside. The ethanol inhibition of the NMDA stimulated fura-2 signal in this preparation would report on the average sensitivity of NMDA receptor subtypes.

Other recent electrophysiological and biochemical studies have added questions concerning whether NMDA-mediated responses are sensitive to very low concentrations of ethanol (5 to 10 mM). Some studies support this level of sensitivity while others do not. For example, electrophysiological studies by White et al. (1990) showed that ethanol inhibited NMDA-activated ion currents at concentrations as low as 2.5 mM and with an

IC$_{50}$ of 10 mM in dorsal root ganglion cells of adult rats. Additional studies by Lovinger et al. (1990) showed that, in whole cell patch clamp experiments using cultured embryonic hippocampal cells, ethanol produced inhibition of NMDA responses at ethanol concentrations as low as 5 mM. In hippocampal slices from adult rats, Lovinger et al. (1990a) reported that ethanol concentrations as low as 1 mM inhibited NMDA responses. However, examination of the data from this report indicates that the apparent inhibition by 1 mM ethanol was not statistically significant. The next higher concentration of ethanol used in this study was 25 mM, which did significantly inhibit NMDA-mediated responses. Woodward and Gonzales (1990) showed that ethanol concentrations as low as 10 mM (ethanol IC$_{50}$ of 21 mM) significantly inhibited endogenous dopamine release from striatal slices isolated from adult rat brain. However, in a later study examining ^{3}H-dopamine release from adult striatal slices, Brown et al. (1991) found that the ethanol threshold for inhibiting NMDA-stimulated release was 30 mM. The reason for the discrepancies between these two reports is not clear. Brown et al. (1991) and Gonzales and Woodward (1990) also reported differing ethanol thresholds for inhibiting NMDA-stimulated ^{3}H-norepinephrine release from cerebrocortical slices of 30 mM and 60 mM, respectively. Similar studies by Fink and Gothert (1990) showed that ^{3}H-norepinephrine release from cortical slices was not inhibited by ethanol concentrations in the range of 5 and 10 mM but was inhibited by 50, 100, and 150 mM ethanol. The IC$_{50}$ for ethanol inhibition of NMDA-stimulated ^{3}H-norepinephrine release reported by Fink and Gothert (1990) was approximately 45 mM.

Electrophysiology studies have also been conducted in which sensitivity to ethanol in the 5 to 10 mM range was not observed. For example, Simon et al. (1991) showed that ethanol produced a potent and dose-dependent inhibition of NMDA-evoked electrophysiological activity in medial septal neurons of adult rats. However, the lowest dose of ethanol, 0.75 g/kg which produces blood levels of approximately 60 mg% (or approximately 15 mM), did not alter the response to iontophoresis of NMDA. The next ethanol dose, 1.25 g/kg which produces blood levels of approximately 125 mg% (or approximately 30 mM), and all subsequent ethanol doses did significantly inhibit NMDA evoked electrophysiological responses in medial septal neurons. In agreement with Lima-Landman and Albuquerque (1989), Simon et al. (1991) also found that the NMDA response in some neurons were sensitive to inhibition by ethanol while others were not.

Teichberg et al. (1984) found that ethanol was more potent in blocking kainate- and quisqualate-induced ^{22}Na$^+$ efflux from brain slices than it was in blocking NMDA-induced ^{22}Na$^+$ efflux. Lodge and Johnson (1991) stated that their initial studies did not show ethanol to be a selective NMDA antagonist, although data to support this statement were not presented. Finally, a recent in vivo microdialysis study examined NMDA-stimulated dopamine release in striatum (Gonzales and Roper, 1992). In vivo dopamine levels were increased in dialysate after NMDA stimulation. However, this response was not inhibited by ethanol (0.5 to 4.0 g/kg ethanol administered intraperitoneally). The inability to demonstrate an inhibitory effect of ethanol in this in vivo microdialysis study is particularly noteworthy since Woodward and Gonzales (1990) had previously shown that ethanol is a potent inhibitor (half-maximal ethanol inhibition of approximately 20 mM) of NMDA-stimulated endogenous dopamine release from striatal slices.

The conclusions that can be drawn from the studies described above are as follows: (1) There is a great deal of agreement from numerous laboratories, using biochemical and electrophysiological techniques, that NMDA receptor mediated responses are highly sensitive to inhibition by ethanol. Ethanol concentrations at or above approximately 25 mM were almost always found to inhibit NMDA receptor function. IC_{50}'s for ethanol in the range of 50 mM are a common but not universal finding. (2) Some reports indicate that very low ethanol concentrations (1 to 10 mM) significantly inhibit NMDA responses. However, other reports do not find statistically significant inhibition at these concentrations. Thus, these discrepancies need to be resolved before it can be concluded that NMDA receptors are sensitive to these very low ethanol concentrations. (3) There is now evidence that different subtypes of NMDA receptors may exist in the brain (Monyer et al., 1992; Sugihara et al., 1992). Some of these NMDA receptor subtypes may be more sensitive to ethanol than others.

Mechanisms of Ethanol Inhibition of NMDA-Mediated Responses

The site on the NMDA receptor which has received the most attention as a target for ethanol interaction is the glycine site. Several reports indicate that high glycine concentrations may reverse ethanol's inhibitory effects on the NMDA receptor. Hoffman et al. (1989) and Rabe and Tabakoff (1990) reported that high concentrations of glycine, in the range of 10 μM, prevented ethanol inhibition of cyclic GMP accumulation and $^{45}Ca^{2+}$ uptake in cerebellar granule cells maintained in cell culture. Similar results were obtained in experiments using dissociated neurons in which glycine at 1, 10, and 100 μM produced a concentration-dependent reversal of ethanol inhibition of NMDA-stimulated calcium uptake (Dildy-Mayfield and Leslie, 1991). Woodward and Gonzales (1990) reported that much lower concentrations of glycine, as low as 0.3 μM, prevented ethanol inhibition of NMDA-stimulated endogenous dopamine release from striatal slices. However, in other studies these investigators (Gonzales and Woodward, 1990) found that the inhibition of NMDA-stimulated ^{3}H-norepinephrine release from cortical slices was not reversed by glycine even at very high concentrations (10 - 500 μM). Finally, Peoples and Weight (1992) reported that in voltage-clamped hippocampal neurons glycine, 0.1 - 100 μM, did not alter ethanol inhibition of NMDA-activated ion currents. Thus, some reports indicate a reversal of ethanol's inhibitory effects of NMDA receptor-mediated responses by glycine and others do not. In studies where glycine does reverse ethanol inhibition, the concentrations of glycine required are quite high. Additional research is needed to clarify the role that the glycine site might have in the actions of ethanol.

Martin et al. (1991) reported that, in adult rat hippocampal CA1 cells, ethanol was more effective as an inhibitor of NMDA responses in the presence of magnesium than in its absence (ethanol IC_{50}'s of 47 and 107 mM, respectively). However, in experiments where magnesium and ethanol were covaried magnesium and ethanol appeared to act at distinct sites. This agrees with the results of recent studies using dissociated neurons in which no interaction was found between ethanol and magnesium inhibition of NMDA-stimulated calcium uptake (Dildy-Mayfield and Leslie, 1991). Furthermore, ethanol did not alter the IC_{50} for magnesium in studies examining NMDA-stimulated calcium entry into cerebellar granule cells grown in culture (Rabe and Tabakoff, 1990). Studies in rat cortical and striatal slices also found that ethanol did not alter the IC_{50}'s for magnesium inhibition of NMDA-stimulated ^{3}H-norepinephrine (Gonzales and Woodward, 1990) or endogenous dopamine release (Woodward and Gonzales, 1990). Thus, the cellular mechanism for inhibition on NMDA receptor function by acute ethanol exposure does not appear to involve an action at the magnesium site on the receptor complex.

The possibility that ethanol may inhibit NMDA receptor function by acting at the NMDA recognition site has also been studied. Ethanol did not alter the percent inhibition of NMDA-stimulated $^{45}Ca^{2+}$ uptake in an NMDA concentration-response study in cerebellar granule cells (Rabe and Tabakoff, 1990), nor did ethanol change the NMDA EC_{50} in acutely dissociated neurons (Dildy-Mayfield and Leslie, 1991). Gonzales and Woodward (1990) also found that ethanol did not alter the EC_{50} for NMDA stimulation of ^{3}H-norepinephrine release. Thus, the competitive NMDA recognition site does not appear to be involved mechanistically in the actions of ethanol on NMDA receptors.

Dildy-Mayfield and Leslie (1991) studied the effects of invitro ethanol exposure on the PCP site of the NMDA receptor complex. These studies were conducted in acutely dissociated neurons isolated from 1 day old rat whole brains. Concentration-response studies were performed with MK-801, a specific antagonist at the PCP site, in the presence of 25 μM NMDA and in the absence or presence of 100 mM ethanol. Analysis of the concentration-response curves showed a statistically significant ethanol x MK-801 interaction suggesting a possible mechanistic involvement of the PCP site in the actions of ethanol. However, ALLFIT analysis of the data in this study showed that ethanol reduced the ability of MK-801 to inhibit NMDA-stimulated calcium uptake into dissociated neurons (IC_{50} values for MK-801 were 133.7 nM and 239.7 nM for controls and ethanol treated preparations, respectively). This observation is opposite to what would be expected if ethanol acts by potentiating PCP-like effects at the NMDA receptor complex.

One possible explanation put forth by Dildy-Mayfield and Leslie (1991) was that ethanol may act at some other site on the NMDA receptor complex (e.g. the glycine site) to inhibit opening of the ion channel. This would restrict MK-801 access to its binding site in the channel and may result in an apparent increase in the MK-801 IC_{50} through this indirect action. Taken together, the results of this study suggest that it is unlikely that ethanol acts at the PCP site to cause its acute inhibitory effects.

In summary, numerous studies have now been conducted on the question of the underlying mechanism of ethanol's inhibitory effects on the NMDA receptor complex. The most promising site of action for the acute effects of ethanol appears to be the glycine site. However, conflicting reports exist in the literature. Furthermore, in studies that have implicated the glycine site, rather large concentrations of glycine are needed to reverse or prevent inhibition of NMDA receptor function by ethanol. The conclusion that can be drawn at this point is that there is no conclusive evidence for a specific mechanism of action that explains how ethanol produces its potent inhibition of NMDA receptor function.

Ethanol and NMDA Receptor Activation during Neuronal Development

Activation of the NMDA receptor complex appears to be of key importance in a variety of physiological and pathological processes, many of which may be impacted by the acute or chronic effects of ethanol. This may be of particular importance in fetal alcohol exposure since NMDA receptor activation is thought to play a role in the development of the central nervous system and in learning and memory (Cotman and Iversen, 1987; Collingridge and Bliss, 1987). The role of NMDA receptors in neuronal development is still at an early stage of understanding. Evidence for such a role of NMDA receptors is strongest in the visual cortex (Rauschecker and Hahn, 1987; Kleinschmidt et al., 1987; Bode-Greuel and Singer, 1989). Recent findings also indicate that activity-dependent early postnatal development in spinal cord may be linked with NMDA receptor activation, which may, in turn, alter the expression of neuronal proteins involved in the development process (Kalb and Kockfield, 1990). Brennaman et al. (1990) suggest that neuronal remodeling and pruning during development requires a critical range of intracellular calcium that may be regulated by NMDA receptors. Increases or decreases in intracellular calcium at key times during development may result in the initiation of cell death; this type of cell death is thought to be responsible for the removal of overproduced neurons during neuronal development (Oppenheim, 1991) and may be controlled, at least in part, through NMDA receptor regulated processes.

There are now several reports that indicate an increased NMDA responsiveness or sensitivity in early ontogeny. This may be derived from an NMDA receptor complex of the developing nervous system that is functionally different from mature neurons. Morrisett et al. (1990) found that NMDA-stimulated excitatory postsynaptic potentials (EPSP's) of hippocampal CA1 slices from immature rats (25 to 35 days) were significantly greater in magnitude than in adult slices from the same region taken from mature rats (75 to 90 days). Magnesium was significantly less potent in immature rats compared to mature rats as an inhibitor of NMDA-stimulated EPSP's. In a study in which mRNA isolated from rat hippocampi of three different age groups (1 to 2, 7 to 8, and 14 to 15 days) was injected into Xenopus oocytes, Kleckner and Dingledine (1991) found that the IC_{50} of magnesium for blocking NMDA-receptor activated currents exhibited age-dependent differences. However, in contrast to the results of Morrisett et al. (1990), currents activated by NMDA receptors expressed in Xenopus oocytes from 14 to 15 day old rats were approximately 2-fold less sensitive to inhibition by magnesium than in the same experiments conducted with NMDA receptors expressed from mRNA isolated from 1 to 2 day old rats. Currents activated by NMDA receptors expressed from mRNA from 7 to 8 day old rats were intermediate in sensitivity to magnesium. Similar experiments conducted with adult (12 week) rats found a magnesium sensitivity similar to that of 14 to 15 day old rats. Kleckner and Dingledine (1991) also conducted studies with the NMDA co-agonist, glycine. NMDA receptors from hippocampi of 1 to 2 day old rats were the most sensitive to stimulation with glycine, 14 to 15 day old rats were intermediate, and adult (12 week) rats were least sensitive to NMDA receptor stimulation by glycine. Thus, the reduced magnesium sensitivity of NMDA receptors from 25 to 35 day old rats compared to adults observed by Morrisett et al. (1990) differs from the findings of Kleckner and Dingledine (1991) who did not observe a difference in magnesium sensitivity of NMDA receptors of 14 to 15 day old rats compared to adult animals. Siviy et al. (1991) showed, using an in vitro slice preparation from cat caudate nucleus, that removal of magnesium from the bathing medium resulted in increased EPSP's in all age groups tested. However, in support of the findings of Morrisett et al. (1990), the response to removal of magnesium from immature preparations resulted in enhanced EPSP's as well as a burst of action potentials. Thus, the magnesium site may be an important contributor to enhanced NMDA-sensitivity during development. Furthermore, the demonstration by Kleckner and Dingledine (1991) of increased sensitivity of neonatal NMDA receptors to glycine is consistent with the concept of increased NMDA sensitivity or responsiveness in neonates.

Taken together, these studies indicate that a period of time may exist in which NMDA receptors of neonates are more prone or sensitive to activation than is the case in adults. These differences in NMDA receptor function may help to explain the cellular basis for neuronal pruning and cell death observed during ontogeny (Mattson et al., 1988; Mattson et al., 1988a). As suggested by Kleckner and Dingledine (1991), such processes may represent a critical period in development of the central nervous system. Interference of these actions by any process (Oppenheim, 1991), including in utero alcohol exposure, may result in abnormalities and/or deficiencies in the formation of the neuronal circuitry.

Prenatal ethanol exposure has been reported to markedly (approximately 50 percent) decrease ^{3}H-glutamate binding in hippocampal formation but not in cerebrocortical or subcortical regions of 45 day old rats (Farr et al., 1988). The reduction in ^{3}H-glutamate binding was observed after both 3 percent and 6 percent ethanol diet exposure administered throughout the gestation period. This produced peak blood ethanol concentrations in pregnant dams of approximately 30 and 70 mM, respectively. In a later study, NMDA-sensitive ^{3}H-glutamate binding site density was found to be reduced by 19 to 29 percent in apical dendritic field regions of dentate gyrus, hippocampal CA1 and subiculum of dorsal hippocampal formation of fetal alcohol exposed rats (Savage et al., 1991). Hippocampal NMDA-sensitive ^{3}H-glutamate binding was not reduced when alcohol was administered to pregnant female rats during the first half of gestation (Savage et al., 1991a). However, prenatal ethanol exposure during the second half or the last third of gestation resulted in reductions in NMDA-sensitive ^{3}H-glutamate binding. Postnatal alcohol exposure did not alter hippocampal ^{3}H-glutamate binding (Savage et al., 1991a). In other studies, prenatal ethanol exposure significantly reduced NMDA-stimulated calcium uptake into dissociated neurons isolated from whole brains of one day old rat pups (Leslie et al., 1992). Morrisett et al. (1989) administered alcohol to pregnant female rats throughout the gestation period and offspring were allowed to mature to 70 to 90 days of age. Hippocampal slices were prepared and NMDA-evoked responses were measured electrophysiologically. In these studies prenatal ethanol exposure caused a reduction in NMDA-stimulated hippocampal responses in 70 to 90 day old offspring.

In conclusion, the results of studies conducted thus far suggest that prenatal ethanol exposure may cause significant reductions in NMDA receptor numbers and/or function which may persist into adulthood. Since NMDA receptors are thought to be involved with development of the nervous system, it may be that neuronal deficits seen in fetal alcohol syndrome and fetal alcohol effects are caused, at least in part, by alcohol-induced abnormalities in NMDA-linked neuronal development. Differential deficits are known to exist in brain regional growth of rats exposed to alcohol from postnatal day 4 to day 10. Growth deficits were most pronounced in the hippocampus and less pronounced in the cerebellum and dentate gyrus (Pierce and West, 1987). This brain regional pattern of alcohol-induced growth deficits fits closely with the regional distribution of NMDA receptors (Cotman et al., 1987). Activation of NMDA receptors is also associated with long-term potentiation (Collingridge and Bliss, 1987), a model of learning and memory. Ethanol blocks long-term potentiation in hippocampal slices (Sinclair and Lo, 1986). Furthermore, early postnatal alcohol exposure impairs the development of spatial navigation learning (Goodlett et al., 1987). This effect of ethanol was greater when a daily dose of 6.6 g/kg was given in a condensed manner than was the case when this dose was given uniformly over a 24 hour period. Thus, high concentrations of ethanol condensed over short but critical periods of neuronal development may be more harmful than low, constant levels of ethanol exposure. It is reasonable to hypothesize that deficits in NMDA function as a consequence of fetal alcohol exposure may lead to cognitive deficiencies during development.

REFERENCES

Bode-Greuel, K.M. and Singer, W., 1989, The development of N-methyl-d-aspartate receptors in cat visual cortex, *Brain Res.* 46:197.

Brenneman, D.E., Forsythe, I.D., Nicol, T., and Nelson, P.G., N-Methyl-d-aspartate receptors influence neuronal survival in developing spinal cord cultures, *Develop. Brain Res.* 51:63.

Brown, L.M., Leslie, S.W., Gonzales, R.A., 1991, The effects of chronic ethanol exposure on N-methyl-d-aspartate-stimulated overflow of [3H]catecholamines from rat brain, *Brain Res.* 547:289.

Collingridge, G.L. and Bliss, T.V.P., 1987, NMDA receptors - their role in long-term potentiation, *TINS*, 10:288.

Cotman, C.W. and Iversen, L.L., 1987, Excitatory amino acids in the brain - focus on NMDA receptors, *TINS* 10:263.

Cotman, C.W., Monaghan, D.T., Ottersen, O.P., and Storm-Mathison, J., 1987, Anatomical organization of excitatory amino acid receptors and their pathways, *TINS* 10:273.

Dildy, J.E. and Leslie, S.W., 1989, Ethanol inhibits NMDA-induced increases in free intracellular Ca^{2+} in dissociated brain cells, *Brain Res.* 499:383.

Dildy-Mayfield, J.E. and Leslie, S.W., 1991, Mechanism of inhibition of N-methyl-d-aspartate-stimulated increases in free intracellular Ca^{2+} concentration by ethanol, *J. Neurochem.* 56:1536.

Dildy-Mayfield, J.E., Machu, T. and Leslie, S.W., 1991, Ethanol and voltage- or receptor-mediated increases in cytosolic Ca^{2+} in brain cells, *Alcohol* 9:63

Farr, K.L., Montano, C.Y., Paxton, L.L., and Savage, D.D., 1988, Prenatal ethanol exposure decreases hippocampal ^{3}H-glutamate binding in 45-day-old rats, *Alcohol* 5:125.

Fink, K. and Gothert, M., 1990, Inhibition of N-methyl-d-aspartate-induced noradrenaline release by alcohols is related to their hydrophobicity, *Eur. J. Pharmacol.* 191:225.

Gonzales, R.A. and Woodward, J.J., 1990, Ethanol inhibits N-methyl-d-aspartate-stimulated [^{3}H]norepinephrine release from rat cortical slices, *J. Pharmacol Exp. Ther.* 253:1138.

Gonzales, R.A. and Roper, L.C., 1992, Ethanol effects on NMDA-stimulated levels of extracellular neurotransmitters by in vivo microdialysis, *Alcohol & Alcoholism* 27 (suppl 1): 21.

Goodlett, C.R., Kelly, S.J., and West, J.R., 1987, Early postnatal alcohol exposure that produces high blood alcohol levels impairs development of spatial navigation learning, *Psychobiol.* 15:64.

Hoffman, P.L., Rabe, C.S., Moses, F., and Tabakoff, B., 1989, N-Methyl-d-aspartate receptors and ethanol: Inhibition of calcium flux and cyclic GMP production, *J. Neurochem.* 52:1937.

Kalb, R.G. and Hockfield, S., 1990, Induction of a Neuronal Proteoglycan by the NMDA receptor in the developing spinal cord, *Science* 250:294.

Kleckner, N.W. and Dingledine, R., 1991, Regulation of hippocampal NMDA receptors by magnesium and glycine during development, *Mol. Brain Res.* 11:151.

Kleinschmidt, A., Bear, M.F. and Singer, W., 1987, Blockade of 'NMDA' receptors disrupts experience-dependent plasticity of kitten striate cortex, *Science* 238:355.

Leslie, S.W., Weaver, M.S., Morris, J.L., Lee, Y-.H. and Randall, P.K., 1992, Glutathione stimulation of NMDA receptor function: Effects of acute and chronic ethanol exposure, *Alcohol and Alcoholism* 27:59.

Lima-Landman, M.T.R. and Albuquerque, E.X., 1989, Ethanol potentiates and blocks NMDA-activated single-channel currents in rat hippocampal pyramidal cells, *FEBS Letters* 247: 61.

Lodge, D. and Johnson, K.M., 1991, Noncompetitive excitatory amino acid receptor antagonists, *TIPS* Special Report p. 13.

Lovinger, D.M., White, G., and Weight, F.F., 1989, Ethanol inhibits NMDA-activated ion current in hippocampal neurons, *Science* 243:1989.

Lovinger, D.M., White, G., and Weight, F.F., 1990, Ethanol inhibition of neuronal glutamate receptor function, *Ann Med.* 22:247.

Lovinger, D.M., White, G., and Weight, F.F., 1990a, NMDA receptor-mediated synaptic excitation selectively inhibited by ethanol in hippocampal slice from adult rat, *J. Neurosci.* 10:1372.

Martin, D., Morrisett, R.A., Bian, X-P., Wilson, W.A., and Swartzwelder, H.S., 1991, Ethanol inhibition of NMDA mediated depolarizations is increased in the presence of Mg^{2+}, *Brain Res.* 546:227.

Mattson, M.P., Dou, P., and Kater. S.B., 1988, Outgrowth-regulating actions of glutamate in isolated hippocampal pyramidal neurons, *J. Neurosci.* 8:2087.

Mattson, M.P., Lee, R.E., Adams, M.E., Guthrie, P.B., and Kater, S.B., 1988a, Interactions between entorhinal axons and target hippocampal neurons: A role for glutamate in the development of hippocampal circuitry, *Neuron* 1: 865.

Monyer, H., Sprengel, R., Schoepfer, R., Herb, A., Higuchi, M., Lomeli, H., Burnashev, N., Sakmann, B., and Seeburg, P.H., 1992, Heteromeric NMDA receptors: Molecular and functional distinctions of subtypes, *Science* 256:1217.

Morrisett, R.A., Martin, D., Wilson, W.A., Savage, D.D., and Swartzwelder, S., 1989, Prenatal exposure to ethanol decreases the sensitivity of the adult rat hippocampus to N-methyl-d-aspartate, *Alcohol* 6:415.

Morrisett, R.A., Mott, D.D., Lewis, D.V., Wilson, W.A., and Swartzwelder, H.S., 1990, Reduced sensitivity of the N-methyl-d-aspartate component of synaptic transmission to magnesium in hippocampal slices from immature rats, *Devel. Brain Res.* 56:257.

Oppenheim, R.W., 1991, Cell death during development of the nervous system, *Annu. Rev. Neurosci.* 14:453.

Pierce, D.R. and West, J.R., 1987, Differential deficits in regional brain growth induced by postnatal alcohol, *Neurotoxicol. Terotol.* 9:129.

Peoples, R.W. and Weight, F.F., 1992, Ethanol inhibition of N-methyl-d-aspartate ion current in rat hippocampal neurons is not competitive with glycine, *Brain Res.* 571:342.

Rabe, C.S. and Tabakoff, B., 1990, Glycine site-directed agonists reverse the actions of ethanol at the N-methyl-d-adpartate receptor, *Mol. Pharmacol.* 38:753.

Rauschecker, J.P. and Hahn, S., 1987, Ketamine-xylazine anaesthesia blocks consolidation of ocular dominance changes in kitten visual cortex, *Nature* 326:183.

Savage, D.D., Queen, S.A., Sanchez, C.F., Paxton, L.L., Mahoney, J.C., Goodlett, C.R., and West, J.R., 1991a, Prenatal ethanol exposure during the last third of gestation in rat reduces hippocampal NMDA agonist binding site density in 45-day-old offspring, *Alcohol* 9:37.

Savage, D.D., Montano, C.Y., Otero, M.A., and Paxton, L.L., 1991, Prenatal ethanol exposure decreases hippocampal NMDA-sensitive [^{3}H]-glutamate binding site density in 45-day-old rats, *Alcohol* 8:193.

Simon, P.E., Criswell, H.E., Johnson, K.B., Hicks, R.E., and Breese, G.B., 1991, Ethanol inhibits NMDA-evoked electrophysiological activity in vivo, *J. Pharmacol. Exp. Ther.* 257:225.

Sinclair, J.G. and Lo, G.F., 1986, Ethanol blocks tetanic and calcium-induced long-term potentiation in the hippocampal slice, *Gen. Pharmacol.* 17:231.

Siviy, S.M., Buchwald, N.A., and Levine, M.S., 1991, Enhanced responses to NMDA receptor activation in the developing cat caudate nucleus, *Neurosci. Letters* 132:77.

Sugihara, H., Moriyoshi, K., Ishii, T., Masu, M., and Nakanishi, S., 1992, Structures and properties of seven isoforms of the NMDA receptor generated by alternative splicing, *Biochem. Biophys. Res. Comm.* 185:826.

Teichberg, V.I., Tal, N., Goldberg, O., and Luini, A., 1984, Barbiturates, alcohols and the CNS excitatory neurotransmission: Specific effects on the kainate and quisqualate receptors, *Brain Res.* 291:285.

White, G., Lovinger, D.M., and Weight, F.F., 1990, Ethanol inhibits NMDA-activated current but does not alter GABA-activated current in an isolated adult mammalian neuron, *Brain Res.* 507:332.

Woodward, J.J. and Gonzales, R.A., 1990, Ethanol inhibition of N-methyl-d-aspartate-stimulated endogenous dopamine release from rat striatal slices: Reversal by glycine, *J. Neurochem.* 54:712.

NEUROTRANSMITTER-GATED ION CHANNELS AS MOLECULAR SITES OF ALCOHOL ACTION

Forrest F. Weight, Robert W. Peoples, Jerry M. Wright, Chaoying Li, Luis G. Aguayo, David M. Lovinger and Geoffrey White

Laboratory of Molecular and Cellular Neurobiology
National Institute on Alcohol Abuse and Alcoholism
National Institutes of Health
Bethesda, MD 20892, USA

INTRODUCTION

The mechanism of alcohol action in the nervous system has long been a subject of great interest. Almost a century ago, two German scientists, Overton (1896; 1901) and Meyer (1899; 1901), found a correlation between the anesthetic potencies of different alcohols and their partition between olive oil and water. The partition between olive oil and water indicates the hydrophobicity of an alcohol; the greater the number of carbons in the alcohol molecule, the greater its hydrophobicity. Subsequent investigations confirmed that, up to a point, as alcohol chain length increases, anesthetic potency increases in proportion to the hydrophobicity of the alcohol (Cole and Allison, 1930; Rang, 1960). More recently, a similar relationship has been reported for the intoxicating properties of alcohols (McCreery and Hunt, 1978). These observations led to the idea that alcohols exert their anesthetic action on the hydrophobic lipid bilayer membrane of cells. This "lipid theory" of alcohol action has had several physico-chemical variants. One proposed that the anesthetic effect was dependent upon the concentration of the anesthetic agent in the lipid (Meyer, 1937). A second suggested that it is the thermodynamic activity of the agent that determines anesthetic potency (Ferguson, 1939). A third hypothesis advanced the idea that a critical volume of the membrane lipid must be occupied by the agent for anesthesia to occur (Mullins, 1954). A fourth proposal advocated the concept that anesthesia results from an expansion of the lipid membrane by the agent (Seeman, 1972). A fifth version attributed the intoxicating and anesthetic properties of alcohols to their lipid-fluidizing action in membranes (Goldstein, 1984).

Recently, evidence has been accumulating that alcohols can affect the function of neurotransmitter-gated ion channels. These membrane proteins are postsynaptic receptors for various excitatory and inhibitory neurotransmitters in the nervous system. The binding of a neurotransmitter to these proteins alters their molecular configuration such that the protein forms a transmembrane pore that permits the flux of particular ions down their concentration gradient. This ion flux in turn generates excitatory or inhibitory postsynaptic potentials (EPSPs or IPSPs) in the postsynaptic cell (cf. Hall, 1992). This paper will review alcohol's effects on various neurotransmitter-gated ion channels, focusing particularly on research in our laboratory. Evidence will be presented that there appear to be differences in the molecular mechanisms involved in the effect of alcohol on the function of different types of neurotransmitter-gated ion channels. Data will also be presented indicating that alcohols can interact directly with the neurotransmitter-gated ion channel protein. On the basis of these observations it is proposed that neurotransmitter-gated ion channels are molecular sites of alcohol action in the nervous system.

Alcohol, Cell Membranes, and Signal Transduction in Brain
Edited by C. Alling *et al.*, Plenum Press, New York, 1993

GLUTAMATE-GATED ION CHANNELS

The major excitatory neurotransmitter in the mammalian central nervous system (CNS) is now recognized to be L-glutamate. This excitatory amino acid activates at least three types of ligand-gated ion channels, which have been designated by the agonists that activate them - NMDA (N-methyl-D-aspartate), kainate and quisqualate (or AMPA). The non-NMDA excitatory amino acid-gated ion channels (*viz.* those activated by kainate and quisqualate) mediate fast excitatory postsynaptic potentials (EPSPs) at the majority of excitatory synapses in the mammalian CNS. NMDA-gated ion channels mediate a somewhat slower EPSP at many central excitatory synapses that requires membrane depolarization to be manifested. NMDA channels have been the subject of considerable recent research, as they are thought to be involved in several types of excitatory neural phenomena, neural plasticity, cognitive function and certain types of learning and memory (*cf.* Mayer and Westbrook, 1987; Collingridge and Lester, 1989). Recent research has also shown that the function of both NMDA- and non-NMDA excitatory amino acid-gated ion channels can be affected by ethanol, as discussed in more detail below.

NMDA Channels

A number of recent studies have shown that ethanol can inhibit the function of NMDA channels in a variety of neural tissues. Figure 1A illustrates the inhibitory effect of 50 mM ethanol on inward current activated by NMDA application in neurons freshly isolated from adult dorsal root ganglia (DRG). The concentration-response curve in Fig. 1B shows that ethanol concentrations from 2.5 to 50 mM produced a concentration-dependent inhibition of NMDA-activated current. In these neurons, maximal inhibition of NMDA-activated current was 83%, which was observed with an ethanol concentration of 50 mM. The IC_{50} for this effect was 10 mM. The inhibition by 100 mM ethanol was not significantly different from the inhibition by 50 mM ethanol.

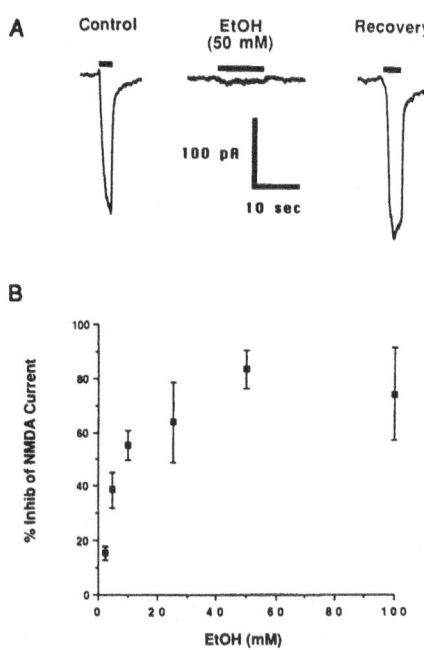

Figure 1. Ethanol inhibition of NMDA-activated ion current. **A.** Effect of 50 mM ethanol (EtOH) on current activated by application of NMDA. **B.** Average percent inhibition of current activated by NMDA, as a function of ethanol concentration. Whole-cell patch-clamp recording from neurons freshly isolated from adult rat dorsal root ganglia. NMDA concentration in A and B was 100 µM. From White *et al.* (1990).

Figure 2A, left, illustrates the effect of 50 mM ethanol on inward current activated by NMDA in cultured hippocampal neurons. In the hippocampal neurons, ethanol produced a concentration-dependent inhibition of NMDA-activated current over the concentration range 5 to 50 mM (Fig. 2A, right). The inhibition by 50 mM ethanol was 61% of control, and the inhibition by 100 mM ethanol was not significantly different from the inhibition by 50 mM. The IC_{50} for inhibition of NMDA-activated current in hippocampal neurons was 30 mM.

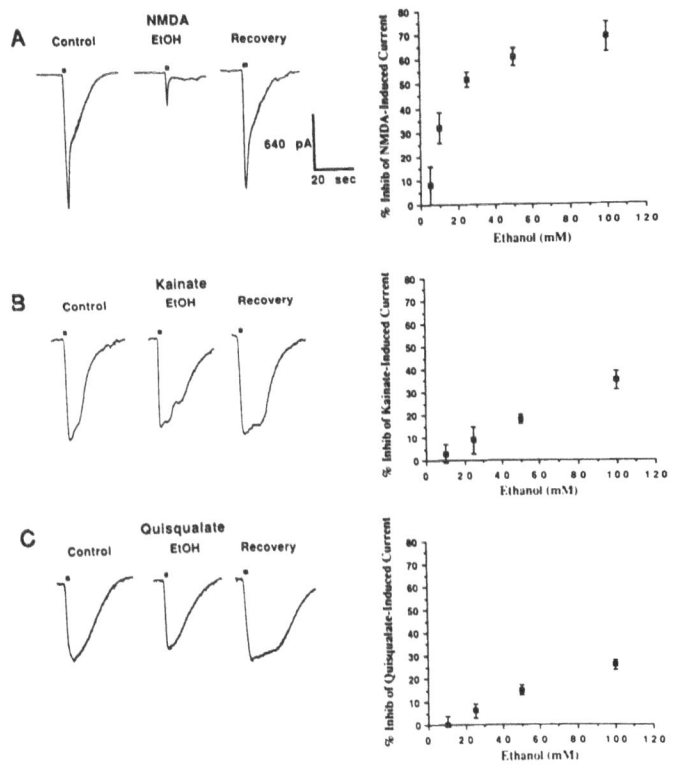

Figure 2. Ethanol inhibition of excitatory amino acid-activated ion currents in hippocampal neurons. **A.** (*Left*) Effect of 50 mM ethanol on current activated by application of 50 μM NMDA. (*Right*) Average percent inhibition of NMDA-activated current as a function of ethanol concentration. **B.** (*Left*) Effect of 50 mM ethanol on current activated by application of 10 μM kainate. (*Right*) Average percent inhibition of kainate-activated current as a function of ethanol concentration. **C.** (*Left*) Effect of 50 μM ethanol on current activated by application of 1 μM quisqualate. (*Right*) Average percent inhibition of quisqualate-activated current as a function of ethanol concentration. Whole-cell patch-clamp recording from cultured mouse hippocampal neurons. From Lovinger *et al.* (1989).

The effect of ethanol on NMDA receptor-mediated synaptic excitation in hippocampal slices is illustrated in Fig. 3. The amplitude of NMDA receptor-mediated population EPSPs (pEPSPS) was decreased by 50 mM ethanol by nearly half, with no apparent effect on the presynaptic volley (Fig. 3A). As shown in Fig. 3B, pEPSP amplitude decreased when ethanol entered the bathing chamber and stabilized at a maximal amplitude within minutes. When ethanol was washed from the recording chamber, pEPSP amplitude gradually returned to control amplitude over several minutes. Although the average reduction of pEPSP amplitude by 50 mM ethanol was less than 50%, synaptic transmission (*viz.* the action potentials triggered by these EPSPs) at this excitatory synapse was almost totally abolished (Fig. 3C). The concentration-response curve for ethanol-induced inhibition of pEPSPs in hippocampal slice was similar to that in cultured hippocampal neurons, with maximal inhibition of 43% observed with a concentration of 50 mM ethanol. As with the cultured neurons, the inhibition by 100 mM ethanol was not significantly different from the inhibition by 50 mM ethanol.

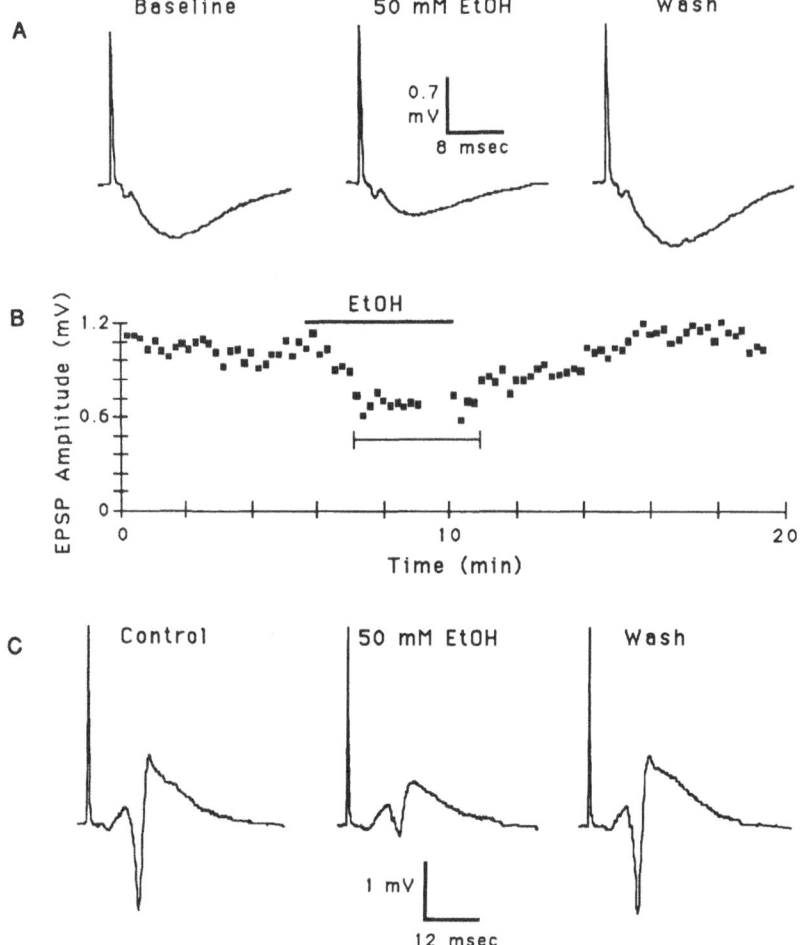

Figure 3. Ethanol inhibition of NMDA receptor-mediated synaptic potentials and synaptically activated action potentials in hippocampus. **A.** Effect of 50 mM ethanol on NMDA receptor-mediated pEPSPs. **B.** Amplitude of individual pEPSPs plotted as a function of time. **C.** Effect of 50 mM ethanol on population spike activated by NMDA receptor-mediated EPSPs. Responses elicited in adult rat hippocampal slice by stimulation of Schaffer collateral-commissural pathway in medium containing 10 μM DNQX and 0.1 mM Mg^{2+}. Responses in A and B recorded in CA1 stratum radiatum, and responses in C recorded in stratum pyramidale. From Lovinger *et al.* (1990a).

Similar ethanol-induced inhibition of NMDA responses has also been observed in other neural tissue, using a variety of experimental methods. In whole-cell patch-clamp experiments, ethanol produced an inhibition of NMDA-activated current in cultured neurons from spinal cord and neocortex that was similar to that observed in hippocampal neurons (Lovinger *et al.*, 1990b). In slices of the amygdala, ethanol suppressed NMDA receptor-mediated synaptic responses (Gean, 1992). In extracellular recording experiments, *in vivo*, the intraperitoneal administration of ethanol inhibited NMDA-activated spike firing of neurons in the medial septal nucleus (Simson *et al.*, 1991) and locus coeruleus (Engberg and Hajos, 1992).

In biochemical experiments, ethanol reduced NMDA-induced intracellular Ca^{2+} elevation in freshly isolated brain cells (Dildy and Leslie, 1989), neurotransmitter release from slices of neocortex (Gothert and Fink, 1989; Gonzales and Woodward, 1990; Fink *et al.*, 1992) and striatum (Woodward and Gonzales, 1990), and Ca^{2+} influx and cyclic GMP elevation in cultured cerebellar granule cells (Hoffman *et al.*, 1989).

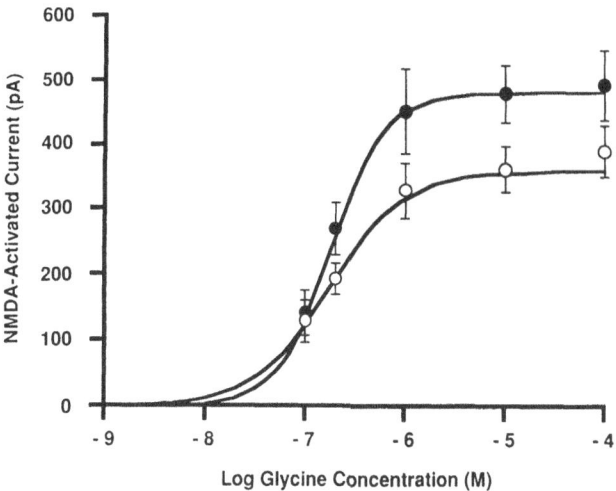

Figure 4. Concentration-response for glycine enhancement of the current activated by 25 μM NMDA in the absence and presence of 50 mM ethanol. Data points are means±S.E. of 12 cells. Ethanol significantly decreased the maximal current activated by glycine (481 pA for control vs 360 pA in the presence of ethanol; analysis of variance, $p<0.01$), but did not alter the potency of glycine (control EC_{50} 183 nM vs 174 nM in the presence of ethanol; analysis of variance, $p>0.7$). Whole-cell patch-clamp recording from cultured rat hippocampal neurons. From Peoples and Weight (1992a).

The mechanism involved in the inhibition of NMDA responses by ethanol has not been established. Several possible mechanisms have been tested in whole-cell patch-clamp experiments on cultured hippocampal neurons (Lovinger *et al.*, 1990c; Peoples and Weight, unpublished observations). Membrane potentials between -60 and +60 mV did not significantly affect the percent inhibition of NMDA-activated current by ethanol, and ethanol did not alter the reversal potential of NMDA-activated current. These observations indicate that the effect of ethanol on NMDA channels is not voltage-dependent, and does not result from an alteration of the ion selectivity of the NMDA channel. Other possible mechanisms include alteration of one of the regulatory sites on the NMDA channel. However, the percent inhibition of NMDA-activated current by ethanol did not differ with different concentrations of NMDA (10-1000 μM), Mg^{2+} (0-500 μM), Zn^{2+} (0-20 μM), ketamine (0-10 μM), spermine (0 & 1 μM), protons (pH 6.0-8.0), or dithiothreitol (0 & 2 mM), indicating that ethanol does not interfere with the regulatory sites on the NMDA channel associated with these agents. It has been suggested that ethanol inhibits NMDA responses by interfering with the glycine modulatory site on the NMDA receptor (Hoffman *et al.*, 1989; Woodward and Gonzales, 1990; Rabe and Tabakoff, 1990; Dildy-Mayfield and Leslie, 1991). However, Fig. 4 illustrates experiments in our laboratory showing that the percent inhibition of NMDA-activated current by ethanol was not altered by glycine concentrations from 100 nM to 100 μM, the concentration range that augments NMDA-activated current. Since glycine concentration in our experiments was less than 100 nM, this observation indicates that interference with the glycine modulatory site would not explain the inhibition of NMDA-activated current by ethanol in our experiments. Activation of protein kinase C (PKC) by phorbol esters has recently been reported to potentiate NMDA-activated current (Kelso *et al.*, 1992). In preliminary experiments in our laboratory, the PKC inhibitor staurosporin does not appear to affect ethanol inhibition of NMDA-activated current in cultured hippocampal neurons (Peoples and Weight, unpublished observations) or in *Xenopus* oocytes expressing NMDA receptors (Masood, Wu, Braunis and Weight, unpublished observations). The effect of ethanol in single channel experiments on NMDA channels using outside-out tear-off patches from cultured cortical neurons is illustrated in Fig. 5. Ethanol concentrations from 10 to 100 mM did not affect the conductance of single NMDA channels. This concentration range did, however, produce a concentration-dependent reduction in channel open probability (Wright and Weight, 1992), similar to that reported previously by Lima-Landman and Albuquerque (1989). This suggests that ethanol may affect the function of NMDA channels by altering gating of the channel.

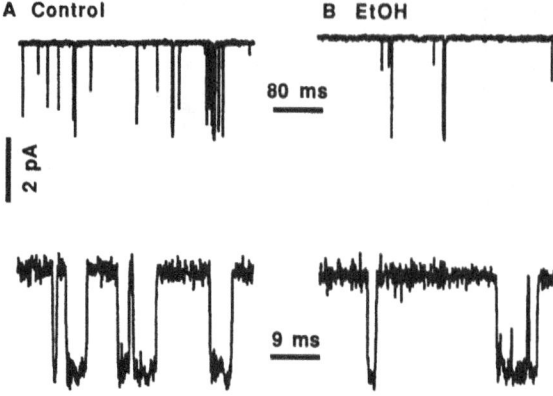

Figure 5. Effect of ethanol on single NMDA channels. **A.** Control single-channel currents activated by 20 μM NMDA in the absence of ethanol. **B.** Effect of 100 mM ethanol on single-channel currents activated by 20 μM NMDA. Note reduced frequency of opening (top) without a significant effect on current amplitude (bottom). Outside-out tear-off patch-clamp recording from cultured cortical neuron. Holding potential was -60 mV in both A and B. Top traces filtered at 200 Hz, lower traces filtered at 2 KHz. From Weight *et al.* (1993).

Figure 6A illustrates the relationship between the hydrophobicity of different alcohols and their effect on NMDA channels. As the hydrophobicity of the alcohols increased, their potency for inhibiting NMDA-activated current increased. In addition, there was a significant linear relation between these parameters, suggesting that the potency of different alcohols for inhibiting NMDA-activated current increases as a function of increasing hydrophobicity. Similar results have also been reported for alcohol-induced inhibition of NMDA-evoked release of [³H]norepinephrine from cortical slices (Gonzales *et al.*, 1991). The relationship between the effect of different alcohols on NMDA channels and their potency for producing intoxication is shown in Fig. 6B. As the potency of different alcohols for inhibiting NMDA-activated current increased, their potency for producing intoxication also increased. Moreover, there was a significant linear relation between these parameters, suggesting that the more potent the alcohol is in inhibiting NMDA-activated current, the greater its potency for producing intoxication.

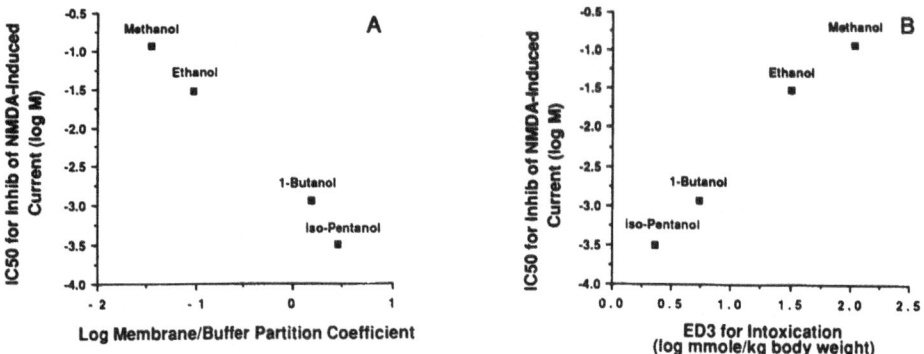

Figure 6. Correlation between the potency of different alcohols for inhibiting NMDA-activated ion current, the hydrophobicity of the alcohols, and the potency of the alcohols for producing intoxication. **A.** Log-log graph plotting IC_{50} of four alcohols tested for inhibition of NMDA-activated current in cultured hippocampal neurons as a function of their membrane-buffer partition coefficients. The linear relation between these parameters has a slope of -1.30±0.055 (*p*<0.01). **B.** Log-log graph plotting ED3 for intoxication by four alcohols versus the IC_{50} for inhibition of NMDA-activated current by alcohols in cultured hippocampal neurons. The linear relation between these parameters has a slope of 1.579±0.099 (*p*<0.025). From Lovinger *et al.* (1989).

Kainate and Quisqualate Channels

The effect of ethanol on kainate- and quisqualate-activated currents in hippocampal neurons is illustrated in Fig. 2. In contrast to NMDA-activated current, kainate- and quisqualate-activated currents were relatively insensitive to low concentrations of ethanol. Thus, athough 50 mM ethanol reduced NMDA -activated current in these neurons by 61% (Fig. 2A), this concentration of ethanol reduced kainate and quisqualate-activated current by only 18% (Fig. 2B) and 15% (Fig. 2C), respectively. Similarly, although 25 mM ethanol inhibited NMDA-activated current by more than 50%, the reduction of kainate- and quisqualate-activated current was less than 10%. Comparison of the concentration-response curves in Fig. 2 reveals that although inhibition of NMDA-activated current did not increase significantly when ethanol concentration increased from 50 to 100 mM, the inhibition of kainate- and quisqualate-activated current continued to increase significantly when ethanol concentration increased from 50 to 100 mM. Moreover, the inhibition of non-NMDA excitatory amino acid-activated current continued to increase at ethanol concentrations greater than 100 mM. As can be seen in Fig. 7, a concentration of 200 mM ethanol inhibited these currents by ~45%. Thus, the IC_{50} for ethanol inhibition of non-NMDA excitatory amino acid-activated ion current was ~220 mM, compared to an IC_{50} for ethanol inhibition of NMDA-activated current of 30 mM.

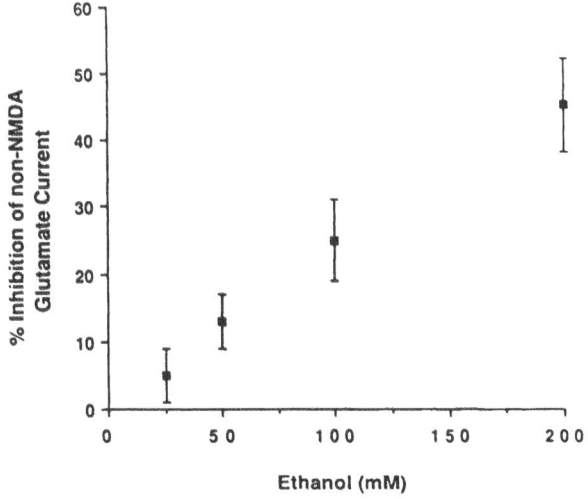

Figure 7. Ethanol inhibition of non-NMDA glutamate-activated ion currents. Graph plots percent inhibition of non-NMDA receptor-mediated glutamate current as a function of ethanol concentration. Data are from cultured mouse hippocampal neurons. The external solution contained 1 mM Mg^{2+} and 50 μM APV. From Lovinger *et al.* (1990b).

GABA-GATED ION CHANNELS

The major inhibitory neurotransmitter in mammalian brain is now recognized to be γ-aminobutyric acid (GABA). This amino acid activates a ligand-gated ion channel that is permeant to chloride ions. Since GABA can also activate signal transduction systems via G-proteins, the ion channels directly activated by GABA have been designated $GABA_A$ receptors. Numerous studies in recent years have shown that $GABA_A$ channels mediate inhibitory postsynaptic potentials (IPSPs) at the majority of inhibitory synapses in mammalian brain (McGeer *et al.*, 1987). In contrast to NMDA responses, which have been reported to sensitive to ethanol in a number of regions of the CNS using a variety of experimental techniques, $GABA_A$ receptor-mediated responses have been reported to be sensitive to ethanol in studies from a number of laboratories (Davidoff, 1973; Nestoros, 1980; Allan and Harris, 1986; Suzdak *et al.*, 1986; Mehta and Ticku, 1988; Celentano *et al.*, 1988; Nakamura *et al.*, 1989;

Takada *et al.*, 1989; Givens and Breese, 1990; Nishio and Narahashi, 1990; Ueha and Kuriyama, 1991; Proctor *et al.*, 1992; Reynolds *et al.*, 1992), but insensitive to the actions of ethanol in studies from a number of other laboratories (Bloom *et al.*, 1984; Gage and Robertson, 1985; Harrison *et al.*, 1987; Mancillas *et al.*, 1986; Siggins *et al.*, 1987; Morelli *et al.*, 1988; Osmanovic and Shefner, 1990; Mihic *et al.*, 1992). Many of the studies in different laboratories used different preparations and different experimental methods, making it difficult to know the extent to which the sensitivity to ethanol results from the preparation or the experimental method used. In our laboratory we used the same experimental method, the whole-cell patch-clamp technique, to study ethanol sensitivity of GABA-activated Cl⁻ current in two different preparations - neurons in tissue culture from cortex or hippocampus, and neurons freshly isolated from adult DRG. Different results were obtained in the two preparations.

The effects of ethanol on GABA-activated Cl⁻ current in cultured neurons from cerebral cortex and hippocampus are illustrated in Fig. 8. In the majority of neurons studied, ethanol potentiated GABA-activated Cl⁻ current in a concentration-dependent manner over the concentration range 1 to 40 mM. Figure 8A illustrates the potentiation of GABA-activated current by 20 mM ethanol in a cortical neuron. The potentiation of the current was not voltage-dependent and ethanol did not alter the reversal potential of the current (Aguayo and Weight, 1990), suggesting that ethanol does not alter the ion selectivity of the channel. Figure 8B shows that in neurons in which GABA-activated current was not sensitive to ethanol, the current was still potentiated by benzodiazepines and barbiturates.

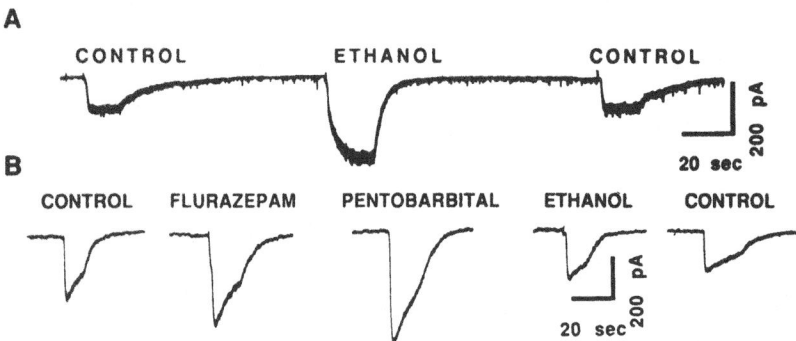

Figure 8. Effects of ethanol on GABA-activated Cl⁻ current in cultured cortical and hippocampal neurons. **A.** Potentiation of current activated by 2.5 μM GABA by the application of 20 mM ethanol in a cultured neuron from mouse cerebral cortex. **B.** Potentiation of ethanol-insensitive GABA (2.5 μM)-activated current by 20 μM flurazepam and 20 μM pentobarbital in a cultured neuron from mouse hippocampus. From Aguayo (1990).

In isolated DRG neurons, on the other hand, the amplitude of GABA-activated Cl⁻ current was not significantly affected by ethanol concentrations from 10 to 100 mM. Figure 9 illustrates the lack of effect ethanol on GABA-activated current in these neurons. The current traces in Fig. 9 show the effect of 50 mM ethanol on current activated by 10 μM GABA (Fig. 9A1) and 50 μM GABA (Fig. 9B1). The bar graphs in Fig. 9 show the average effect of 10, 50 and 100 mM ethanol on current activated by 10 μM GABA (Fig. 9A2) and 50 μM GABA (Fig. 9B2). No group was significantly different from control ($p > 0.1$; paired Student's *t*-test). Similar results were obtained when current was activated by 100 μM GABA. Ethanol also had no effect on GABA-activated current when the internal (pipette) solution contained 10 mM BAPTA instead of EGTA, Ca²⁺ was not added or ATP was added to the internal solution. In addition, ethanol did not alter GABA-activated current in the presence of the benzodiazepine agonist flurazepam or the inverse agonist DMCM.

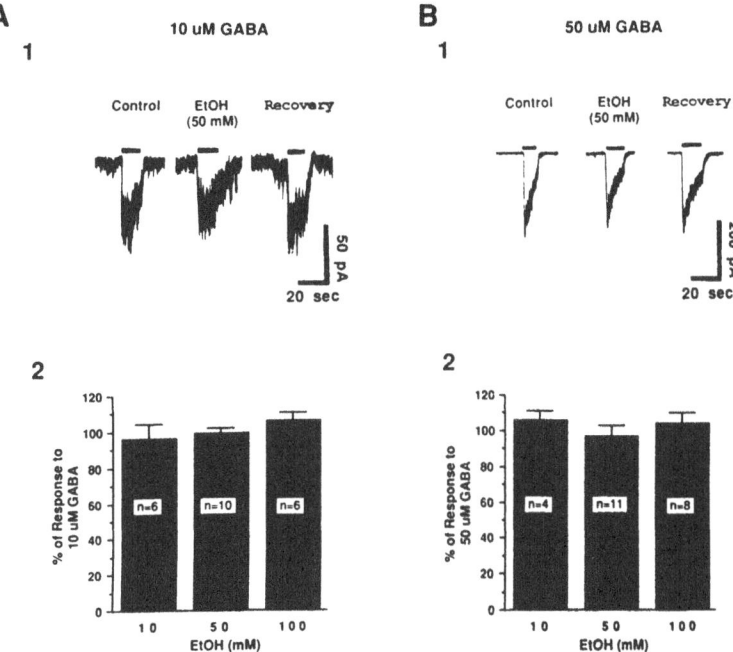

Figure 9. Lack of effect of ethanol on GABA-activated Cl⁻ current in isolated DRG neurons. Current traces illustrate effect of 50 mM ethanol (EtOH) on current activated by 10 μM GABA (**A1**) and 50 μM GABA (**B1**). Bar graphs show average effect of 10, 50 and 100 mM ethanol on the amplitude of current activated by 10 μM GABA (**A2**) and 50 μM GABA (**B2**). Neurons were freshly dissociated from adult rat dorsal root ganglia. From White *et al.* (1990).

The preceding results from our laboratory, using essentially the same experimental technique, suggest that the sensitivity of GABA$_A$ receptors to ethanol is different in different types of neurons. One explanation for this difference is that the ethanol sensitivity of GABA$_A$ receptors results from differences in the subunits that form GABA$_A$ receptors in different types of neurons (*cf*. Olsen and Tobin, 1990). Wafford *et al.* (1991) suggested that the presence of the γ2L subunit in the GABA$_A$ receptor is necessary for potentiation of GABA-activated current by ethanol. Presumably this hypothesis would explain the observation that when mRNA from brains of long-sleep(LS) and short-sleep (SS) mice was expressed in *Xenopus* oocytes, ethanol facilitated GABA responses in oocytes injected with mRNA from LS mice, but inhibited responses in oocytes injected with mRNA from SS animals (Wafford *et al.*, 1990). However, in an analysis of subunit mRNAs in LS and SS mouse brain using polymerase chain reaction (PCR) amplification, similar levels of γ2L and γ2S were found in LS and SS mice (Zahniser *et al.*, 1992). In addition, although GABA-activated current in DRG neurons neurons from adult rat are not sensitive to ethanol (White *et al.*, 1990), PCR experiments indicate that these neurons contain the γ2L but not the γ2S subtype (Hartnett, 1991).

Another possible explanation for differences in ethanol sensitivity is regulation of the GABA$_A$ receptor. The γ2L, but not the γ2S, subunit of the GABA$_A$ receptor contains a protein kinase C (PKC) phosphorylation site (Whiting *et al.*, 1990), and it has been proposed that phosphorylation of this site is essential for ethanol sensitivity of GABA$_A$ receptors (Wafford and Whiting, 1992). However, in preliminary experiments in our laboratory, we have been unable to induce ethanol sensitivity of GABA-activated currents in adult rat DRG neurons using phorbol esters to activate PKC (Li and Weight, unpublished observations), even though these neurons apparently contain the γ2L subunit. On the other hand, ethanol sensitivity of GABA responses has been reported to be induced by activation of β-adrenergic receptors in cerebellar Purkinje neurons (Lin *et al.*, 1991); however, β-adrenergic receptor activation would be expected to activate protein kinase A (PKA), rather than PKC. Clearly, further work is needed to establish the physiological and/or molecular basis of ethanol sensitivity of GABA$_A$ receptors.

SEROTONIN-GATED ION CHANNELS

Serotonergic mechanisms have been suggested to be important in alcohol abuse and alcoholism (cf. Sellers et al., 1992). Serotonin-containing nerve fibers arise predominantly from neurons in the raphe nucleus and project throughout the CNS. Many serotonin (5-hydroxytryptamine, 5-HT) receptors in the CNS involve G-protein mediated signal transduction systems (cf. Julius, 1991). In addition, serotonin activates a ligand-gated ion channel that has been designated the 5-HT$_3$ receptor (Derkach et al., 1989; Maricq et al., 1991). Binding experiments indicate differential distribution of 5-HT$_3$ receptors throughout the brain with high concentrations in brain stem, cortical and limbic areas (Kilpatrick et al., 1987; 1989).

The observations discussed above on glutamate- and GABA-gated ion channels raised the question of whether the function of other neurotransmitter-gated ion channels are also sensitive to ethanol. This qustion was investigated on 5-HT$_3$ channels in NCB-20 neuroblastoma cells using the whole-cell patch-clamp technique (Lovinger, 1991). Ethanol, in concentrations from 25 to 100 mM, produced a concentration-dependent increase in the amplitude of 5-HT$_3$ receptor-mediated current. The current activated by 1 µM 5-HT was increased 59% in amplitude by an ethanol concentration of 100 mM. Ethanol concentrations greater than 100 mM also potentiated 5-HT$_3$ current, but the potentiation was not greater than that observed with 100 mM ethanol. With 5-HT concentrations of 1 or 2 mM, potentiation of 5-HT$_3$ current by ethanol concentrations of 50 mM or greater was observed in 86% of the cells studied.

Further investigation of ethanol effects on 5-HT$_3$ channels in NCB-20 cells showed that the magnitude of the potentiation decreased with increasing 5-HT concentration (Lovinger and White, 1991). With 1 µM 5-HT, the potentiation by 50 mM ethanol was 45%, with 2 µM 5-HT the potentiation was 18%, and with 10 µM 5-HT no potentiation of 5-HT$_3$ current was observed. Membrane voltage did not affect the potentiation of 5-HT$_3$ current by ethanol, and the augmented current had an increased rate of decay. In addition, ethanol concentrations from 25 to 100 mM were also shown to potentiate 5-HT$_3$ current in neurons freshly dissociated from rat nodose ganglia. In these neurons, 100 mM ethanol potentiated the current activated by 1 µM 5-HT by 39%. Binding experiments on NCB-20 and rat cerebral membranes suggest that the potentiation of 5-HT$_3$ current by ethanol does not result from an increased affinity of the 5-HT$_3$ receptor for agonist (Hellevuo et al., 1991). The effects of ethanol on 5-HT-activated currents are discussed in more detail in the chapter by Lovinger in this volume.

ATP-GATED ION CHANNELS

Adenosine-5'-triphosphate (ATP) is generally considered to be an intracellular energy donor; however, in recent years, evidence has accumulated that ATP can also function as a extracellular mediator (cf. Burnstock, 1990). ATP is released from a variety of cell types including nerve fibers upon stimulation, and acts via two types of transduction systems - activation of ligand-gated ion channels and coupling to G-proteins (cf. Illes and Norenberg, 1993). The ATP-gated ion channels have been studied most extensively in the peripheral nervous system, where it appears that ATP can be an excitatory neurotransmitter or cotransmitter (Bean and Friel, 1990; Bean, 1992; Evans et al., 1992). It has also recently been shown that ATP can excite neurons in the CNS by activation of ligand-gated ion channels (Edwards et al., 1992; Shen and North, 1993). However, the distribution of ATP-gated channels throughout the CNS and their functional significance are not yet known.

Since several types of neurotransmitter gated ion channels are sensitive to ethanol, we have recently investigated whether ATP-gated channels are also sensitive to ethanol (Li et al., 1991). In whole-cell patch-clamp experiments on neurons freshly dissociated from bullfrog DRG, ethanol concentrations from 6 to 250 mM produced a concentration-dependent decrease in the amplitude of ATP-activated current. The IC$_{50}$ for this effect was 160 mM ethanol. The average inhibition by 100 mM ethanol of current activated by 5 µM ATP was 31%. The effect of ethanol on the concentration-response curve for ATP differed from the effect of ethanol on the concentration-response curve for NMDA, observed previously in hippocampal neurons by Lovinger et al. (1989). Ethanol had been found to inhibit the maximal response to NMDA but it did not affect the EC$_{50}$ for NMDA, whereas ethanol produced a parallel shift to the right of the concentration-response curve for ATP, shifting the EC$_{50}$ from 2.4 to 5.3

μM. In addition, as observed for NMDA-activated current, methanol was less potent and 1-propanol was more potent than ethanol in inhibiting ATP-activated current. However, much to our surprise, 1-butanol and isopentanol were without effect on the ATP-activated current. In addition, the ATP-activated current was not affected by trichloroethanol or dichloroethanol, but was inhibited by monochloroethanol and trifluroethanol (Li and Weight, unpublished observations).

DISCUSSION

The studies discussed above indicate that the function of several types of neurotransmitter-gated ion channels can be affected by ethanol in a pharmacologic concentration range (1 to 100 mM). Neurotransmitter-gated ion channels underlie the generation of fast postsynaptic excitation and inhibition in the nervous system, and thus play an important role in intraneuronal communication and the regulation of neuronal excitability. Consequently, it seems likely that the effects of ethanol on the function of neurotransmitter-gated ion channels may contribute significantly to the behavioral effects of ethanol. Although the behavioral effects of ethanol may result from a combination of actions on different ion channels and other molecular sites in the CNS, it is of interest to consider some of the behavioral effects that may result from ethanol actions on different types of neurotransmitter-gated ion channels.

Ethanol inhibited NMDA-activated current in hippocampal neurons in a concentration-dependent manner over the concentration range 5-50 mM. This corresponds to the concentration range over which intoxication occurs in nontolerant humans. Impairment of fine motor control and delayed reaction time can be detected at blood-ethanol concentrations as low as 5 mM (Ritchie, 1985). Increasing blood-ethanol concentrations up to 50 mM are associated with increasing impairment of mental ability and motor coordination that are generally recognized as signs of intoxication (Kissen, 1988). Since NMDA channels are thought to be involved in complex excitatory neural phenomena and cognitive function, the correspondence between ethanol inhibition of NMDA-activated current and ethanol-induced intoxication suggests that the inhibition of NMDA-activated current by ethanol may contribute to intoxication. This possibility is supported by the observation that the potency of different alcohols for inhibiting NMDA-activated current is correlated with their potency for producing intoxication. This possibility is also supported by the observation that the intracerebroventricular injection of APV, a specific NMDA receptor antagonist, produces behavioral effects that appear to be similar to those observed during ethanol intoxication (Woods et al., 1987).

Although ethanol concentrations in the intoxicating range had a relatively small effect on kainate- and quisqualate-activated currents in hippocampal neurons, concentrations of ethanol greater than 50 mM produced significant concentration-dependent inhibition of these currents. Blood-ethanol concentrations greater than 50 mM are associated with signs of general anesthesia. Clinically, as blood-ethanol concentration increases above 50 mM there is increasing CNS depression manifested as progression from sedation to stupor to coma, and with blood-ethanol concentrations greater than 100 mM, death can occur from respiratory depression (Kissen, 1988). Because non-NMDA excitatory amino acid receptors mediate fast EPSPs at the majority of excitatory synapses in the CNS, inhibition of EPSPs at these excitatory synapses would be expected to result in general CNS depression. This suggests that ethanol inhibition of kainate and quisqualate responses may contribute to the general anesthetic effects of ethanol. Consistent with this hypothesis, we have found that the general anesthetic agents, pentobarbital (Peoples et al., 1990), trichloroethanol, the active metabolite of chloral anesthetics such as chloral hydrate (Peoples and Weight, 1991), and the volatile anesthetics, halothane, enflurane and isoflurane (Peoples and Weight, 1992b), all inhibit kainate- and quisqualate-activated currents.

There are still many uncertainties regarding the effect of ethanol on GABA-gated ion channels. Because there are reports of both ethanol-sensitive and ethanol-insensitive GABA$_A$ responses in different brain regions and in different types of neurons, in different laboratories, using different experimental methods, it is still not established what brain regions or types of neurons have GABA$_A$ receptors that are sensitive to ethanol. It is also not established whether the reported differences in ethanol sensitivity result from differences in the regulation of the GABA$_A$ receptors, differences in the molecular composition of GABA$_A$ receptors or other factors yet to be determined. Without knowing the inhibitory synaptic pathways in brain that are affected by ethanol, it is difficult to speculate on the behavioral significance of ethanol effects on $GABA_A$ receptors. Nevertheless, if these complexities are ignored, and it is assumed that ethanol has a general potentiating action on $GABA_A$ responses in brain, some insight may be gained from the effects of benzodiazepines. Benzodiazepine agonists, such as diazepam (Valium), potentiate most $GABA_A$ responses in the CNS (Rall, 1990). They are widely used clinically, primarily for their anxiolytic actions; moreover, in their clinical concentration range, benzodiazepines do not produce an intoxicated state (Baldessarini, 1990). This suggests that ethanol potentiation of $GABA_A$ responses may contribute primarily to the anxiolytic effects of ethanol, rather than its intoxicating actions. It goes without saying that clinical concentrations of benzodiazepines do not produce a general anesthetic state. In high concentrations, benzodiazepines can have sedative/hypnotic actions; however, such effects are usually not considered to be of sufficient depth for general surgery. In addition, in most studies, ethanol concentrations in the anesthetic range (*viz.* greater than 50 mM) do not produce concentration-dependent potentiation of $GABA_A$ responses. Taken together, these observations suggest that ethanol potentiation of $GABA_A$ responses may not contribute significantly to the general anesthetic effects of ethanol.

There have been a number of recent studies related to the question of whether $5-HT_3$ receptors may be involved in the rewarding properties of drugs of abuse. The actions of several drugs of abuse, such as ethanol, morphine, and nicotine, are associated with an elevation of mesolimbic dopamine concentration (Imperato and Di Chiara, 1986; Di Chiara and Imperato, 1988), and $5-HT_3$ antagonists have been found to block this dopamine elevation and associated behavioral effects (Carboni *et al.*, 1989; Wozniak *et al.*, 1990; Carboni *et al.*, 1988; Costal *et al.*, 1990). In this context, the potentiation of $5-HT_3$ current by ethanol is of interest because it may provide an explanation for the elevation of mesolimbic dopamine concentration by ethanol. For example, if dopaminergic neurons that innervate mesolimbic areas have $5-HT_3$ receptors that are potentiated by ethanol, this would augment 5-HT-mediated synaptic excitation of these neurons and hence increase the release of dopamine in the mesolimbic area. It is of interest to note that the ability of pigeons to discriminate between ethanol and water can be blocked by the $5-HT_3$ antagonists ICS205-930 and MDL 72222 (Grant and Barrett, 1991), suggesting that ethanol potentiation of $5-HT_3$ responses may contribute to the ability to recognize the action of ethanol. In addition, MDL 72222 has been reported to suppress voluntary ethanol consumption in alcohol-preferring rats (Fadda *et al.*, 1991). However, further work is needed to determine whether ethanol effects on central $5-HT_3$ receptors are involved in the rewarding properties of ethanol.

Although it is not yet known whether ethanol affects the function of ATP-gated channels in the CNS and if it does what the functional significance of such effects might be, the inhibition of ATP-gated ion channel function by ethanol is of interest in several respects. First, the ethanol inhibition of ATP channel function differs from that of NMDA channels. The concentration-response curve for NMDA-activated current was reduced in amplitude by ethanol but the EC_{50} for agonist was not significantly affected, whereas ethanol shifted the concentration-response curve for ATP to the right, including the EC_{50} for agonist. This suggests that the mechanism of ethanol's action may be different on NMDA-gated channels and ATP-gated channels. Second, the parallel shift to the right of the ATP concentration-response curve by ethanol is consistent with either a competitive interaction or a decrease in the affinity of the agonist binding site. Since structurally ethanol seems unlikely to compete with the binding of ATP, it appears more likely that ethanol induces an allosteric interaction in the channel protein that decreases the affinity of the ATP-binding site. By contrast, ethanol appears to alter the gating of the NMDA receptor. Third, the increasing inhibition of ATP-activated current by one, two and three carbon alcohols is consistent with a hydrophobic interaction, as had been found for NMDA channels. If the effect of alcohols on the function of ATP-gated channels were secondary to an interaction of the alcohols with membrane lipids, the potency of the alcohols would continue to increase in proportion to the number of carbons in the alcohol. However, our observation that alcohols consisting of four carbons or more did not affect ATP-activated current, suggests that the site of alcohol action is either a small hydrophobic pocket in the channel protein or there may be spatial limitations in accessing the hydrophobic site, for example via the channel pore. These interpretations are supported by the observation that trichloroethanol and dichloroethanol did not affect ATP-activated current, while monochloroethanol and trifluroethanol both inhibited the current. These observations indicate that alcohols produce their effect on the function of

ATP-gated channels by an interaction with a hydrophobic region of the channel protein, rather than with membrane lipids.

SUMMARY AND CONCLUSIONS

The "lipid theory" has dominated thinking on the mechanism of alcohol action in the nervous system for almost a century. Research studies in recent years have shown that alcohol can affect the function of neurotransmitter-gated ion channels in mammalian neurons. Since these ion channels mediate fast excitatory and inhibitory postsynaptic potentials (EPSPs and IPSPs) in the nervous system, alteration of their function by alcohol may underlie many of alcohol's behavioral effects. Recent molecular biological studies indicate that neurotransmitter-gated ion channels are heterooligomeric membrane proteins, and each subunit of these proteins appears to have four hydrophobic membrane-spanning domains. Data are presented for ATP-gated channels that alcohols can interact with a hydrophobic site on the channel protein. It is suggested that alcohols may also affect the function of other neurotransmitter-gated ion channels by interaction with a hydrophobic site (or sites) on the channel protein. This "ion channel hypothesis" is of heuristic value, as it suggests experiments using molecular biological techniques, such as antisense oligonucleotides or site-directed mutagenesis, to determine the hydrophobic sites of alcohol interaction on different types of neurotransmitter-gated ion channels.

REFERENCES

Aguayo, L.G., 1990, Ethanol potentiates the GABA$_A$-activated Cl⁻ current in mouse hippocampal and cortical neurons, *Eur. J. Pharmacol.* 187:127.

Aguayo, L.G. and Weight, F.F., 1990, Ethanol potentiates the whole-cell GABA$_A$- and glycine-activated Cl⁻ currents in mouse hippocampal neurons, *FASEB J.* 4:A741.

Allan, A.M., and Harris, R.A., 1986, Gamma-aminobutyric acid and alcohol actions: neurochemical studies of long sleep and short sleep mice, *Life Sci.* 39:2005.

Baldessarini, R.J., 1990, Benzodiazepines, *in*:"The Pharmacological Basis of Therapeutics," A.G. Gilman, T.W. Rall, A.S. Nies, and P. Taylor, eds., Pergamon, New York, pp. 424-427.

Bean, B.P., 1992, Pharmacology and electrophysiology of ATP-activated ion channels, *Trends Pharmacol. Sci.*, 13:87.

Bean, B.P. and Friel, D.D., 1990, ATP-activated channels in excitable cells, *in*: "Ion Channels, Vol. 2," T. Narahashi, ed., Plenum, New York.

Bloom, F.E., Siggins, G.R., Foote, S.L., Gruol, D., Aston-Jones, G., Rogers, J., Pittman, Q., and Staunton, D., 1984, Noradrenergic involvement in the cellular actions of ethanol, *in*: "Catecholamines: Neuropharmacology and Central Nervous System - Theoretical Aspects," E. Usdin *et al.*, eds., Alan R. Liss, New York, pp. 159-167.

Burnstock, G., 1990, Purinergic Mechanisms, *Ann. N.Y. Acad. Sci.* 603:1.

Carboni, E., Acquas, E., Frau, R., and Di Chiara, G., 1989, Differential inhibitory effects of a 5-HT$_3$ antagonist on drug-induced stimulation of dopamine release, *Europ. J. Pharmacol.*164:515.

Carboni, E., Acquas, E., Leone, P., Perezzani, L., and Di Chiara, G., 1988, 5-HT$_3$ receptor antagonists block morphine- and nicotine-induced place-preference conditioning, *Europ. J. Pharmacol.* 151:159.

Celentano, J.J., Gibbs, T.T., and Farb, D.H., 1988, Ethanol potentiates GABA- and glycine-induced chloride currents in chick spinal neurons, *Brain Res.* 455:377.

Cole, W.H., and Allison, J.B., 1930, Chemical stimulation by alcohols in the barnacle, the frog, and planaria, *J. Gen. Physiol.* 14:71.

Collingridge, G.L., and Lester, R.A.J., 1989, Excitatory amino acid receptors in the vertebrate central nervous system, *Pharmacol. Rev.* 41:143.

Costal, B., Naylor, R.J., and Tyers, 1990, The psychopharmacology of 5-HT$_3$ receptors, *Pharmac. Ther.* 47:181.

Davidoff, R.A., 1973, Alcohol and presynaptic inhibition in an isolated spinal cord preparation, *Arch. Neurol.* 28:60.

Derkach, V., Surprenant, A., and North, R.A., 1989, 5-HT$_3$ receptors are membrane ion channels, *Nature* 339:706.

Di Chiara, G., and Imperato, A., 1988, Drugs abused by humans preferentially increase synaptic dopamine concentrations in the mesolimbic system of freely moving rats, *Proc. Natl. Acad. Sci. USA* 85:5274.

Dildy, J.E., and Leslie, S.W., 1989, Ethanol inhibits NMDA-induced increases in free intracellular Ca^{2+} in dissociated brain cells, *Brain Research* 499:383.

Dildy-Mayfield, J.E., and Leslie, S.W., 1991, Mechanism of inhibition of *N*-methyl-D-aspartate-stimulated increases in free intracellular Ca^{2+} concentration by ethanol, *J. Neurochem.* 56:1536.

Edwards, F.A., Gibb, A.J., and Colquhoun, D., 1992, ATP receptor-mediated synaptic currents in the central nervous system, *Nature* 359:144.

Engberg, G., and Hajos, M., 1992, Ethanol attenuates the response of locus coeruleus neurons to excitatory amino acid agonists in vivo, *Naunyn-Schmiedebergs Arch. Pharmacol.* 345:222.

Evans, R.J., Derkach, V., and Surprenant, A., 1992, ATP mediates fast synaptic transmission in mammalian neurons, *Nature* 357:503.

Fadda, F., Garau, B., Marchei, F., Colombo, G., and Gessa, G.L., 1991, MDL 72222, a selective 5-HT$_3$ receptor antagonist, suppresses voluntary ethanol consumption in alcohol-preferring rats, *Alcohol Alcoholism* 26:107.

Ferguson, J., 1939, The use of chemical potentials as indices of toxicity, *Proc. Roy. Soc. Lond., B* 127:387.

Fink, K., Schultheiss, R., and Gothert, M., 1992, Inhibition of N-methyl-D-aspartate- and kainate-evoked noradrenaline release in human cerebral cortex slices by ethanol, *Naunyn-Schmiedebergs Arch. Pharmacol.* 345:700.

Gage, P.W., and Robertson, B., 1985, Prolongation of inhibitory postsynaptic currents by pentobarbitone, halothane and ketamine in CA1 pyramidal cells in rat hippocampus, *Brit. J. Pharmacol.* 85:675.

Gean, P.W., 1992, Ethanol inhibits epileptiform activity and NMDA receptor-mediated synaptic transmission in rat amygdaloid slices, *Brain Res. Bull.* 28:417.

Givens, B.S., and Breese, G.R., 1990, Site-specific enhancement of γ-aminobutyric acid-mediated inhibition of neural activity by ethanol in the rat medial septal area, *J. Pharmacol. Exp. Ther.* 254:528.

Goldstein, D.B., 1984, The effects of drugs on membrane fluidity, *Annu. Rev. Pharmacol. Toxicol.* 24:43.

Gonzales, R.A., Westbrook, S.L., and Bridges, L.T., 1991, Alcohol-induced inhibition of N-methyl-D-aspartate-evoked release of [^{3}H]norepinephrine from brain is related to lipophilicity, *Neuropharmacol.* 30:441.

Gonzales, R.A., and Woodward, J.J., 1990, Ethanol inhibits N-methyl-D-aspartate-stimulated [^{3}H]norepinephrine release from rat cortical slices, *J. Pharmacol. Exp. Ther.* 253:1138.

Gothert, M., and Fink, K., 1989, Inhibition of N-methyl-D-aspartate (NMDA)- and L-glutamate-induced noradrenaline and acetylcholine release in the rat brain by ethanol, *Naunyn-Schmiedebergs Arch. Pharmacol.* 340:516.

Grant, K.A., and Barrett, J.E., 1991, Blockade of the discriminative stimulus effects of ethanol with 5-HT$_3$ receptor antagonists, *Psychopharmacol.*, 104:451.

Hall, Z.W., 1992, "Molecular Neurobiology," Sinauer Assoc., Sunderland, MA.

Harrison, N.L., Majewska, M.D., Harrington, J.W., and Barker, J.L., 1987, Structure-activity relationships for steroid interaction with gamma-aminobutyric acidA receptor complex, *J. Pharmacol. Exp. Ther.* 241:346.

Hartnett, C., 1991, Detection by PCR of GABA$_A$ receptor subtypes expressed in dorsal root ganglia of the adult rat, *Soc. Neurosci. Abst.* 17:797.

Hellevuo, K., Hoffman, P.L., and Tabakoff, B., 1991, Ethanol fails to modify [^{3}H]GR65630 binding to 5-HT$_3$ receptors in NCB-20 cells and in rat cerebral membranes, *Alcoholism: Clin. Exptl. Research*, 15:775.

Hoffman, P.L., Rabe, C.S., Moses, F., and Tabakoff, B., 1989, N-methyl-D-aspartate receptors and ethanol: inhibition of calcium flux and cyclic GMP production, *J. Neurochem.* 52:1937.

Illes, P., and Norenberg, W., 1993, Neuronal ATP receptors and their mechanism of action, *Trends Pharmacol. Sci.* 14:50.

Imperato, A., and Di Chiara, G., 1986, Preferential stimulation of dopamine release in the nucleus accumbens of freely moving rats by ethanol, *J. Pharmacol. Exp. Ther.* 239:219.

Julius, D., 1991, Molecular biology of serotonin receptors, *Annu. Rev. Neurosci.* 14:335.

Kelso, S.R., Nelson, T.E., and Leonard, J.P., 1992, Protein kinase C-mediated enhancement of NMDA currents by metabotropic glutamate receptors in *Xenopus* oocytes, *J. Physiol.* 449:705.

Kilpatrick, G.J., Jones, B.J., and Tyers, M.B., 1987, Identification and distribution of 5-HT$_3$ receptors in rat brain using radioligand binding, *Nature* 330:746.

Kilpatrick, G.J., Jones, B.J., and Tyers, M.B., 1989, Binding of the ligand, [^{3}H]GR65630, to rat area postrema, vagus nerve and the brains of several species, *Eur. J. Pharmacol.* 159:157.

Kissin, B., 1988, Alcohol abuse and alcohol-related illnesses, *in*: "Cecil Textbook of Medicine," J.B. Wyngaarden, and L.H. Smith, eds., Saunders, Philadelphia, pp. 48-51.

Li, C., Aguayo, L., and Weight, F.F.,1991, Ethanol inhibits ATP-activated currents in sensory neurons, *Third IBRO World Congress Neurosci. Abst.*, 64.

Lima-Landman, M.T., and Albuquerque, E.X., 1989, Ethanol potentiates and blocks NMDA-activated single-channel currents in rat hippocampal pyramidal cells, *FEBS. Lett.* 247:61.

Lin, A.M.-Y., Freund, R.K., and Palmer, M.R., 1991, Ethanol potentiation of GABA-induced electrophysiological responses in cerebellum: requirement for catecholamine modulation, *Neurosci. Lett.* 122:154.

Lovinger, D.M., 1991, Ethanol potentiation of 5-HT$_3$ receptor-mediated ion current in NCB-20 neuroblastoma cells, *Neurosci. Lett.* 122:57.

Lovinger, D.M., and White, G., 1991, Ethanol potentiation of 5-hydroxytryptamine$_3$ receptor-mediated ion current in neuroblastoma cells and isolated adult mammalian neurons, *Mol. Pharmacol.* 40:263.

Lovinger, D.M., White, G., and Weight, F.F., 1989, Ethanol inhibits NMDA-activated ion current in hippocampal neurons, *Science* 243:1721.

Lovinger, D.M., White, G., and Weight, F.F., 1990a, NMDA receptor-mediated synaptic excitation selectively inhibited by ethanol in hippocampal slice from adult rat, *J. Neurosci.* 10:1372.

Lovinger, D.M., White, G., and Weight, F.F., 1990b, Ethanol inhibition of neuronal glutamate receptor function, *Ann. Medicine* 22:247.

Lovinger, D.M., White, G., and Weight, F.F. 1990c, Ethanol inhibition of NMDA-activated ion current is not voltage-dependent and ethanol does not interact with other binding sites on the NMDA receptor/ionophore complex, *FASEB J.* 4:A678.

Mancillas, J.R., Siggins, G.R., and Bloom, F.E., 1986, Systemic ethanol: selective enhancement of responses to acetylcholine and somatostatin in the rat hippocampus, *Science* 231:161.

Maricq, A.V., Peterson, A.S., Brake, A.J., Myers, R.M., and Julius, D., 1991, Primary structure and functional expression of the 5HT$_3$ receptor, a serotonin-gated ion channel, *Science* 254:432.

Mayer, M.L., and Westbrook, G.L., 1987, The physiology of excitatory amino acids in the vertebrate central nervous system, *Prog. Neurobiol.* 288:197.

Mehta, A.K., and Ticku, M.K., 1988, Ethanol potentiation of GABAergic transmission in cultured spinal cord neurons involves γ-aminobutyric acid$_A$-gated chloride channels, *J. Pharmacol. Exp. Ther.* 246:558.

Meyer, H., 1899, Welche eigenschaft der anasthetica bedingt ihre narkitische wirkung?, *Naunyn-Schmiedebergs Arch. Exp. Path. Pharmakol.* 42:109.

Meyer, H., 1901, Zur theorie der alkolnarkose: der einfuss wechselnder temperatur auf wirkungsstarke und theilungscoefficient der narcotica, *Naunyn-Schmiedebergs Arch. Exp. Path. Pharmakol.* 46:338.

Meyer, K.H., 1937, Contributions to the theory of narcosis, *Transac. Faraday Soc.* 33:1062.

McCreery, M.J., and Hunt, W.A., 1978, Physico-chemical correlates of alcohol intoxication, *Neuropharmacol.* 17:451.

McGeer, P.L., Eccles, J.C., and McGeer, E.G., 1987, "Molecular Neurobiology of the Mammalian Brain," Plenum, New York, pp. 197-234.

Mihic, S.J., Wu, P.H., and Kalant, H., 1992, Potentiation of γ-aminobutyric acid-mediated chloride flux by pentobarbital and diazepam but not ethanol, *J. Neurochem.* 58:745.

Morelli, M., Deidda, S., Garau, L., Carboni, E., and Di Chiara, G., 1988, Ethanol: lack of stimulation of chloride influx in rat brain synaptoneurosomes, *Neurosci. Res., Commun.* 2:77.

Mullins, L.J., 1954, Some physical mechanisms of narcosis, *Chem. Rev.* 54:289.

Nakamura, J., Sasa, M., and Takaori, S., 1989, Ethanol potentiates the effect of gamma-aminobutyric acid on medial vestibular neurons responding to horizontal rotation, *Life Sci* 45:971.

Nestoros, J.N., 1980, Ethanol specifically potentiates GABA-mediated neurotransmission in feline cerebral cortex, *Science* 209:708.

Nishio, M., and Narahashi, T.,1990, Ethanol enhancement of GABA-activated chloride current in rat dorsal root ganglion neurons, *Brain Res.* 518:283.

Olsen, R.W., and Tobin, A.J., 1990, Molecular biology of GABA$_A$ receptors, *FASEB J.* 4:1469.

Osmanovic, S.S., and Shefner, S.A., 1990, Enhancement of current induced by superfusion of GABA in locus coeruleus neurons by pentobarbital, but not ethanol, *Brain Res.* 517:324.

Overton, E., 1896, Uber die osmotischen eigenschaften der zelle in ihrer betdeutung fur die toxikologie und pharmakologie, *Zeitschrift. Physikal. Chem.* 22:189.

Overton, E., 1901, "Studien uber die Narkose Zugleich ein Beitrag zur Allgemeinen Pharmakologie," Verlag von Gustav Fischer, Jena.

Peoples, R.W., Lovinger, D.M., and Weight, F.F., 1990, Inhibition of excitatory amino acid currents by general anesthetic agents, *Soc. Neurosci. Abst.* 16:1017.

Peoples, R.W., and Weight, F.F., 1991, Modulation of amino acid-activated ion currents by trichloroethanol, *Third IBRO World Congress Neurosci. Abst.* 63.

Peoples, R.W., and Weight, F.F., 1992a, Ethanol inhibition of N-methyl-D-aspartate-activated ion current in rat hippocampal neurons is not competitive with glycine, *Brain Res.* 571: 342.

Peoples, R.W., and Weight, F.F. 1992b, Inhibition of excitatory amino acid-activated ion currents by inhalational anesthetics, *Soc. Neurosci. Abst.* 18:248.

Proctor, W.R., Soldo, B.L., Allan, A.M. and Dunwiddie, T.V., 1992, Ethanol enhances synaptically evoked GABA$_A$ receptor-mediated responses in cerebral cortical neurons in rat brain slices, *Brain Res.* 595:220.

Rabe, C.S., and Tabakoff,B., 1990, Glycine site-directed agonists reverse the actions of ethanol at the N-methyl-D-aspartate receptor, *Mol. Pharmacol.*, 38:753.

Rall, T.W., 1990, Benzodiazepines, *in*:"The Pharmacological Basis of Therapeutics," A.G. Gilman, T.W. Rall, A.S. Nies, and P. Taylor, eds., pp. 346-358, Pergamon, New York.

Rang, H.P., 1960, Unspecific drug action. The effects of a homologous series of primary alcohols, *Brit. J. Pharmacol.* 15:185.

Reynolds, J.N., Prasad, A., and MacDonald, J.F., 1992, Ethanol modulation of GABA receptor-activated Cl$^-$ currents in neurons of the chick, rat and mouse central nervous system, *Eur. J. Pharmacol.* 224:173.

Ritchie, J.M., 1985, Ethyl Alcohol, *in*: "The Pharmacological Basis of Therapeutics," A.G. Goodman, L.S. Gilman, T.W. Rall, and F. Murad, eds., Macmillan, New York, pp. 372-381.

Seeman, P., 1972, The membrane actions of anesthetics and tranquilizers, *Pharmacol. Rev.* 24:583.

Sellers, E.M., Higgins, G.A., and Sobell, M.B., 1992, 5-HT and alcohol abuse, *Trends Pharmacol. Sci.* 13:69.

Shen, K.-Z., and North, R.A., 1993, Excitation of rat locus coeruleus neurons by adenosine 5'-triphosphate: ionic mechanism and receptor characterization, *J. Neurosci.* 13:894.

Siggins, G.R., Pittman, Q.J., and French, E.D., 1987, Effects of ethanol on CA$_1$ and CA$_3$ pyramidal cells in the hippocampal slice preparation: an intracellular study, *Brain Res.* 414:22.

Simson, P.E., Criswell, H.E., Johnson, K.B., Hicks, R.E., and Breese, G.R., 1991, Ethanol inhibits NMDA-evoked electrophysiological activity in vivo, *J. Pharmacol. Exp. Ther.*, 257:225.

Suzdak, P.D., Schwartz, R.D., Skolnick, P., and Paul, S.M.,1986, Ethanol stimulates γ-aminobutyric acid receptor-mediated chloride transport in rat brain synaptoneurosomes, *Proc. Natl. Acad. Sci. USA* 83:4071.

Takada, R., Saito, K., Matsuura, H., and Inoki, R., 1989, Effect of ethanol on hippocampal GABA receptors in the rat brain, *Alcohol* 6:115.

Ueha, T., and Kuriyama, K., 1991, Ethanol-induced alterations in the function of cerebral GABA$_A$ receptor complex: effect on GABA-dependent ^{36}Cl$^-$ influx into cerebral membrane vesicles, *Alcohol Alcoholism* 26:17.

Wafford, K.A., Burnett, D.M., Dunwiddie, T.V.; and Harris, R.A., 1990, Genetic differences in the ethanol sensitivity of GABA$_A$ receptors expressed in *Xenopus* oocytes, *Science* 249:291.

Wafford, K.A., Burnett, D.M., Leidenheimer, N.J., Burt, D.R., Wang, J.B., Kofuji, P., Dunwiddie, T.V., Harris, R.A., and Sikela, J.M., 1991, Ethanol sensitivity of the GABA$_A$ receptor expressed in Xenopus oocytes requires 8 amino acids contained in the γ2L subunit, *Neuron* 7:27.

Wafford, K.A., and Whiting, P.J., 1992, Ethanol potentiation of GABA$_A$ receptors requires phosphorylation of the alternatively spliced variant of the γ2 subunit, *FEBS Lett.* 313:113.

Weight, F.F., Peoples, R.W., Wright, J.M., Lovinger, D.M., and White, G., 1993, Ethanol action on excitatory amino acid activated ion channels, *Alcohol Alcoholism*, in press.

White, G., Lovinger, D.M., and Weight, F.F., 1990, Ethanol inhibits NMDA-activated current but does not alter GABA-activated current in an isolated adult mammalian neuron, *Brain Res.* 507:332.

Whiting, P., McKernan, R.M., and Iversen, L.L., 1990, Another mechanism for creating diversity in γ-aminobutyrate type A receptors: RNA splicing directs expression of two forms of γ2 subunit, one of which contains a protein kinase C phosphorylation site, *Proc. Natl. Acad. Sci. USA* 87:9966.

Woods, J. H., Koek, W., and Ornstein, P., 1987, A preliminary study of PCP-like behavioral effects of 2-amino-5-phosphonovalerate in rhesus monkeys, *in*: "Excitatory Amino Acid Transmission," T. P. Hicks, D. Lodge, and H. McLennan, eds., Liss, New York, pp. 205-212.

Woodward, J.J., and Gonzales, R.A., 1990, Ethanol inhibition of N-methyl-D-aspartate-stimulated endogenous dopamine release from rat striatal slices: reversal by glycine, *J. Neurochem.* 54:712.

Wozniak, K. M., Pert, A., and Linnoila, M., 1990, Antagonism of 5-HT$_3$ receptors attenuates the effects of ethanol on extracellular dopamine, *Europ. J. Pharmacol.* 187:287.

Wright, J.M., and Weight, F.F., 1992, Effects of ethanol on NMDA receptor-channels in single channel recordings, *Soc. Neurosci. Abst.* 18:257.

Zahniser, N.R., Buck, K.J., Curella, P., McQuilkin, S.J., Wilson-Shaw, D., Miller, C.L., Klein, R.L., Heidenreich, K.A., Keir, W.J., Sikela, J.M., and Harris, R.A., 1992, GABA$_A$ receptor function and regional analysis of subunit mRNAs in long-sleep and short-sleep mouse brain, *Molec. Brain Res.* 14:196.

MOLECULAR MECHANISMS OF ALCOHOL NEUROTOXICITY

Fulton T. Crews, Hunter Newsom, Mark Gerber,
Colin Sumners, L. Judson Chandler, and Gerhard Freund

Center for Alcohol Research
Department of Pharmacology
University of Florida College of Medicine
Gainesville, Florida 32610-0267

Neuropathology of Chronic Alcohol Abuse

Heavy alcohol consumption over a period of years can lead to cognitive and neurological impairments. Studies have indicated that alcoholics have greatly reduced brain volume. Computer tomographic studies of the brains of alcoholics have indicated that chronic alcoholism leads to enlargement of the ventricles, decreases in tissue volume and increases in cerebral spinal fluid volume, i.e. brain shrinkage (Pfefferbaum et al. 1992). Although changes are somewhat similar to those that occur during normal aging, the increase in size of the lateral ventricles and the increase in the cortical space between sulci clearly indicate that the brain of alcoholics have a decrease in cellular mass above that found in age matched non-alcoholics (Pfefferbaum et al., 1992; Jernigan et al., 1992). The increases in cerebral spinal fluid spaces are particularly associated with loss of the gray matter with some reduction also occurring in white matter. These losses in cortical tissue are possibly an acceleration of age-induced effects as well as the cumulative toxicity that occurs during a lifetime of chronic alcohol exposure.

The reduction in brain volume observed in imaging studies is consistent with human postmortem histopathological studies of chronic alcoholics which have reported neuronal cell body shrinkage and neuronal loss (Harper and Krill, 1990). A number of studies have shown that chronic alcoholics (Carlen et al., 1978; Harper et al., 1985; Harper et al., 1987) and heavy social drinkers (Cala et al., 1983) exhibit cerebral cortical and cerebellar vermal atrophy (Pfefferbaum et al., 1990). The gross brain histopathology of alcoholics is characterized by a uniformly diffuse atrophy of the cerebral cortex (Freund, 1985). Microscopic changes consist of a diffuse and patchy loss of neurons with dendritic and axonal regression in many of the remaining neurons. These neuronal changes are not specific for alcohol abuse nor are they distinguishable from those that occur during aging and/or from hypoxia (Lynch, 1960). Furthermore, ethanol intoxication frequently precedes symptoms of ischemic brain damage, particularly in patients under 40 years of age (Hillbom, 1978). Although alcohol associated atrophy in the central nervous system is well documented in humans, these findings are complicated by problems of nutrition. However, Walker's studies with rats receiving nutritionally complete diets have provided definitive evidence that chronic ethanol consumption has neurotoxic effects even in the presence of adequate nutrition. Chronic ethanol treatment of rats produces a significant loss of hippocampal CA1 pyramidal cells, dentate granule cells and interneurons. In addition, the surviving neurons have attenuated dendrites and decreased density of dendritic spines. These changes are associated with memory deficits in animals comparable to the dementia often found in alcoholics (Walker and Hunter,

1987; Walker et al., 1981). Thus, chronic alcohol abuse is clearly associated with neurotoxic processes that lead to neuronal cell damage and reductions in brain cellular mass.

Although ethanol is clearly neurotoxic, the exact mechanism(s) are not known. The neuropathological similarities of chronic ischemia and ethanol abuse may be due to overlapping pathological etiologies. Epidemiologic studies have associated ethanol consumption with strokes, particularly ischemic stroke where more than 3 drinks per day has been suggested as a threshold level (Gorelick, 1990). Cerebral ischemia is the largest category of stroke and both the World Health Organization and the Stroke Council list chronic ethanol as a risk factor for stroke (Gorelick, 1990). Hypoxic/ischemic brain damage is strongly linked to excitatory amino acid (EAA) mediated excitotoxicity (Choi, 1990; Clark, 1989). Ischemia has been shown to increase extracellular levels of glutamate more than 10 fold (Clark, 1989) which secondarily excites neurons causing them to release more glutamate. The excitotoxic hypothesis of brain injury proposes that glutamate is a principle cause of damage in ischemia. Three components of this hypothesis have been tested and largely proved in experimental studies. First, elevated concentrations of glutamate cause excessive excitation at a subtype of glutamate receptors, the NMDA receptor. Second, excitation at this receptor leads to excessive influx of sodium, chloride and water, which cause acute neuronal damage, and calcium, which causes delayed and more permanent damage. Third, pharmacologic blockade at the NMDA receptor-ion channel complex prevents ischemic neuronal damage (Choi, 1990; Clark, 1989). The similarities between ischemic brain damage and that found in the chronic alcoholic is consistent with common mechanisms of action. Ethanol could sensitize neurons to neuronal damage due to changes in glutamate mediated excitation of neurons making them more susceptible to excitotoxicity. This chapter will review various aspects of the effects of ethanol on neurotransmission and excitotoxicity and present the hypothesis that the neuropathological actions of ethanol are due to sensitization of neurons to excitotoxicity during chronic ethanol exposure.

Effects of Ethanol on Neurotransmission

Ethanol has been found to have a variety of acute effects on neurotransmission. Acute ethanol tends to reduce neuronal excitation (Figure 1). A variety of biochemical and electrophysiological studies suggest that *in vitro* ethanol, at pharmacologically relevant concentrations, is an inhibitor of NMDA receptors (Hoffman et al., 1989; Hoffman et al., 1989; Lovinger et al., 1989; Dildy and Leslie, 1989; Woodward and Gonzales, 1990). In addition to inhibition of calcium influx through NMDA channels, ethanol has also been shown to inhibit calcium influx through voltage sensitive calcium channels (VSCC). In concentrations as low as 25 mM, ethanol was shown to significantly inhibit fast-phase $^{45}Ca^{2+}$ influx into cerebral cortical synaptosomes (Leslie et al., 1983) and to inhibit fast-phase $^{45}Ca^{2+}$ uptake into PC-12 cells (Skattebol and Rabin, 1987). Electrophysiologically, ethanol reduces the calcium component of action potentials (Oakes and Pozos, 1982) and inhibits calcium spikes in cultured neurons (Triestman et al., 1985). Thus, inhibition of voltage sensitive calcium channels by acute ethanol would tend to reduce $[Ca^{2+}]_i$ and excitation. Similarly, ethanol has been shown to potentiate GABA stimulated Cl^- flux through $GABA_A$ receptor operated ion channels. Ethanol has been found to potentiate the 5-HT$_3$ receptor ion channel and inhibit the adenosine transporter. A variety of studies have also suggested that ethanol may directly alter the coupling of guanine nucleotide proteins to enzymes, particularly adenylate cyclase. Taken together, these actions likely contribute to the intoxicating and sedating actions of ethanol. Adaptive changes in the presence of ethanol leads to tolerance and physical dependence. The hyperexcitability syndrome associated with ethanol withdrawal may be related to adaptive changes in these components of neurotransmission which could reduce ethanol's ability to blunt neuronal excitation. This hyperexcitable state may be directly related to the neurotoxic effects of ethanol.

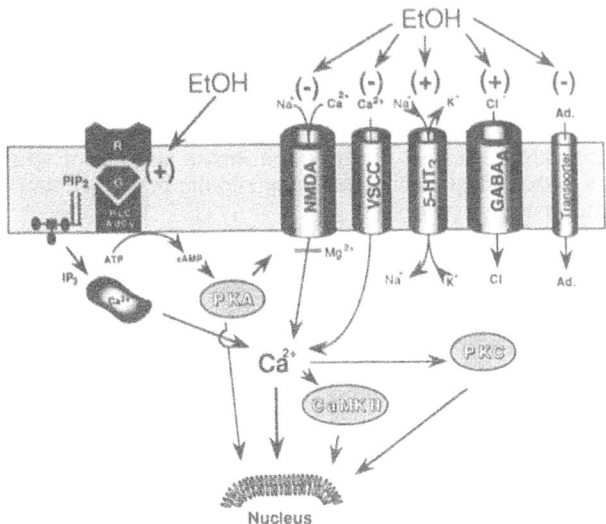

Figure 1. Ethanol sensitive components of neurotransmission. Pharmacological concentrations of ethanol have been found to selectively modify a variety of neurotransmitter signals. NMDA receptor operated and the voltage sensitive calcium channels (VSCC) are both inhibited by ethanol which reduce calcium influx and reduce neuronal excitation (Gonzales and Hoffman, 1991). In addition, ethanol potentiates GABA mediated CL⁻ flux (Allan and Harris, 1986) which further reduces neuronal excitation. Ethanol has also been shown to potentiate the 5-hydroxytryptamine3 (5HT3) cation channel (Lovinger and White, 1990) and to increase the accumulation of adenosine by increasing release and blocking adenosine transport (Clark and Dar, 1989; Nagy et al., 1990). Increased extracellular adenosine can activate adenosine A2 receptors stimulating cyclic AMP and reducing neuronal excitation. Ethanol may increase cyclic AMP by promoting the activity of guanine nucleotide proteins (Hoffman and Tabakoff, 1990). Although these actions of ethanol vary in magnitude between different neuronal types, changes in intracellular calcium and other second messenger regulated protein kinases likely lead to modifications in the nuclear transcription of proteins which lead to tolerance and dependance to ethanol.

Excitotoxicity and Excitatory Amino Acid Receptors

Glutamate and aspartate are the major neurotransmitters responsible for rapid excitatory neurotransmission as well as excitotoxicity. EAA mediated excitotoxicity has been divided into two components: a rapid component associated with osmotic swelling and a delayed component that occurs as a progressive degenerative process over a period of hours to days. Both processes are likely due at least in part to the excessive accumulation of calcium (Meldrum and Garthwaite, 1990; Clark, 1989; Olney, 1990). In addition to Ca^{2+}, the early process likely involves Na^+ flux through activated membrane channels, an increase in intracellular volumes and a depletion of intracellular energy stores. The slow progressive neuronal degeneration, i.e. delayed neuronal cell death, appears to involve the following events. Glutamate depolarizes neurons leading to increased $[Ca^{2+}]_i$ through receptor-gated ion channels and VSCC (Choi et al., 1988; Clark, 1989). Both glutamate and the increase in $[Ca^{2+}]_i$ activate phospholipases generating Ins 1,4,5P3 which releases intracellular calcium, diacylglycerol which activates protein kinase C (PKC), and arachidonic acid which is metabolized by prostaglandin synthetases generating free radicals and prostaglandins. Ultimately, additional proteases and endonucleases are activated leading to cell death (Orrenius et al., 1989). Although a brief exposure to glutamate initially increases $[Ca^{2+}]_i$ even after removal of glutamate, $[Ca^{2+}]_i$ remains increased throughout the progression to delayed neuronal death. This sustained increase in $[Ca^{2+}]_i$ appears to involve activation of PKC since PKC inhibitors or down-regulation of PKC have been reported to prevent the sustained increase in $[Ca^{2+}]_i$ and block delayed neuronal cell death (Favaron, et al., 1990; Manev et al., 1989). Studies *in vivo* and *in vitro* have indicated that the properties of EAA receptors are reflected in the excitotoxic actions of glutamate. Thus, excitotoxicity is a pathological mechanism unique to the brain that may be involved in a number of neurodegenerative processes.

Pharmacological studies have been used to define four classes of glutamate receptors that molecular biology studies have indicated are extremely heterogeneous. Three of the receptor classes are receptor operated ion channels, e.g. N-methyl-D-aspartic acid (NMDA), kainate, and amino-3-hydroxy-5-methyl-4-isoxalone proprionic acid (AMPA). The fourth group of receptors are referred to as metabotropic receptors since they couple through guanine nucleotide proteins to stimulate phosphoinositide hydrolysis and/or inhibit adenylate cyclase activity. The excitatory potency of glutamate analogs generally parallel the order of toxicity with NMDA and kainate being the most excitatory receptors that are linked to neurotoxicity. Each of the pharmacological EAA receptor subtypes are heterogeneous (Monaghan et al., 1988; Cotman et al., 1989). NMDA receptors are characterized by voltage-dependent block by extracellular magnesium, high Ca^{2+}/Na^+ permiability ratio, requirement for glycine as a co-agonist, and by slow onset and offset time courses of channel-gating. The high influx of calcium during activation of NMDA receptors is critical for activity dependent synaptic plasticity as well as playing a dominant role in excitotoxicity. Several subtypes of the NMDA receptor are likely to exist (Nakanishi, 1992; Michaelis, et al., 1992). Molecular cloning has identified several different mRNA species which can form heteromeric ion channel structures (Monyer et al., 1992b). These subunits exhibit differential expression patterns in brain regions suggesting that different heteromeric NMDA channels exist in various brain regions. Michaelis et al. (1992) has suggested that 4 to 5 proteins of different molecular weights combine to form functional NMDA channels. Although the exact diversity of NMDA receptors remains to be fully elucidated, it is clear that NMDA receptors play a key role in excitotoxicity.

Kainate and AMPA receptors are a heterogeneous group of ion channels composed of multiple subunits each with a unique pharmacology. There are multiple genes (GluR1-GluR7) that encode subunits which combine to form multisubunit AMPA-kainate receptors (Boulter et al., 1992). In general, GluR1-4 subunits group into AMPA-kainate receptors whereas GluR5-7 tend to be predominantly kainate receptors. These ion channels are permiant to sodium and potassium (Boulter et al., 1992; Monyer et al., 1992a) and in some instances, conduct Ca^{2+} (Gilbertson et al., 1990). A major difference is that AMPA receptors readily desensitize during stimulation whereas the response to kainate is prolonged with little or no desensitization. There are differences in rank order of potencies, ligand affinities, and distribution of binding sites in brain which likely reflect multiple combinations of various subunits which form heteromeric ion channels with different agonist responses (Barnard et al., 1992). GluR1 and GluR2 are the predominant mRNA's in cortex and hippocampus. GluR2 has the unique property in that one amino acid out of approximately 900 has been shown to regulate Ca^{2+} conductance for these receptor channels (Monyer et al., 1992a). Since Ca^{2+} flux is a major determinant of excitotoxicity, changes in this subunit could impact tremendously on glutamate stimulated excitotoxicity for neurons not containing GluR2 subunits within its kainate-AMPA ion channel complex. The subunit composition of various subtypes of AMPA and kainate receptors that are particularly excitotoxic and their brain regional distribution are not clearly delineated at this time.

Cerebral cortical cell culture studies have found that glutamate neurotoxicity depends largely on activation of NMDA receptors. Direct exposure of cortical neurons in culture to NMDA for 3-5 min leads to massive and progressive neuronal death whereas kainate exposure requires several hours of treatment to induce excitotoxicity (Choi et al., 1988). However, this may depend on a number of factors. Certain neurons found to be relatively resistant to NMDA appear to be particularly sensitive to kainate neurotoxicity (Koh and Choi, 1988).

Voltage sensitive calcium channels are likely to play a role in certain types of excitotoxicity. Studies in cultured cortical neurons have found that the dihydropyridine calcium channel antagonist nifedipine has little affect on the marked neuronal degeneration produced by brief exposure to high concentrations of NMDA, whereas nifedipine markedly reduces the neurotoxicity produced by prolonged exposure to quinolinate, kainate and AMPA (Weiss et al., 1990). Quinolinate is a weak NMDA agonist that requires prolonged exposure for neurotoxicity. Other studies in striatal neurons have shown that the dihydropyrindine antagonist nitrendipine markedly reduced the rise in intracellular free calcium produced by kainate (Murphy and Miller, 1989). Thus, VSCC are important for neurotoxicity of non-NMDA glutamate receptors and submaximal stimulation of NMDA receptors that requires prolonged stimulation.

Nitric Oxide and Excitotoxicity

Nitric oxide (NO) may play an important role in excitotoxicity. Nitric oxide is formed from arginine by the action of NO synthase (NOS). In neurons, NOS is Ca^{2+}/calmodulin dependent and appears to be activated by increased intracellular calcium (Bredt and Snyder, 1990). Although the physiological actions of NO in brain have not yet been clearly defined, formation of NO has been strongly linked to NMDA receptor activation and may be involved in synaptic plasticity and long-term potentiation (Bohme et al., 1991; Schuman and Madison, 1991; O'Dell et al., 1992). NMDA receptor activation has been shown to increase NO formation in a variety of cell culture systems via the influx of calcium through NMDA channels (Garthwaite et al., 1989). Recent evidence suggests that NO may be an important mediator of NMDA receptor stimulated excitotoxicity. Inhibition of NOS activity *in vitro* has been reported to protect against NMDA receptor neurotoxicity in primary neuronal cultures (Dawson et al., 1991) while inhibition of NOS *in vivo* has been shown to dramatically reduce the volume of cortical infarcts (a measure of neuronal cell death which is strongly linked to excitotoxicity) following irreversible focal ischemia in the mouse (Nowicki et al., 1991). Thus, an important factor in excitotoxicity is likely to be the formation of NO subsequent to NMDA receptor activation.

Free radical formation has been implicated in various forms of neurotoxicity. NO is itself a highly reactive free radical and the cytotoxic properties of NO most likely relate to its high chemical reactivity or to that of metabolic products of NO metabolism such as peroxynitrate anions ($ONOO^-$) which are formed when NO reacts with superoxide ions. Precedent for NO as a mediator of cytotoxicity is found in cells such as macrophages and neutrophils which utilize NO as a bacteriocidal and tumoricidal agent. These cells contain a form of NOS that is not dependent upon Ca^{2+}/calmodulin and is transcriptionally regulated. Activation of the Ca^{2+}/calmodulin dependent form of NO synthase found in neurons results in transient NO formation due to the transient nature of Ca^{2+} flux, whereas formation of NO following induction of the Ca^{2+}/calmodulin independent form is sustained and prolonged resulting in a much higher concentration of NO in being produced. In brain, it appears that at low concentrations, NO plays a neurotransmitter or neuromodulatory role. However, when produced in high concentrations, such as might occur in response to prolonged or excessive stimulation of NMDA receptors following stroke or hypoxia/ischemia, NO exhibits neurotoxic actions. In addition, very recent studies have shown that brain microglia, which are frequently associated with reactivity to neuronal insult such as that following hypoxia/ischemia, also possess an inducible form of NOS (Zielasek et al., 1992; Chandler et al., 1992a). It is not known what role, if any, formation of NO by microglia may play in the excitotoxic process. In any case, it is clear that glutamate stimulated Ca^{2+} flux will activate NO formation which generates free radicals and other actions which may play a key role in excitotoxicity.

Ethanol and Excitotoxicity

Exposure of cerebral cortical neurons in culture to NMDA for a little as 5 minutes results in a progressive neuronal death over the next 20 hours. When ethanol is added during NMDA treatment it reduces the neurotoxic effects of NMDA (Chandler et al., 1992; Greenberg, 1992). At maximal concentrations of NMDA, ethanol can block excitotoxicity in a dose dependent manner with approximately 40 % inhibition at 25 mM ethanol and complete inhibition of excitotoxicity at 200 mM ethanol (Chandler et al., 1992b). Ethanol acts as a non-competitive antagonist of excitotoxicity by reducing the excitotoxic effects of NMDA at all concentrations of NMDA that cause excitotoxicity (Greenberg, 1992). Thus, acute *in vitro* ethanol can actually protect neurons from excitotoxicity.

Although acute ethanol treatment may protect against EAA excitotoxicity, tolerance to ethanol is remarkably rapid. Ethanol withdrawal represents a hyperexcitable state that can include seizures after prolonged drinking episodes. Seizures have been associated with EAA neurotransmission and excitotoxic neuronal lesions at the site of the seizure focus. During ethanol withdrawal, neuronal hyperexcitability is likely due at least in part to increased EAA neurotransmission. Both NMDA antagonists (Grant et al., 1990) and VSCC antagonists (Little et al., 1986) reduce ethanol withdrawal symptoms. These data suggest that ethanol withdrawal hyperexcitability involves increased neuronal calcium flux which could sensitize neurons to excitotoxicity. The loss of cortical pyramidal neurons (Harper and Kril, 1990) and decreased cognitive function (Oscar-Berman and Ellis, 1987) are consistent with chronic ethanol sensitizing cortical neurons to excitotoxicity. Increased sensitivity to excitotoxicity during chronic ethanol could underlie the neuropathology of chronic ethanol. Studies in cell culture have clearly indicated that chronic ethanol sensitizes neurons to NMDA mediated excitotoxicity (Figure 3; Chandler et al., 1993).

The mechanism by which chronic ethanol increases excitotoxicity is likely due to adaptive changes in neuronal ion channels which regulate excitability and Ca^{2+} flux during chronic ethanol abuse (Figure 4). Michaelis has pioneered studies for more than a decade indicating increases in L-glutamate binding sites during chronic ethanol treatment (Michaelis, 1989; Michaelis et al., 1990; Michaelis et al., 1978). These studies have lead to the cloning of the cDNA for a 70-kD glutamate binding protein that forms a heteromeric protein channel with approximately 3 other proteins. This channel is one form of the NMDA receptor which has been found to play an important role in excitotoxicity (Mattson et al., 1989; Michaelis et al., 1992). Thus, the increased L-glutamate sites reported in human alcoholics and rats treated chronically with ethanol appears to represent an increase in the agonist binding site on NMDA ion channels. Other studies have found that chronic ethanol treatment increases the density of MK-801 sites in the hippocampus (Grant et al., 1990). MK-801 is a ligand used to identify the NMDA ion channel itself. Studies in cultured cerebellar neurons have found that chronic ethanol can significantly increase NMDA stimulated calcium influx (Iorio et al., 1992). Increases in NMDA receptor ion channels and NMDA mediated calcium flux are not the only changes that would sensitize neurons to excitotoxicity. Chronic ethanol treatment of cell cultures (Brennan et al., 1989; Messing et al., 1986) and animals (Brennan et al., 1990) results in an increase in membrane calcium channels, particularly the L-type VSCC. In hippocampal CA1 pyramidal neurons, the L-type voltage-gated channels are clustered at the base of major dendrites (Westenbroek et al., 1990) where they may play a role in the high sensitivity of this group of neurons to excitotoxicity. Chronic ethanol exposure has also been found to decrease GABA stimulated Cl⁻ flux with little change in receptor binding (Harris and Allan, 1989). The reduction in GABA feedback inhibition would reduce homeostatic mechanisms preventing excessive excitation. In addition to receptor gated channels, chronic ethanol increases the maximum velocity of the Na^+/Ca^{2+} antiporter (Michaelis, 1989). This change in antiporter might protect or it could play a role in excitotoxic increases in calcium as the influx in Na^+ might actually reverse the antiporter leading to further increases in $[Ca^{2+}]_i$. Thus, a variety of data suggest that chronic ethanol treatment may disrupt calcium homeostatic mechanisms enhancing EAA excitotoxicity. Recent studies have suggested that chronic ethanol may increase receptor stimulated production of NO (Davda et al., 1992). Although these studies were not done in neurons, an increase in NMDA stimulated NO production would

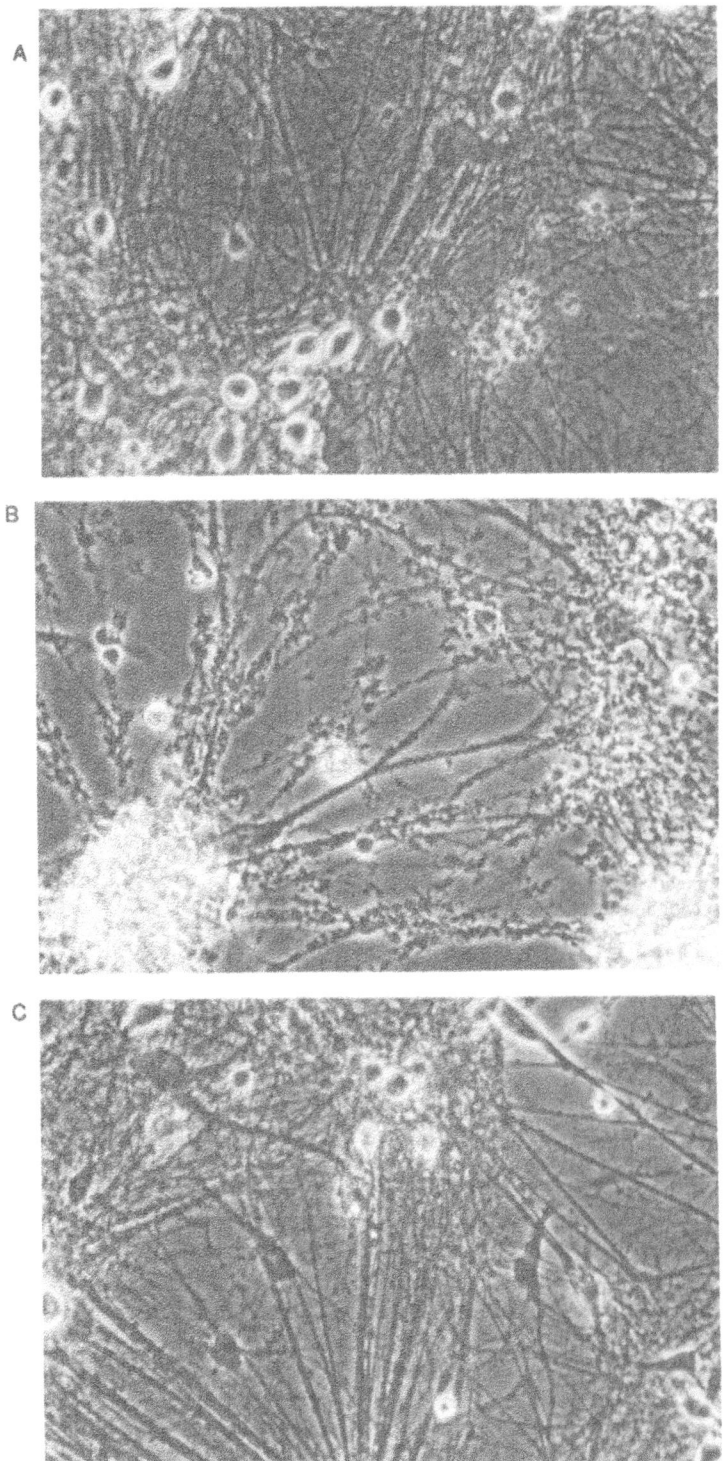

Figure 2. Phase contrast microscopy (X200) of neuronal cultures after 12 days in vitro. A: control neurons, B: neurons 20 hrs after 25 min of exposure to 100 μM NMDA, C: neurons 20 hrs after 25 min of exposure to 100 μM NMDA in the presence of 100 mM ethanol. Note extensive granulation and disintegration of neuronal cell bodies and neuritic processes following treatment with NMDA (A versus B) and its protection by ethanol (C).

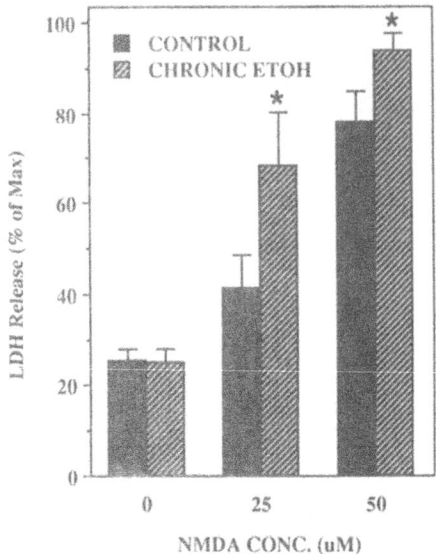

Figure 3. Effect of chronic ethanol exposure on NMDA stimulated neurotoxicity in cultured rat cerebral cortical neurons. Cultures were exposed to buffer or the indicated concentration of NMDA for 25 min in Mg^{2+}-free buffer and cell damage quantitated by measuring the accumulation of lactate dehydrogenase (LDH) in the culture media 20 hrs after NMDA washout (Chandler and Crews, 1992). The effects of chronic ethanol on cultures were determined after incubation with 100 mM ethanol for 96 hrs followed by removal of ethanol prior to NMDA exposure (Chandler et al., 1993). Values represent the means ± s.e.m. of four separate representative experiments each performed in triplicate. Asterisks indicate significant increase ($P < 0.05$) in NMDA stimulated LDH release in cells chronically exposed to ethanol compared to controls. Note that at the lower concentration of NMDA (25 μM), there was little increase in LDH release in controls, whereas a marked excitotoxic response was observed in ethanol treated cells. This suggests that neurons previously resistant to excitotoxic damage from low levels of stimulation fall within the excitotoxic range following chronic ethanol treatment.

be expected to increase excitotoxicity. Thus, a variety of data suggest that chronic ethanol treatment may disrupt calcium homeostatic mechanisms enhancing EAA excitotoxicity.

Studies in Human Brain

Studies of human alcoholics have clearly indicated that there is neuronal loss and a decrease in brain volume. Since chronic ethanol can sensitize to excitotoxicity, the brains of human alcoholics may degenerate secondary to excitotoxic loss of neurons. Repeated cycles of intoxication and withdrawal could repeatedly prune neurons from particular brain regions susceptible to excitotoxicity. Although acute ethanol has been shown to increase the density VSCC and NMDA receptors, these calcium channels return to control levels within 24 hours after ethanol withdrawal (Gulya et al., 1991). The effects of chronic alcohol on human brain EAA receptors have been investigated by several groups. Dodd et al. (1992) found no change in the binding of [3H]MK-801 to membranes of frontal superior cortex. Most of these patients did not show neuropathology. In studies of hippocampus and other brain regions there appear to be decreases in binding sites associated with EAA receptors and particularly the NMDA receptor. Michaelis et al. (1990) reported that [3H]glutamate receptor binding was increased in the

CONTROL

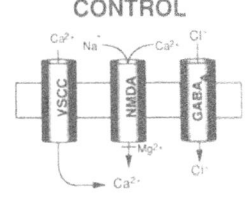

CHRONIC ALCOHOL

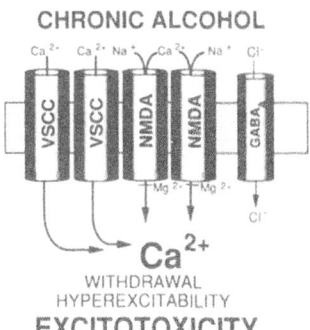

Ca²⁺
WITHDRAWAL
HYPEREXCITABILITY
EXCITOTOXICITY

Figure 4. Schematic diagram of the interaction of acute and chronic ethanol with particular ion channels. Under **control** conditions, neuronal excitability is enhanced by NMDA receptor mediated and voltage sensitive calcium channel (VSCC) mediated sodium and calcium influx. The GABA$_A$ channel reduces neuronal excitability by stimulating Cl⁻ influx which tends to hyperpolarize neurons. During **chronic ethanol**, adaptive mechanisms increase NMDA and VSCC density. During **withdrawal** from ethanol, the increased Ca^{2+} flux secondary to the presence of more NMDA and VSCC channels causes hyperexcitability often associated with ethanol withdrawal and sensitizes neurons to excitotoxicity by increasing [Ca^{2+}]$_i$. This sensitization to excitotoxicity may underlie the neurodegeneration associated with chronic alcohol abuse.

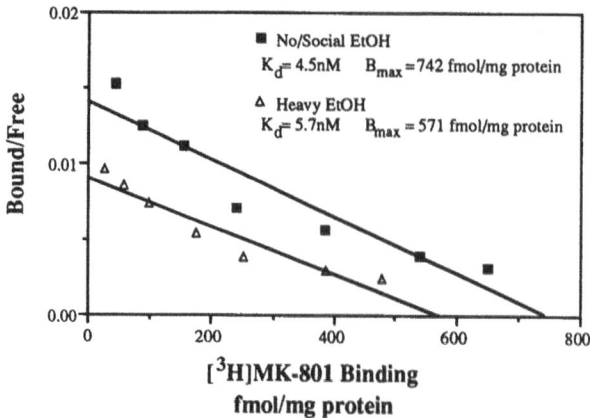

Figure 5. Scatchard plot of [3H]MK-801 binding to human hippocampal homogenates pooled from contols (n = 4) and alcoholics (n = 4). Each point represents the mean of ligand binding conducted in triplicate. Using the EBDA computer program (Biosoft), the controls (■) yielded a K_d = 4.52 nM with a B_{max} = 745 fmol/mg protein. The alcoholic patients (Δ) yielded a K_d = 5.69 nM with a B_{max} = 571 fmol/mg protein. The concentration of [3H]MK-801 ranged from 0.2 to 15 nM. Although the Scatchard plots hint at the possibility of multiple binding sites, the Hill coefficients were 1.038 and 1.026 in controls and alcoholics, respectively.

hippocampus of alcoholics, some of whom had measurable blood alcohol levels at the time of death. This binding represented several subtypes of EAA receptors. In addition, the binding of the NMDA antagonist [3H]CPP was found to be decreased in the hippocampus of alcoholics. Other investigators have reported a 9% decrease in NMDA specific [3H] glutamate binding in hippocampus and a 20% decrease in striatal binding in alcoholics (Cummins et al., 1990). These findings are consistent with our studies on [3H]MK-801 binding in hippocampal membranes prepared from the brains of alcoholics which had not consumed alcohol for several weeks prior to death (Figure 5). We have also found a 26% decrease in the binding of [3H]CGS-19755, an NMDA antagonist, to frontal cortex membranes compared to matched controls (Freund et al., 1993). Decreases in NMDA receptor density are consistent with the loss of neurons enriched with NMDA receptors. Thus, studies on human alcoholic brain are consistent with the loss of neurons containing EAA receptors and the hypothesis that the neurotoxic effects of ethanol are related to excitotoxicity.

Conclusions

Ethanol has a distinct interaction with a number of elements related to excitotoxicity. Excitotoxicity occurs through the activation of a heterogeneous group of EAA receptors which are likely to have a differential sensitivity to ethanol. The NMDA family of EAA receptors are the most excitotoxic and the most sensitive to acute inhibition by ethanol. This inhibition by ethanol is likely to contribute to the sedative and cognitive disrupting effects of ethanol. During chronic ethanol exposure, there are adaptive changes related to tolerance to ethanol that result in increases in NMDA and VSCC channel density that contribute to the sensitization of neurons to excitotoxicity. During ethanol withdrawal hyperexcitablility, it is likely that certain neurons reach a threshold of excitation that triggers neuronal death. Chronic ethanol clearly increases the sensitivity to NMDA mediated excitotoxicity. Further, alcoholics show decreases in the density of calcium channels consistent with the loss of neurons containing these channels in chronic alcoholics. The similarity of the neuropathology of hypoxic/ischemic brain damage and chronic ethanol abuse are consistent with excitotoxic neuronal death occurring over a prolonged period leading to diffuse neuronal loss and decreases in brain function and size. Thus, increased sensitivity to excitotoxicity is a likely mechanism of ethanol's neurotoxicity.

Acknowledgments. Supported by NIAAA grants AA06069, AA00127, AA09267 and Public Health Service Grant NS-19441. Hunter Newsom's work was supported by National Institutes of Health Program Fellowship Grant # P35-8L07489.

References

Allan, A. M. and Harris, R. A., 1986, GABA and alcohol actions: Neurochemical studies of long sleep and short sleep mice, *Life Sci.* 39:2005.

Barnard, E. A., Ambrosini, A., Bateson, A. N., Darlison, M. G., Harvey, R. J., Henley, J. M., Hutton, M. L., Ishimaru, H., Rodriguez-Ithurralde, D., Sudan, H., and Usherwood, P. N. R., 1992, Purification, reconstitution and DNA cloning of novel excitatory animo acid receptors, *in:* "Excitatory Amino Acids", R.P. Simon, ed.,Thieme Medical Publishers, Inc., Stuttgart, N.Y., pp:3-7.

Bohme, G. A., Bon, C., Stutzman, J. M., Doble, A. and Blanchard, J. C., 1991, Possible involvement of nitric oxide in long-term potentiation, *Eur. J. Pharmacol.* 199:379.

Boulter, J., Bettler, B., Dingledine, R., Duvoisin, R., Egebjerg, J., Gasic, G., Hartley, M., Hermans-Borgmeyer, I., Hollmann, M., Hughes, T. E., Hume, R. I., Moll, C., Rogers, S., and Heinemann S., 1992, Molecular biology of the glutamate receptors, *in:* "Excitatory Amino Acids", R. P. Simon, ed., Thieme Medical Publishers, Inc., Stuttgart, N.Y., pp:9-13.

Bredt, D. S., and Snyder, S. H., 1990, Isolation of nitric oxide synthetase, a calmodulin-requiring enzyme, *Proc. Natl. Acad. Sci. USA.* 87:682.

Brennan, C. H., Crabbe, J., and Littleton, J. M., 1990, Genetic regulation of dihydropyridine-sensitive calcium channels in brain may determine susceptibility to physical dependence on alcohol, *Neuropharmacol.* 29:429.

Brennan, C. H., Lewis, A., and Littleton, J. M., 1989, Membrane receptors, involved in up-regulation of calcium channels in bovine adrenal chromaffin cells, chronically exposed to ethanol, *Neuropharmacol.* 28:1303.

Cala, L. A., Jones, B., Burns, P., Davis, R. E., Stenhouse, N., and Mastaglia, F. L., 1983, Results of computerized tomography, psychometric testing and dietary studies in social drinkers, with emphasis on reversibility after abstinence, *Med. J. Aust.* 2:264.

Carlen, P. L., Wortzman, G., Holgate, R. C., Winkinson, D. A., and Rankin, J. G., 1978, Reversible cerebral atrophy in recently abstinent chronic alcoholics measured by computer tomographic scans, *Science.* 200:1076.

Chandler, L. J., Guzman, N., Crews, F. T,. and Sumners, C., 1992(a), Induction of nitric oxide synthase in rat brain microglia, Soc. Neurosci. *(Abst).* 18: 633.

Chandler, L. J., Sumners, C., and Crews, F. T., 1992(b), Ethanol inhibits NMDA receptor-mediated excitotoxicity in rat primary neuronal cultures, *Alcohol. Clin. Exp. Res.* 16:(in press).

Chandler, L.J., Newsom, H., Sumners, C., and Crews, F.T., 1993, Chronic ethanol exposure potentiates NMDA excitotoxicity in cerebral cortical neurons, submitted to *J. Neurochem.*

Choi, D. W., 1990, The Role of glutamate neurotoxicity in hypoxic-ischemic neuronal death. *Annu. Rev. Neurosci.* 13:171.

Choi, D. W., Koh, J., and Peters, S., 1988, Pharmacology of glutamate neurotoxicity in cortical cell culture: Attenuation by NMDA antagonists, *J. Neurosci.* 8:185.

Clark, G. D., 1989, Role of excitatory amino acids in brain injury caused by hypoxia-ischemia, status epilepticus, and hypoglycemia, *Clin. Perinatol.* 16:459.

Clark, M., and Dar, M. S., 1989, Effect of acute ethanol on release of endogenous adenosine from rat cerebellar synaptosomes, *J. Neurochem.* 52:1859.

Cotman, C. W., Bridges, R. J., Taube, J. S., Clark, A. S., Geddes, J. W., and Monaghan, D. T., 1989, The role of the NMDA receptor in central nervous system plasticity and pathology, *J.NIH Res.* 1:65.

Cummins, J. T., Sack, M., von-Hungen, K., 1990, The effect of chronic ethanol on glutamate binding in human and rat brain, *Life Sci.* 47(10):877.

Davda, R. K., Chandler, L. J., Crews, F. T., and Guzman, N. J., 1992, Ethanol enhances the endothelial nitric oxide synthase response to agonists, *Hypertension* , (In press)

Dawson, V. L., Dawson, T. M., London, E. D., Bredt, D. S., and Snyder, S. H., 1991, Nitric oxide mediates glutamate neurotoxicity in primary cortical cultures, *Proc. Natl. Acad. Sci. USA.* 88:6368.

Dildy, J. E., and Leslie, S. W., 1989, Ethanol inhibits NMDA-induced increases in free intracellular Ca^{2+} in dissociated brain cells, *Brain Res.* 499: 383.

Dodd, P.R. , Thomas, G. J., Harper, C. G., and Kril, J. J., 1992, Amino acid neurotransmitter receptor changes in cerebral cortex in alcoholism: effect of cirrhosis of the liver, J. Neurochem. 59:1506.

Favaron, M., Manev, H., Siman, R., Bertolino, M., Szekely, A. M., and DeErausquin, G., 1990, Down-regulation of protein kinase C protects cerebellar granule neurons in primary culture from glutamate-induced neuronal death, *Proc. Natl. Acad. Sci. U.S.A..* 87:1983.

Freund, G., 1985, Neuropathology of alcohol abuse, *in:* "Alcohol and the Brain Chronic Effects", R.E. Tarter, P.D. and D.H.V. Thiel, M.D. ed., Plenum Publishing Company, New York, N.Y.

Garthwaite, J., Garthwaite, G., Palmer, R. M. J., and Moncada, S., 1989, NMDA receptor activation induces nitric oxide synthesis from arginine in rat brain slices, *Eur. J. Pharmacol.* 172:413.

Gilbertson, T. A., Scobey, R., and Wilson, M.,1990, Permeation of calcium ions through non-NMDA glutamate channels in retinal bipolar cells, *Science.* 251:1613.

Gonzales, R. A., and Hoffman, P. L., 1991, Receptor-gated ion channels may be selective CNS targets for ethanol, *TIPS.* 12:1.

Gorelick, P. M., 1990, Stroke from alcohol and drug abuse, *Postgrad. Med.* 88:171.

Grant, K. A., Valverius, P., Hudspith, M., and Tabakoff, B., 1990, Ethanol withdrawal seizures and the NMDA receptor complex, *Eur. J. Pharmacol.* 176:289.

Greenberg, D. A., Sampson, H. A., and Chan, J., 1992, Ethanol protection from excitotoxic neural injury in vitro. *in:* "Excitatory Amino Acids", Simon R.P., ed., Thieme Medical Publishers, Inc., Stuttgart, N.Y.

Gulya, K., Grant, K.A., Valverius, P., Hoffman, P. L., and Tabakoff, B., 1991, Brain regional specificity and time-course of changes in the NMDA receptor-ionophore complex during ethanol withdrawal, *Brain Res.* 547:129.

Harper, C. B., and Kril, J. J., 1990, Neuropathology of alcoholism, *Alcohol.* 25:207.

Harper, C. G., Kril, J. J., and Daly, J., 1987, Are we drinking our neurons away? *Br. Med. J.* 294:534.

Harper, C. G., Kril, J. J., and Holloway, R. L., 1985, Brain shrinkage in chronic alcoholics, *Br. Med. J.* 290:501.

Harris, R. A., and Allan, A. M., 1989, Alcohol intoxication: ion channels and genetics, *FASEB J.* 3:1689.

Hillbom, 1978, Does ethanol intoxication promote brain infarction in young adults? *Lancet.* 2:1181.

Hoffman, P. L., Moses, F., and Tabakoff, B., 1989(a), Selective inhibition by ethanol of glutamate-stimulated cyclic GMP production in primary cultures of cerebellar granule cells, *Neuropharmacology.* 28: 1239.

Hoffman, P. L., Rabe, C. S., Moses, F., and Tabakoff, B., 1989(b), N-Methyl-D-Aspartate receptors and ethanol: inhibition of calcium flux and cyclic GMP production, *J. Neurochem.* 52:1937.

Hoffman, P.L., and Tabakoff, B., 1990, Ethanol and guanine nucleotide binding proteins: A selective interaction, *FASEB J.* 4:2612.

Iorio, K. R., Reinlib, L., Tabakoff, B., and Hoffman, P. L., 1992, Chronic exposure of cerebellar granule cells to ethanol results in increased N-methyl-D-aspartate receptor function, *Mol. Pharmacol.* 41:1142.

Jernigan, T. L., Butter, N., and Cermak, L. S., 1992, Studies of brain structure in chronic alcoholism using magnetic resonance imaging, *in:* "Imaging in Alcohol Research"--NIAAA Research Monograph-21, S. Zakhari and E. Witt, ed., U.S. Dept. of Health and Human Services, Rockville, M.D.

Koh, J., and Choi, D. W., 1988, Vulnerability of cultured cortical neurons to damage by excitotoxins: differential susceptibility of neurons containing NADPH-Diaphorase, *J. Neurosci.* 8:2153.

Leslie, S. W., Barr, E., Chandler, L. J., and Farrar, R. P., 1983, *J. Pharmacol. Exp. Ther.* 225:571.

Little, H. J., Dolin, S. J., and Halsey, M. J., 1986, Calcium channel antagonists decrease the ethanol withdrawal syndrome, *Life Sci.* 39:2059.

Lovinger, D.M., 1991, Ethanol potentiation of 5-HT$_3$, receptor-mediated ion currents in NCB-20 neuroblastoma cells, *Neurosci. Lett.* 122:57.

Lovinger, D. M., and White, G., 1991, Ethanol potentiation of 5-HT receptor-mediated ion currents in neuroblastoma cells and isolated adult mammalian neurons, *Mol.Pharmacol.* (in press)

Lovinger, D. M., White, G., and Weight, F. F., 1989, Ethanol inhibits NMDA-activated ion current in hippocampal neurons, *Science.* 243: 1721.

Lovinger, D. M., White, G., and Weight, F. F., 1990, NMDA receptor-mediated synaptic excitation selectively inhibited by ethanol in hippocampal slice from adult rat, *J. Neurosci.* 10:1372.

Lynch, M. J., 1960, Brain lesions in chronic alcoholism, *AMA Arch. Path.* 69:342.

Manev, H., Favaron, M., Guidotti, A., and Costa, E., 1989, Delayed increase of Ca^{2+} influx elicited by glutamate: role in neuronal death, *Mol. Pharmacol.* 36:106.

Mattson, M. P., Guthrie, P. B., and Kater, S. B., 1989, A role for Na^+-dependent CA^{2+} extrusion in protection against neuronal excitotoxicity, *The FASEB Journal.* 3:2519.

Meldrum, B., and Garthwaite, J., 1990, Excitatory amino acid neurotoxicity and neurodegenerative disease, *Trends Pharmacol. Sci.* 11:379.

Messing, R. O., Carpenter, C. L., Diamond, I., and Greenberg, D. A., 1986, Ethanol regulates calcium channels in clonal neural cells, *Proc. Natl. Acad. Sci. USA.* 83:6213.

Michaelis,, E. K., Freed, W. J., Galton, N., Foye, J., Michaelis, M. L., Phillips, I., and Kleinman J. E., 1990, Glutamate receptor changes in brain synaptic membranes from human alcoholics, *Neurochem. Res.* 15:1055.

Michaelis, E. K., Kumar, K. N., Tilakaratne, N., Johnson, P. S., Allen, A. E., Aistrup, G., and Schowen, R., 1992, Purification and reconstitution of an NMDA receptor-ion channel complex, and cloning of a cDNA for the glutamate binding subunit, *in:* "Excitatory Amino Acids", R. P. Simon, ed., Thieme Medical Publishers, Inc., Stuttgart, N.Y., pp:23-27.

Michaelis, E. K., Mulvaney, M. J., and Freed, W. J., 1978, Effects of acute and chronic ethanol intake on synaptosomal glutamate binding activity, *Biochem. Pharmacol.* 27:1685.

Monaghan, D. T., Olverman, H. J., Nguyen, L., Watkins, J. C., and Cotman, C. W., 1988, Two classes of NMDA recognition sites: differential distribution and differential regulation by glycine, *Proc. Natl. Acad. Sci. U.S.A.* 85:9836.

Monyer, H., Sommer, B., Wisden, W., Verdoorn, T. A., Burnashev, N., Sprengel, R., Sakmann, B., and Seeburg, P. H., 1992(a), Glutamate-gated ion channels in the brain: genetic mechanisms for generating molecular and functional diversity. *in:* "Excitatory Amino Acids", R. P. Simon, ed.,Thieme Medical Publishers, Inc., Stuttgart, N.Y., pp:29-33.

Monyer, H., Sprengel, R., Schoepfer, R., Herb, A., Higuchi, M., Lomeli, H., Burnashev, N., Sakmann, B. and Seeburg, P. H., 1992(b), Heteromeric NMDA receptors: molecular and functional distinction of subtypes, *Science.* 256:1217.

Monyer, H., and Choi, D. W., 1990, Glucose deprivation neuronal injury in vitro is modified by withdrawal of extracellular glutamine, *J. Cereb. Blood Flow Metab.* 10:337.

Murphy, S. N., and Miller, R. J., 1989, Regulation of Ca^{2+} influx into striatal neurons by kainic acid, *J. Pharmacol. Exp. Ther.* 249:184.

Nagy, L.E., Diamond, I., Casso, D. J., et al., 1990, Ethanol increases extracellular adenosine by inhibiting adenosine uptake via the nuceloside transporter, *J. Biol.Chem.* 265:1946.

Nakanishi, S., 1992, Molecular characterization of the family of metabotropic glutamate receptors. *in:* "Excitatory Amino Acids", R. P. Simon, ed., Thieme Medical Publishers, Inc., Stuttgart, N.Y.. pp:21-22.

Nowicki, J. P., Duval, D., Poignet, H., and Scatton, B., 1991, Nitric oxide mediates neuronal death after focal cerebral ischemia in the mouse, *Eur. J. Pharmacol.* 204:339.

O'Dell, T. J., Hawkins, R. D., Kandel, E. R., and Arancio, O., 1992, Tests of the roles of two diffusible substances in long-term potentiation: evidence for nitric oxide as a possible early retrograde messenger, Proc. Natl. Acad. Sci. USA. 88:11285.

Oakes, G., and Pozos, R. S., 1982, Electrophysiologic effects of acute ethanol exposure. I. Alterations in the action potentials of dorsal root ganglia neurons in dissociated culture. *Dev. Brain Res.* 5:243.

Olney, J. W., 1990, Excitotoxin-mediated neuron death in youth and old age, *Prog. Brain Res.* 86:37.

Orrenius, S., McConkey, D. J., Bellomo, G., and Nicotera, P., 1989, Role of Ca^{2+} in toxic cell killing, *Trends Pharmacol. Sci.* 10:281.

Oscar-Berman, M, and Ellis, R. J., 1987, Cognitive deficits related to memory impairments in alcoholism, *in:* "Recent Developments in Alcoholism", M. Galanter, ed., Plenum, New York, N.Y.

Pfefferbaum, A., Lim, K. O., and Rosenbloom, M. A., 1992, Structural imaging of the brain in chronic alcoholism, *in:* "Imaging in Alcohol Research"--NIAAA Research Monograph-21, S. Zakhari and E. Witt, ed., U.S. Dept. of Health and Human Services, Rockville, M.D.

Pfefferbaum, A., Lim, K. O., Ha ,C. N. and Zipursky, R. B., 1990, Changes in brain white and gray matter volume in alcoholics: an MRI study. *Alcohol. Clin. Exp. Res.* 14:328.

Pfefferbaum, A., Rosenbloom, M. J., Crusan, K., and Jernigan, T. L., 1988, Brain CT changes in alcoholic. The effects of age and alcohol consumption, *Alcohol. Clin. Exp. Res.* 12:81.

Schuman, K. and Madison, D. V., 1991, A requirement for the intercellular messenger nitric oxide in long-term potentiation, *Science.* 254:1503.

Skattebol, A., and Rabin, R. A., 1987, Effects of ethanol on $^{45}Ca^{2+}$ uptake in synaptosomes and in PC12 cells, *Biochem. Pharmacol.* 36:2227.

Tecoma, E. S., and Choi, D. W., 1989, GABAergic neocortical neurons are resistant to NMDA receptor-mediated injury, *Neurology.* 39:676.

Triestman, S. N., Comacho-Nasi, P., and Wilson A., 1985, Alcohol effects on voltage-dependent currents in identified cells, *Alcoholism Clin. Exp. Res.* 9:201.

Walker, D. W., and Hunter B. E., 1987, Neuronal adaptation in the hippocampus induced by long-term ethanol exposure, *in:* "The Role of Neuroplasticity in the Response to Drugs.", D.P. Friedman and D.H. Clouet, ed., U.S. Government Printing Office, Washington, D.C.

Walker, D. W., Hunter, B. E., and Abraham W. C., 1981, Neuroanatomical and functional deficits subsequent to chronic ethanol administration in animals, *Alcoholism Clin. Exp. Res.* 5:267.

Weiss, J. H., Hartley, D. M., Koh, J., and Choi, D. W., 1990, The calcium channel blocker nifedipine attenuates slow excitatory amino acid neurotoxicity, *Science.* 247:1474.

Westenbroek, R. E., Ahlijanian, M. K., and Catterall, W. A., 1990, Clustering of L-type Ca^{2+} channels at the base of major dendrites in hippocampal pyramidal neurons, *Nature.* 347:281.

Woodward, J. J., and Gonzales, R. A., 1990, Ethanol inhibition of N-Methyl-D-Aspartate-stimulated endogenous dopamine release from rat striatal slices: reversal by glycine, *J. Neurochem.* 54:712.

Zielasek, J., Tausch, M., Toyka, K. V., and Hartung H. P., 1992, Production of nitrite by neonatal rat microglial cells/brain macrophages, *Cell. Immun.* 141:111.

MOLECULAR PROPERTIES OF AN NMDA RECEPTOR COMPLEX AND EFFECTS OF ETHANOL ON THE EXPRESSION OF THIS COMPLEX

Elias K. Michaelis, Mary L. Michaelis and Keshava N. Kumar

Department of Pharmacology and Toxicology
and the Center for Biomedical Research
University of Kansas, Lawrence, KS 66045

INTRODUCTION

One of the currently held beliefs about ethanol is that some of the acute and chronic effects of this alcohol on the central nervous system (CNS) of humans and experimental animals are the result of changes in the activity of glutamate neurotransmission in the brain and spinal cord. This belief is based on the recent demonstrations that ethanol inhibits the activity of a subset of neuronal receptors for the excitatory amino acid L-glutamic acid, the class of receptors known as the N-methyl-D-aspartate receptor (NMDAR). These acute effects of ethanol on excitatory neurotransmitter function have been demonstrated for neurons grown *in vitro* in primary neuronal cultures as well as for neurons in brain slices (Hoffman *et al.*, 1989; Lima-Landman and Albuquerque, 1989; Lovinger *et al.*, 1989, 1990; White *et al.*, 1990). The NMDAR is a subtype of the ion-channel forming receptors in the brain that are activated by the excitatory neurotransmitter L-glutamic acid (Watkins *et al.*, 1990).

Since L-glutamate forms the most widespread excitatory transmitter network in the mammalian CNS, inhibition by ethanol of neuronal NMDAR's probably leads to the suppression of some aspects of neuronal excitability. This suppressive effect of ethanol on neuronal excitability may be the initiating step in a cascade of events that alters many processes in addition to interneuronal communication along certain CNS pathways. Thus, chronic exposure of an organism to ethanol could affect synaptic plasticity, long-term potentiation of synaptic activity, memory formation, or neuronal development and survival. The excitation produced by L-glutamic acid through activation of NMDAR's is important in the development of the enhanced synaptic efficacy seen in long-term potentiation and memory formation, and it plays a crucial role in synaptogenesis and enhanced neuronal survival during development of the nervous system (Balazs *et al.*, 1988; Cotman and Monaghan, 1988; Mattson *et al.*, 1988; Brewer and Cotman, 1989).

EFFECTS OF CHRONIC ETHANOL ADMINISTRATION ON GLUTAMATE RECEPTORS IN THE CNS

Glutamate receptors fall into two major classes in terms of effect or coupling systems. Certain receptors are coupled to the enzyme phospholipase C and lead to the formation of 1,4,5-inositol trisphosphate, a metabolic response (Recasens et al., 1988). The other class of receptors form ligand-gated ion channels (Watkins et al., 1990). The receptors that form ion channels are further subclassified according to their sensitivity to specific agonists. There are receptors that are activated by the glutamate analogs kainate, quisqualate, and alpha-3-hydroxy-5-methyl-4-isoxazolepropionic acid (AMPA), and those that are activated by N-methyl-D-aspartate (NMDA). The NMDAR's contain a number of distinct ligand binding domains through which their function is regulated. These include ligand binding sites for the agonists L-glutamate and NMDA, for the competitive antagonists (±)3-carboxypiperazin-4-yl-propyl-phosphonic acid (CPP), 2-amino-5-phosphonopentanoic acid (2-AP5) and 2-amino-7-phosphonoheptanoic acid (2AP7), for the co-activator glycine, and for the ion channel blockers phencyclidine (PCP), 1-[1-(2-thienyl)cyclohexylpiperidine (TCP), and dizocilpine (MK-801) (Wong and Kemp, 1991).

Several years ago we suggested that one way in which the brain might adapt to the continuous exposure to ethanol during chronic alcohol abuse is to increase the number of receptors for L-glutamic acid (Freed and Michaelis, 1978; Michaelis et al., 1978, 1980). Recent evidence obtained in pharmacological studies fits with the idea that neuronal adaptations that involve the glutamate/NMDA receptor system are contributing to the development of physiological dependence on ethanol. For example, [^{3}H]MK-801 was used to label NMDAR's in brain neuronal membranes isolated from mice that had been chronically exposed to ethanol and from controls (Grant et al., 1990). The estimated number of [^{3}H] MK-801 binding sites was increased in the neuronal membranes from chronically treated mice compared with the respective controls. Furthermore, MK-801 was shown to be an effective inhibitor of the post-ethanol withdrawal seizures in mice exposed chronically to this alcohol (Grant et al., 1990; Morrisett et al., 1990). Thus, there is substantial evidence to support the idea that at least one of the sites of action of ethanol in the mammalian CNS is the glutamate/NMDA receptor.

In early studies from our laboratories, we demonstrated that a period of chronic ethanol administration to experimental animals resulted in a significant increase in the sensitivity of the animals to the convulsant effects of kainic acid (Freed and Michaelis, 1978), a potent glutamate analog that activates the kainate/AMPA receptors, and an increase in the binding sites for L-[^{3}H]glutamate (Michaelis et al., 1978, 1980). Chronic ethanol intake (14 days) in rats led to the appearance of a 50% increase in the density of glutamate binding sites in synaptic plasma membranes without a change in their affinity for the ligand L-[^{3}H] glutamate (Michaelis et al., 1987).

The binding studies in which L-[^{3}H]glutamic acid is used as the ligand to label glutamate receptor sites in synaptic membranes do not allow one to discern which subpopulation of glutamate receptors is increased as a result of the chronic exposure of CNS neurons to ethanol. In order to achieve such discrimination among the glutamate receptors affected by ethanol, we have explored the definition of the changes in subtypes of glutamate receptors that may result from chronic alcohol intake in synaptic membranes isolated from the brains of human alcoholics and non-alcoholics (Michaelis et al., 1990). The ligand binding studies performed with human brain tissue included measurements of both L-[3H] glutamate and [3H]CPP binding site densities and affinities in a brain region in which glutamate-mediated excitation is the dominant type of excitatory neurotransmission, i.e., the hippocampus.

The B_{max} for L-[3H]glutamate binding to synaptic membranes from the groups of alcoholics was at least twice that for glutamate binding to the membranes from the groups of non-alcoholics. When the more selective ligand for NMDAR's [3H]CPP was used, a threefold decrease in the density of NMDAR's was detected in the membranes from the hippocampus of chronic alcoholics. These observations would lead one to conclude that NMDAR density decreased while the density of other subtypes of glutamate receptors increased following chronic alcoholic intake. There was also an apparent shift to a higher affinity in the sites for [3H]CPP binding to the synaptic membranes from the hippocampal region in brains from alcoholics.

There are several explanations that might be advanced to account for the changes in glutamate and NMDAR's detected in synaptic membranes from the brains of human alcoholics. The first is that the studies performed in experimental animals which were fed ethanol-containing diets for a short period of time, for example 14 days, may not be an appropriate model for determining the types of changes brought about by ethanol in CNS neurons of human alcoholics. The second is that human brain neurons may adapt to the effects of exposure to ethanol over a prolonged period of time by increasing the affinity of NMDAR's for the competitive antagonists, and presumably also for the agonists, but decreasing the total density of such receptors. Such a change may enhance the probability of activation of NMDAR's by L-glutamic acid released from neurons following depolarization, but may decrease the overall contribution of these receptors to the excitatory response initiated by glutamate activation of post-synaptic receptors. Such an adaptive response would indicate that neurons in the brains of chronic alcoholics become increasingly more reliant on an enhanced density of other types of glutamate receptor, such as the kainate/AMPA and the metabotropic receptors.

A final possibility is that the adaptive response is even more complex than can be accounted for in either of the schemes presented above. It is possible that the NMDAR's are made of multiple subunits and that different subunits contain the various ligand-binding sites associated with the NMDA receptors. Therefore, the adaptive response of neurons may be a reflection of differential regulation of gene expression for specific subunits following chronic exposure of neurons to ethanol. It is obvious that ligand binding measurements represent only one view into the possible changes in structure and function of glutamate and NMDA receptors in the CNS following chronic alcohol intake. A more precise understanding of the molecular changes that may be produced in neurons as a result of chronic exposure to ethanol can only be achieved by probing the molecular structure of glutamate receptors in both normal and alcoholic brain tissue. In order to accomplish this, specific probes need to be developed that will allow the exploration of changes in receptor protein structure and composition that might be associated with an adaptation of neurons to chronic ethanol exposure.

MOLECULAR CHARACTERISTICS OF TWO FAMILIES OF NMDAR'S

There are two structures that have been presented in the literature for glutamate/NMDA receptor-ion channel complexes. One is a large protein which has all expected characteristics of an NMDA receptor-ion channel when it is expressed in frog oocytes but exhibits relatively low conductance of the ion channel activated by NMDA (Moriyoshi *et al.*, 1991). The inferred structure of this protein based on cDNA sequence is that of a 105 kDa protein that has a fairly high degree of homology to some of the previously cloned kainate/AMPA receptor proteins (Hollman *et al.*, 1989; Keinanen *et al.*, 1990). This protein is called NMDAR1. Three genes that predict proteins homologous to the 105 kDa NMDAR were recently cloned and found to have none of the characteristics of an NMDAR when expressed by themselves in frog oocytes but form functional NMDAR complexes when co-expressed with the 105 kDa protein (Katsuwada *et al.*, 1992; Meguro *et al.*, 1992: Monyer *et al.*, 1992). These complexes exhibit higher NMDA receptor-ion channel conductances than the complexes formed from the homomeric association of the 105 kDa NMDAR protein, and they resemble the receptor-associated ion channels found in neuronal membranes. In addition, these heteromeric complexes have more complex patterns of regulation of ion channel conductances by agonists, antagonists and ion channel blockers. There is very little doubt that these cloned cDNA's represent the genes for neuronal NMDAR's. The three proteins which form complexes with NMDAR1 have predicted molecular weights of 120,000 to 150,000 and are called NMDAR2A, B, and C. Presumably the NMDAR1 and R2 complexes form pentamers of molecular weight approximating 600,000, as has been observed for the kainate/AMPA proteins (Wenthold *et al.*, 1992).

The second structure of an NMDAR1 identified is that of a hetero-oligomeric complex composed of protein subunits with molecular weights ranging between 30,000 and 60,000 (Ikin *et al.*, 1990; Kumar *et al.*, 1991; Ly and Michaelis, 1991; Michaelis *et al.*, 1992). The structure of this putative NMDAR complex appears to include four proteins with molecular sizes determined by SDS electrophoresis to be equal to 67–70, 57–60, 42–46 and 31–36 and has an estimated weight of approximately 230,000 (Michaelis *et al.*, 1992). This complex has NMDA-sensitive L-$[^3H]$glutamate binding sites, strychnine-insensitive $[^3H]$glycine recognition sites, 2-AP5-sensitive $[^3H]$CPP binding sites, and MK-801-sensitive $[^3H]$TCP binding sites (Ikin *et al.*, 1990; Michaelis *et al.*, 1992). In addition, the NMDAR agonists glutamate and NMDA, the co-agonist glycine, and the modulators spermidine and spermine kinetically activated the binding of $[^3H]$TCP to the isolated protein complex (Ikin *et al.*, 1990; Kumar *et al.*, 1992; Michaelis *et al.*, 1992). These observations are indicative of the presence of functional NMDA receptor-ion channels in this protein complex isolated from brain synaptic membranes since the activation of the binding of channel blockers such as TCP to NMDAR's is considered to be an analog of the agonist and co-agonist activation of the ion channel of NMDAR's (Kloog *et al.*, 1988).

Reconstitution of a partially purified preparation of this complex into liposomes formed L-glutamate and NMDA-sensitive ion channels that were detected by rapid-kinetic measurements of cation fluxes into the liposomes (Ly and Michaelis, 1991). More highly purified preparations of the same protein complex have also been reconstituted in planar lipid bilayer membranes and, following such reconstitution, L-glutamic acid and NMDA-activated ion channels with conductance characteristics similar to those of neuronal NMDAR's (Minami *et al.*, 1991; Michaelis *et al.*, 1992). The estimated K_{act} for glutamate-induced channel conductance is 0.8 μM, a value that is very similar to the K_D for glutamate binding to this protein complex. The glutamate and NMDA-activated conductances exhibit a linear voltage-current relationship and the characteristics of a channel that opens to three different conductance states. The estimated slope condutances for the three types of channel open states were 23, 47, and 67 pS, with the 47 pS state the predominant one.

NEURONAL DISTRIBUTION AND FUNCTION OF THE HETEROMERIC, SMALL MOLECULAR SIZE NMDAR

The properties of the heteromeric, small molecular size protein complex described above suggest that this complex functions as an NMDA receptor-ion channel. If this suggestion is correct, then one would have to accept the possibility that there are two distinct families of the ion-channel forming NMDAR's in the CNS. Since recent reports from other laboratories indicate of yet another class of NMDAR's, those that are coupled in a negative fashion to adenylate cyclase (Itano et al., 1992), it is likely that there are several families of NMDAR's in the mammalian CNS. This apparent multiplicity of receptor families and classes requires careful scrutiny and ample documentation before it is accepted as an operational scheme. This is particularly necessary for the ion channel-forming NMDAR's.

As indicated above, there is little doubt that the cloned 105–150 kDa proteins represent functional receptor-ion channels for glutamate and NMDA. On the other hand, the protein complex of smaller protein subunits identified by us and other investigators may represent either a new family of NMDAR's or proteins derived through proteolytic breakdown of the 105–150 kDa subunits of the NMDAR. The latter possibility has recently been excluded by cloning and sequencing of the cDNA's for the 67–70 subunit (Kumar et al., 1991) and for the 57–60 and 42–46 kDa subunits (unpublished observations). Neither the cloned cDNA for the 67–70 kDa subunit, nor the ones for the other two subunits exhibit any significant homology to the 105 kDa NMDAR1.

The 67–70 kDa subunit of the heteromeric complex was identified by immunochemical procedures as the protein that we had previously isolated from rat and bovine brain synaptic membranes and called the glutamate-binding protein (Chen et al., 1988; Kumar et al., 1991; Ly and Michaelis, 1991; Wang et al., 1992). Transformation of E. coli with the phagemid that contains the cDNA insert for this protein leads to the expression of a fusion protein of approximately 60 kDa that is recognized by the antibodies raised against the brain glutamate-binding protein. The protein that is expressed in bacteria following transformation with the phagemid described above also binds L-[^{3}H]glutamic acid with the same affinity and selectivity as the brain protein does (Kumar et al., 1991).

The 57–60 kDa protein has also been identified by immunochemical procedures (unpublished observations) as the synaptic membrane protein we had previously isolated and called the CPP-binding protein (Cunningham and Michaelis, 1990). Since CPP, 2-AP5, and 2-AP7 selectively interact with this protein, we assume that this protein is a component of an NMDA receptor complex and that it represents the antagonist binding subunit. The cDNA for this protein has been inserted into a phagemid and bacterial cells have been transformed with this phagemid. The transformants express a protein that is recognized by the antibodies raised against the brain protein, and the protein expressed in bacteria has ligand recognition sites that are selective for the phosphonoaminocarboxylic acid antagonists of NMDA receptors (unpublished observations).

The results of the cloning studies described above are indicative of the presence of distinct protein entities that have the characteristics of specific ligand-recognizing subunits of a heteromeric NMDAR complex. Therefore, there is a high probability that there are two families of ion channel-forming NMDAR's, those composed of subunits of 105–150 kDa and those that consist of subunits of 30–60 kDa molecular size. The presence of such diversity in the structure of the NMDAR's of the brain may be viewed as a molecular index of the importance of this receptor for neuronal function and viability in the CNS. It is quite possible that the different families of receptors are under different genetic controls. Thus, neuronal adaptations to environmental stimuli, including those produced by acute and chronic exposure to ethanol, might involve differential regulation of gene expression of the receptor subunits in these two families of NMDAR's.

REGULATION OF NEURONAL EXPRESSION OF THE
SMALL MOLECULAR SIZE HETEROMERIC NMDAR

Specific antibodies raised against one of the protein subunits of the smaller, heteromeric complex, the glutamate-binding protein, have been used to examine the distribution and expression of this protein in CNS neurons. Light microscopic examination revealed that it is localized in dendritic processes of neurons in the hippo-campus and that, at the electron microscopic level, these immunoreactive proteins are most heavily represented in the post-synaptic membranes of neurons (Eaton *et al.*, 1990). The protein is expressed abundantly in cell bodies and dendrites of hippo-campal neurons in primary neuronal cultures but not in their axons (Mattson *et al.*, 1991). Pre-exposure of these neurons to the anti-glutamate binding protein antibodies protects hippocampal neurons from NMDA-induced, but not from kainate or quisqualate-induced, cell degeneration (Mattson *et al.*, 1991). Finally, developmental expression of this protein in hippocampal neurons and in granule cells of the cere-bellum in primary cultures correlates most closely with the appearance of functional NMDA receptors and not with those for kainate or AMPA (Mattson *et al.*, 1991; Balazs *et al.*, 1992).

The studies described above indicate that for at least one of the components of the heteromeric complex of the small molecular size NMDAR, the expression and localization of the protein detected by immunocytochemical studies correlates with the presence and function of NMDAR's. In collaboration with M. Mattson and B. Cheng, we have recently determined that through the introduction of antisense oligonucleotides into the neurons in culture a reduction in the expression of this protein as determined by immunochemical approaches and a significant reduction in the mRNA for this protein measured by Northern blot hybridization analyses is achieved. Correlated with the decreases in the expression of this protein following treatment of neurons with antisense oligonucleotides is a marked attenuation in the response of the neurons to NMDA in terms of increases in intracellular calcium accumulation and resistance to the toxic effects of high concentrations of NMDA (Kumar *et al.*, 1992).

ENHANCED EXPRESSION OF THE GLUTAMATE-BINDING
PROTEIN FOLLOWING CHRONIC EXPOSURE OF
NEURONS TO ETHANOL

We have previously used polyclonal antibodies raised against the glutamate-binding protein to quantify changes in the amounts of this protein in synaptic mem-branes isolated from the brains of rats treated chronically with ethanol (Michaelis *et al.*, 1987). The approximately 50% increase in the B_{max} of glutamate binding sites associ-ated with the synaptic membranes isolated from ethanol-treated animals as compared with the membranes from dietary controls was matched by a nearly identical increase in the immunochemically detected amount of the glutamate binding protein in the membranes purified from alcohol-treated animals (Michaelis *et al.*, 1987). In these studies we did not have any measures of changes in the functional characteristics of either the NMDA or the kainate/AMPA receptors. However, in more recent studies performed in collaboration with P. Hoffman, K. Iorio and B. Tabakoff, we have attempted to explore the alterations in NMDAR function and the expression of different glutamate receptor molecular species in cerebellar granule cells exposed to ethanol *in vitro*. In neurons exposed to 100 mM ethanol for 4 days *in vitro*, there was an approximately 100% increase in NMDAR response measured as NMDA-induced calcium influx into the neurons, and this was correlated with an approximately 100% increase in the expression of the glutamate binding protein measured by Western blot analysis (Iorio *et al.*, 1992). The increase in the expression of the glutamate-binding protein was also correlated with an increase in the mRNA for this protein detected by Northern blot analysis. Following 4 days exposure to ethanol, the increase in mRNA in treated cells was approximately 45% above that in the control cells. On the other hand, in the same neurons there was only a 12% increase in the mRNA for the 105 kDa NMDAR1 protein and a 3% increase in the kainate/AMPA receptor protein mRNA (GluR1).

CONCLUSIONS

The studies described above represent the first attempts to probe the responses of neurons to chronic ethanol treatment in terms of expression of various kainate/AMPA and NMDA receptor subunits. From the studies performed thus far in experimental animals (rats) and in cerebellar granule cells in primary neuronal cultures, it appears that the glutamate-binding protein subunit marker of the heteromeric small molecular size NMDAR may reveal the best correlation between changes in neuronal NMDAR function and receptor protein subunit expression. The work carried out thus far represents only an initial attempt at exploring the molecular changes that may occur in NMDAR subunits during chronic ethanol treatment, and further detailed studies will be necessary to characterize fully the multiplicity of neuronal molecular adaptations likely to occur in this glutamatergic system in the presence of ethanol.

Future studies should focus on an analysis of the mechanisms for changes in the expression of the small molecular size NMDAR complex following chronic exposure of neurons to ethanol. The steps of molecular transduction from surface receptor activation by the agonists to gene expression are not well understood but are thought to involve the activation of response elements in the promoter region of immediate early genes and possibly of other genes (Sonnenberg *et al.*, 1989; Szekely *et al.*, 1989). Since activation of NMDA receptors causes substantial influx of calcium through receptor-ion channels as well as through voltage-sensitive channels, the molecular responses in neurons may be linked to either phosphorylation of cellular proteins following activation of calcium/ calmodulin kinase and kinase C or dephosphorylation of existing cell proteins following activation of calcium/calmodulin-sensitive phosphatases such as calcineurin (Halpain and Greengard, 1990; Halpain *et al.*, 1990). One would have to assume that under normal conditions of activation of NMDA receptors, expression of the proteins that form these receptor-ion channels is under continuous regulation by some factor or factors. However, upon inhibition of the receptors by ethanol, such regulation is removed and enhanced gene transcription occurs (Fig. 1). This hypothetical scheme of molecular signal transduction that involves ethanol's effects at the cellular level in the CNS should be the focus of future studies.

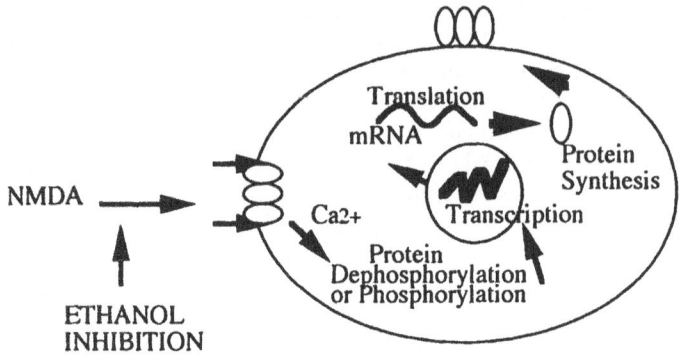

Fig. 1. A scheme for the possible regulation of gene transcription and protein synthesis in neurons induced by activation of NMDA receptors. It is assumed that continuous activation of NMDA receptors would cause down-regulation of the expression of these receptor proteins and that ethanol inhibition of NMDA receptors would have the opposite effect, i.e., an increase in the expression of NMDA receptor proteins.

Acknowledgements. This work was supported by grants AA04732 from NIAAA and DAAL03-88-K-0017 and DAAL03-91-G-0167 from the ARO to E.K.M. We acknowledge the support of the Center for Biomedical Research and we thank Kim Bland and Nancy Harmony for their assistance in the preparation of this paper.

REFERENCES

Balazs, R., Jorgensen, O.S. and Hack, N., 1988, N-Methyl-D-aspartate promotes the survival of cerbellar granule cells in culture, *Neurosci.* 27:437-451.

Balazs, R., Resink, A., Hack, N., Van der Valk, J.B.F., Kumar, K.N. and Michaelis, E.K., 1992, NMDA treatment and K$^+$-induced depolarization selectively promote the expression of an NMDA-preferring class of the ionotropic glutamate receptors in cerebellar granule neurones, *Neurosci. Lett.* 137:109-113.

Bannon, M.J., Poosch, M.S., Xia, Y., Goebel, D.J., Cassin, B. and Kapatos, G., 1992, Dopamine transporter mRNA content in human substantia nigra decreases precipitously with age, *Proc. Natl. Acad. Sci. USA* 89:7095-7099.

Brewer, G.J. and Cotman, C.W., 1989, NMDA receptor regulation of neuronal morphology in cultured hippocampal neurons, *Neurosci. Lett.* 99:268-273.

Chen, J.-W., Cunningham, M.D., Galton, N. and Michaelis, E.K., 1988, Immune labelling and purification of a 71-kDa glutamate-binding protein from brain synaptic membranes, *J. Biol. Chem.* 263:417-426.

Cotman, C.W. and Monaghan, D.T., 1988, Excitatory amino acid neurotransmission: NMDA receptors and Hebb-type synaptic plasticity, *Ann. Rev. Neurosci.* 11:61-80.

Cunningham, M.D. and Michaelis, E.K., 1990, Solubilization and partial purification of a 3-((±)-2-carboxypiperazine-4-yl)-[1,2-^{3}H]propyl-1-phosphonic acid recognition proteins from rat brain synaptic membranes, *J. Biol. Chem.* 265:7768-7778.

Eaton, M.J., Chen, J.-W., Kumar, K.N., Cong, Y. and Michaelis, E.K., 1990, Immunochemical characterization of brain synaptic membrane glutamate-binding proteins, *J. Biol. Chem.* 265:16195-16204.

Freed, W.J. and Michaelis, E.K., 1978, Glutamic acid and ethanol dependence, *Pharmacol. Biochem. Behav.* 8:509-514.

Grant, K.A., Valverius, P., Hudspith, M. and Tabakoff, B., 1990, Ethanol withdrawal seizures and the NMDA receptor complex, *Eur. J. Pharmacol.* 176:289-296.

Halpain, S. and Greengard, P., 1990, Activation of NMDA receptors induces rapid dephosphorylation of the cytoskeletal protein MAP2, *Neuron* 5:237-246.

Halpain, S., Girault, J-A. and Greengard, P., 1990, Activation of NMDA receptors induces dephosphorylation of DARPP-32 in rat striatal slices, *Nature* 343:369-372.

Hoffman, P.L., Rabe, C.S., Moses, F. and Tabakoff, B., 1989, N-Methyl-D-Aspartate receptors and ethanol: Inhibition of calcium flux and cyclic GMP production, *J. Neurochem.* 52:1937-1940.

Hollman, M., O'Shea-Greenfield, A., Rogers, S. W. and Heinemann, S., 1989, Cloning by functional expression of a member of the glutamate receptor family, *Nature* 342:643-648.

Ikin, A.F., Kloog, Y. and Sokolovsky, M., 1990, N-Methyl-D-aspartate/phencyclidine receptor complex of rat forebrain: Purification and biochemical characterization, *Biochemistry* 29:2290-2295.

Iorio, K., Hoffman, P.L., Tabakoff, B., Kumar, K.N. and Michaelis, E.K., 1992, Increased NMDA receptor function in cerebellar granule cells exposed chronically to ethanol, *Neurosci. Abs.* 18:978.

Itano, Y., Murayama, T., Kitamura, Y. and Nomura, Y., 1992, Glutamate inhibits adenylate cyclase activity in dispersed rat hippocampal cells directly via an N-methyl-D-aspartate-like metabotropic receptor, *J. Neurochem.* 59:822-828.

Keinanen, K., Wisden, W., Sommer, B., Werner, P., Herb, A., Verdoorn, T.A., Sakman, B. and Seeburg, P.H., 1990, A family of AMPA-selective glutamate receptors, *Science* 249:556-560.

Kloog, Y., Nadler, V. and Sokolovsky, M., 1988, Mode of binding of [^{3}H]dibenzocycloalkenimine (MK-801) to the N-methyl-D-aspartate (NMDA) receptor and its therapeutic implication, *FEBS Lett.* 230:167-170.

Kumar, K.N., Tilakaratne, N., Johnson, P.S., Allen, A.E. and Michaelis, E.K., 1991, Cloning of cDNA for the glutamate-binding subunit of an NMDA receptor complex, *Nature* 354:70-73.

Kumar, K.N., Mattson, M.P., Wang. H., Cheng, B. and Michaelis, E.K., 1992, Antisense oligonucleotides to a 71 kDa glutamate-binding protein decrease expression of functional NMDA receptors, *Neurosci. Abs.* 18:258.

Kutsuwada, T., Kashiwabuchi, N., Mori, H., Sakimura, K., Kushiya, E., Araki, K., Meguro, H., Masaki, H., Kumanishi, T., Arakawa, M. and Mishina, M., 1992, Molecular diversity of the NMDA receptor channel, *Nature* 358:36-41.

Lima-Landman, M.T.R. and Albuquerque, E.X., 1989, Ethanol potentiates and blocks NMDA-activated single-channel currents in rat hippocampal pyramidal cells, *FEBS Lett.* 247:61-67.

Lovinger, D.M., White, G. and Weight. F.F., 1989, Ethanol inhibits NMDA-activated ion current in hippocampal neurons, *Science* 243:1721-1724.

Lovinger, D.M., White, G. and Weight, F.F., 1990, NMDA receptor-mediated synaptic excitation selectively inhibited by ethanol in hippocampal slice from adult rat, *J. Neurosci.* 10:1372-1379.

Ly, A.M. and Michaelis, E.K., 1991, Solubilization, partial purification, and reconstitution of glutamate- and N-methyl-D-asparate-activated cation channels from brain synaptic membranes, *Biochemistry* 30:4307-4316.

Mattson, M.P., Ping, D. and Kater, S.B., 1988, Outgrowth-regulating actions of glutamate in isolated hippocampal pyramidal neurons, *J. Neurosci.* 8:2087-2100.

Mattson, M.P., Wang, H. and Michaelis, E.K., 1991, Developmental expression, compartmentalization, and possible role in excitotoxicity of a putative NMDA receptor protein in cultured hippocampal neurons, *Brain Res.* 565:94-108.

Meguro, H., Mori, H., Araki, K., Kushiya, E., Kutsuwada, T., Yamazaki, M., Kumanishi, T., Arakawa, M., Sakimura, K. and Mishina, M., 1992, Functional charaterization of a heteromeric NMDA receptor channel expressed from cloned cDNAs, *Nature* 357:70-74.

Michaelis, E.K., Mulvaney, J.J. and Freed, W.J., 1978, Effects of acute and chronic ethanol intake on synaptosomal glutamate binding activity, *Biochem. Pharmacol.* 27:1685-1691.

Michaelis, E.K., Michaelis, M.L. and Freed, W.J., 1980, Chronic ethanol intake and synaptosomal glutamate binding activity, *in* "Biological Effects of Alcohol-Advances in Experimental Medicine and Biology." H. Begleiter, ed., Vol. 126, Plenum, New York, pp. 43-56.

Michaelis, E.K., Roy, S., Galton, N., Cunningham, M., LeCluyse, E. and Michaelis, M.L., 1987, Correlation of glutamate binding activity with glutamate-binding protein immunoreactivity in the brain of control and alcohol-treated rats, *Neurochem. Int.* 11:209-218.

Michaelis, E.K., Freed, W.J., Galton, N., Foye, J., Michaelis, M.L., Phillips, I. and Kleinman, J.E., 1990, Glutamate receptor changes in brain synaptic membranes from human alcoholics, *Neurochem. Res.* 15:1055-1063.

Michaelis, E.K., Michaelis, M.L., Kumar, K.N., Tilakaratne, N., Joseph, D.B., Johnson, P.S., Babcock, K.T., Aistrup, G.L. and Schowen, R.L., 1992, Purification, reconstitution, and cloning of an NMDA-ion channel complex from rat brain synaptic membranes: Implications for neurobiological changes in alcoholism, *Ann. N.Y. Acad. Sci.* 654:7-18.

Minami, H., Sugawara, M., Odashima, K. and Umezawa, Y., 1991, Ion channel sensors for glutamic acid, *Anal. Chem.* 63:2787-2795.

Monyer, H., Sprengel, R., Schoepfer, R., Herb, A., Higuchi, M., Lomeli, H., Burnashev, N., Sakmann, B. and Seeburg, P.H., 1992, Heteromeric NMDA receptors: Molecular and functional distinction of subtypes, *Science* 256:1217-1221.

Morrisett, R.A., Rezvani, A.H., Overstreet, D., Janowsky, D.S., Wilson, W.A. and Swartzwelder, H.S., 1990, MK-801 potently inhibits alcohol withdrawal seizures in rats, *European J. Pharmacol.* 176:103-105.

Moriyoshi, K., Masu, M., Ishii, T., Shigemoto, R., Mizuno, N. and Nakanishi, S., 1991, Molecular cloning and characterization of the rat NMDA receptor, *Nature* 354:31-37.

Recasens, M., Guiramand, J., Nourigat, A., Sassetti, I., and Devilliers, G., 1988, A new quisqualate receptor subtype (sAA$_2$) responsible for the glutamate induced inositol phosphate formation in rat brain synaptoneurosomes, *Neurochem. Int.* 13:463-467.

Sonnenberg, J.L., Mitchelmore, C., Macgregor-Leon, P.F., Hempstead, J., Morgan, J.I. and Curran, T., 1989, Glutamate receptor agonists increase the expression of Fos, Fra and Ap1-DNA binding activity in the mammalian brain, *J. Neurosci. Res.* 24:72-80.

Szekely, A.M., Barbaccia, M.L., Alho, H. and Costa, E., 1989, In primary cultures of cerebellar granule cells the activation of NMDA-sensitive glutamate receptors induces *c-fos* mRNA expression, *Mol. Pharmacol.* 35:401-408.

Watkins, J.C., Krogsgaard-Larsen, P. and Honore, T., 1990, Structure activity relationships in the development of excitatory amino acid receptor agonists and competitive antagonists, *Trends Pharmacol. Sci.* 11:25-33.

Wenthold, R.J., Yokotani, N., Doi, K. and Wada, K., 1992, Immunochemical characterization of the non-NMDA glutamate receptor using subunit-specific antibodies, *J. Biol. Chem.* 267:501-507.

White, G., Lovinger, D.M. and Weight, F.F., 1990, Ethanol inhibits NMDA-activated current but does not alter GABA-activated current in an isolated adult mammalian neuron, *Brain Res.* 507:332-336.

Wang, H., Kumar, K.N. and Michaelis, E.K., 1992, Isolation of glutamate-binding proteins from rat and bovine brain synaptic membranes and immunochemical and immunocytochemical characterization, *Neuroscience* 46:793-806.

Wong, E.H.F. and Kemp, J.A., 1991, Sites for antagonism on the N-methyl-D-aspartate receptor channel complex, *Ann. Rev. Pharmacol.* 31:401-425.

Xia, Y., Goebel, D.J., Kapatos, G. and Bannon, M.J., 1992, Quantitation of rat dopamine transporter mRNA: Effects of cocaine treatment and withdrawal, *J. Neurochem.* 59:1179-1182.

IN VIVO STUDIES OF ETHANOL INTERACTIONS WITH N-METHYL-D-ASPARTATE SYSTEMS IN RAT BRAIN

Lynn C. Roper and Rueben A. Gonzales

Department of Pharmacology
University of Texas
Austin, Texas 78712

INTRODUCTION

Several electrophysiological and biochemical studies have indicated that N-methyl-D-aspartate (NMDA) receptor functions may be altered by ethanol (Gonzales, 1990). Many have suggested that ethanol potently and selectively inhibits the actions of agonists at the NMDA receptor (Hoffman et al.,1989; Lovinger et al., 1989; Gothert and Fink, 1989; Hoffman et al., 1990; Woodward and Gonzales, 1990). Others have implicated the involvement of the NMDA receptor in acute behavioral effects of ethanol such as memory dysfunction (Hoffman et al., 1990) and in certain aspects of ethanol withdrawal symptoms, such as seizures (Grant et al., 1990).

Both in vitro (Clow and Jhamandas, 1988; Woodward and Gonzales, 1990; Wang, 1991) and in vivo (Carter et al., 1988; Moghaddam et al., 1990; Imperato et al., 1990; Keefe et al., 1990) pharmacological studies have shown that glutamate and / or NMDA stimulates the release of striatal dopamine (DA). There are, however, discrepencies as to whether NMDA, non-NMDA, or both receptor subtypes are involved in this process.

It has been shown that ethanol inhibits the NMDA-induced release of endogenous DA from striatal slices (Woodward and Gonzales, 1990). This NMDA stimulation was also completely blocked by 1 mM magnesium (Mg^{++}), suggesting that activation of the NMDA receptor is responsible for stimulating the release of DA from dopaminergic neurons since non-NMDA receptor activation is not blocked by this cation (Nowak et al., 1984; Mayer and Westbrook, 1987).

Because the interaction between ethanol and the NMDA receptor has so many important implications, the aims of this study were 1) to characterise the effect of NMDA on extracellular striatal DA in vivo and 2) to observe the effect of ethanol on NMDA-induced increases in striatal DA using microdialysis in awake animals.

METHODS

One week after their arrival at the university, male Sprague-Dawley rats were stereotaxically implanted with a guide cannula into the left striatum. The coordinates used were +0.8 AP, +3.0 ML, -4.0 DV (Paxinos and Watson, 1986). Surgery was performed under equithesin anaesthesia. The guide cannula was fixed to the skull with dental cement, using microscrews as stabilizing anchors. After surgery, rats were individually housed and had free access to food and water. They were allowed one week to recover before being

Alcohol, Cell Membranes, and Signal Transduction in Brain
Edited by C. Alling *et al.*, Plenum Press, New York, 1993

used in a dialysis experiment. Prior to the dialysis experiment the probe, which has a tip length of 3 mm (concentric design from Carnegie Medicin), was calibrated with an ascorbic acid-containing standard solution of DA to establish the percentage of DA exchanged through the membrane. The probe was flushed with filtered artificial cerebrospinal fluid (ACSF: 147 mM NaCl, 4 mM KCl, and 2.3 mM $CaCl_2$) at 15 µl/min using a microinfusion pump. DA samples were collected at 15 min intervals with a flow rate of 2 µl/min and compared to corresponding DA reservoir samples. On the day of dialysis, the probe was connected to a liquid switch which allows different solutions to be delivered into the striatum during the same experiment. The flow rate was set at 2 µl/min, and the rat lightly anaesthetized with halothane to facilitate probe insertion. The animal was then hooked up to a free arm apparatus and placed into a containment bowl. Neurotransmitter levels were allowed to stabilize for 45 min, and basal DA samples then collected every 15 min for an hour. After 4 basal DA samples had been collected, drugs were administered as specified in the figure legends. Samples were collected into vials containing 1 M perchloric acid with 6.5 mg/l EDTA and 6.5 mg/l sodium bisulphite. These were placed on dry ice until analyzed with HPLC. Each rat was perfused with 10% formalin solution and it's brain sliced and stained with cresyl violet to verify probe location according to Paxinos and Watson (1986).

HPLC Analysis

The HPLC-EC system consisted of a Beckman 100A pump and Rheodyne 7125 injector valve with a reverse phase, 3 µm C18 column (0.32 X 10 cm) and a 100 µl sample loop. The column eluant is monitored with a BAS amperometric detector (LC4B; sensitivity set at 0.5 nA full scale) with the detector potential set at +0.7 V (relative to Ag/AgCl) using a glassy carbon working electrode. The detector was connected to a Hewlett Packard 3390A integrator for initial peak processing. The mobile phase (pH 3.1) consists of 0.15 M MCA, 0.9687 mM sodium octyl sulfate, 2.015 mM Na_2EDTA, and 0.1175 M NaOH. With a 0.8 ml/min flow rate, the DA retention time was approximately 8.0 min. In some cases a 1 x 100 mm 3 µm C18 column was used along with a Rheodyne 8125 injector and a 20 µl sample loop.

RESULTS

Average DA recoveries from probe calibrations were 19-25%. The infusion of NMDA (0.3-10 mM) resulted in a concentration-dependent increase in extracellular striatal DA release (Table 1).

Table 1. NMDA concentration-effect on extracellular striatal DA

NMDA concentration (mM)	DA (pg/µl)	N
0.0	0.53 ± 0.06	14
0.3	1.75 ± 0.61	5
1.0	8.99 ± 3.92	2
3.0	12.8 ± 4.97	6
10.0	16.01	1

Effect of NMDA concentrations on striatal dialysate DA levels. Samples were collected every 15 min before and after a 1 hour NMDA infusion. The control value represents the mean of the basal DA obtained from all the rats prior to NMDA infusion. Experimental values represent the means ± sem of the DA samples collected during the 1 hour NMDA infusion of each dose. ANOVA showed significant differences for each concentration from the control, ($p < 0.05$).

The addition of 1 mM MgCl$_2$ to the ACSF (figure 1) caused a 56% decrease in basal dialysate DA, but the NMDA stimulation in the presence of Mg^{++} (1134% above its basal) was actually greater than that in the magnesium-free ACSF (736% above its basal).

Further characterisation of the striatal DA response to NMDA is represented in figure 2. The 1 mM NMDA-induced increase in striatal dialysate DA was blocked by the coinfusion of NMDA with each of the following antagonists: 10 µM 2-amino-5-

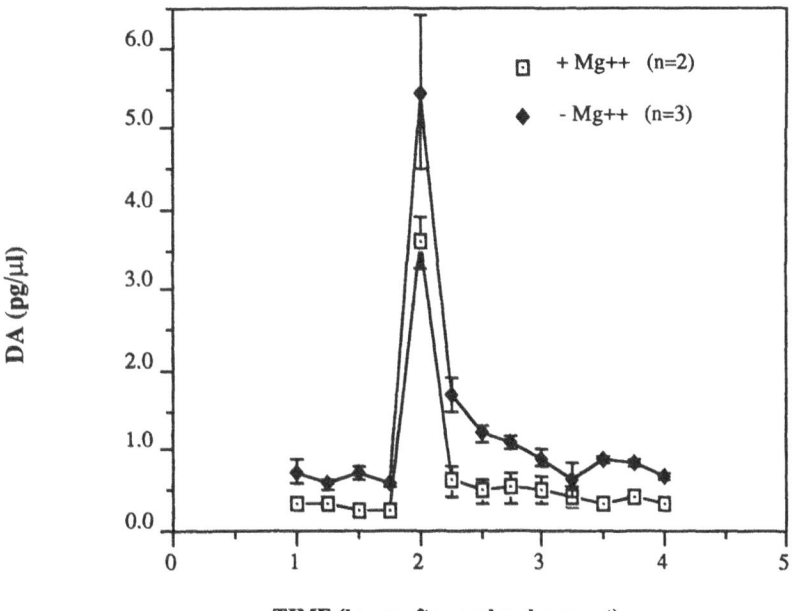

TIME (hours after probe placement)

Figure 1. Effect of 1 mM MgCl$_2$ on the NMDA-induced increase in striatal dialysate DA. Samples were collected every 15 min before and after a 10 min NMDA 1 mM infusion. MgCl$_2$, 1 mM, was included in the ACSF of a set of experiments to observe its effect on the NMDA stimulation of dialysate DA. Analysis of variance showed a significant time and group effect but no interaction, (p < 0.05). Shown are the means ± sem.

phosphopentanoic acid (AP5) - a selective competitive NMDA antagonist, 25 µM 10,11-dihydro-5-methyl-5H-dibenzo[a,c]cyclohepten-5,10-imine (MK-801) - a selective non-competitive NMDA antagonist, and 10 µM 6,7-dinitroquinoxaline-2,3-dione (DNQX) - a selective non-NMDA receptor blocker. The concentration of antagonists used had no effect on striatal DA when infused into the striatum by themselves (data not shown).

In the last part of our study, we saw neither a significant effect of intraperitoneal (ip) administration of ethanol on basal dialysate DA nor on the 1 and 3 mM NMDA-induced stimulation of dialysate DA (figure 3). We have found a similar result with 0.5 g/kg and 3.0 g/kg ip ethanol injection on a 1 mM NMDA-induced response (manuscript in preparation).

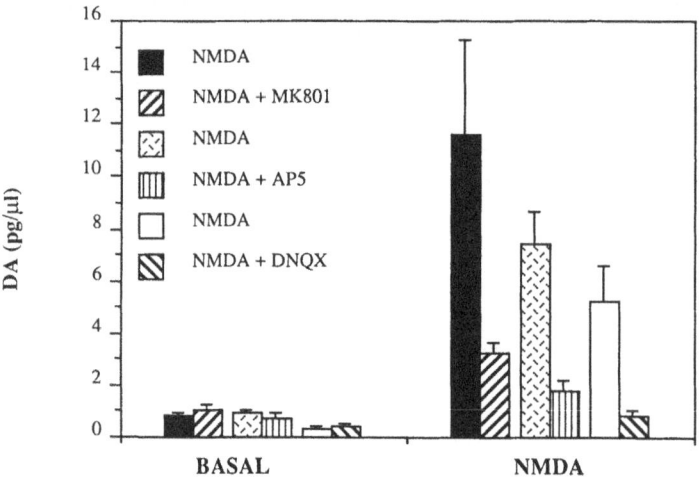

Figure 2. Characterization of striatal DA response to NMDA. In the MK-801 experiment, MK-801, NMDA or a combination of both was infused into the striatum for 10 min. To observe the effects of AP5 and DNQX, the antagonist was infused into the striatum for 1 hour prior to a 10 min infusion of a combination of NMDA and the antagonist. All three antagonists significantly depressed the dialysate DA response to NMDA (ANOVA, $p < 0.05$).

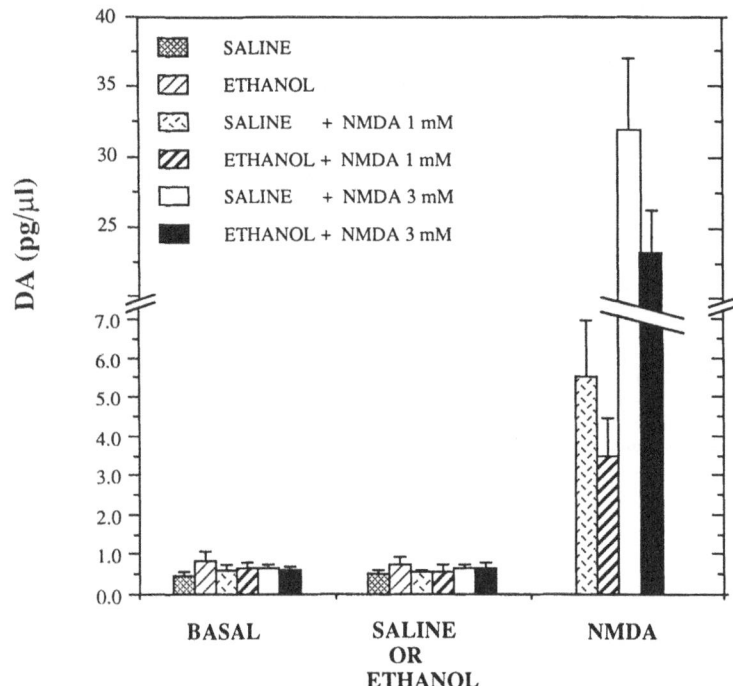

Figure 3. Effect of ethanol on striatal dialysate DA and on the response of DA to an NMDA infusion. Saline or ethanol was injected ip before the beginning of either a 10 min infusion of 1 mM NMDA or a 15 min infusion of the 3 mM NMDA (i.e. in those animals which received the NMDA treatment). One post-injection sample was collected prior to the NMDA-stimulated sample. This served as a measure of the effect of saline or ethanol on basal DA dialysate content. Samples were collected every 15 min throughout the experiment. Data represents the ·mean ± the sem (N = 5, except in the effect of ethanol on 3 mM NMDA, where N = 4). Analysis of variance showed a significant time effect in the NMDA 1 and 3 mM experiments, but no effect of ethanol on basal efflux nor on the DA response to either concentration of NMDA, ($p < 0.05$).

DISCUSSION

Although a 0.3 mM NMDA infusion was sufficient to increase striatal dialysate DA, we characterised the 1 mM NMDA effect since it is easier to see the effect of antagonists on a larger, more reliable control (in this case, NMDA) response. AP5 and MK-801 inhibition of the NMDA-induced response suggests that the stimulated DA levels are a consequence, in part, of NMDA receptor activation. The blockade attained with DNQX may or may not implicate the involvement of non-NMDA receptors since, although known to be a non-NMDA receptor antagonist, it's also been shown to inhibit postsynaptic NMDA responses by interacting at the glycine site of the NMDA receptor (Birch and Hayes, 1988; Patel et al., 1990).

Magnesium, by acting at the NMDA channel, has been shown to be a specific non-competitive antagonist of NMDA receptors (Nowak et al., 1984; Mayer and Westbrook, 1987). The lack of inhibition of the NMDA-induced response by 1 mM magnesium lends itself also to the possibility of a non-NMDA receptor component in the NMDA-induced stimulation of dialysate DA, i.e., stimulation of a non-NMDA channel could partially depolarise a nerve terminal which could relieve the magnesium blockade of the NMDA channel. In 1991, Wang showed that 100 μM NMDA resulted in a significant release of [^{3}H]DA from rat striatal synaptosomes. This effect was completely inhibited by the inclusion of 1 mM Mg^{++} in the buffer. He then showed that 10 μM glycine potentiated the NMDA response, and that this potentiation was reduced by the addition of 1 mM Mg^{++} to the buffer, but not completely inhibited as had been the case with the NMDA stimulation in the abscence of glycine. This suggests that if glycine were present, the NMDA receptor could still be activated, even in the presence of Mg^{++}. In an in vivo stuation, it is probable that sufficient extracellular glycine may still be present in the striatum which could also account for this lack of Mg^{++} inhibition on the NMDA-induced increase in dialysate DA.

The lack of an ethanol effect on basal dialysate DA parallels that seen in some in vitro preparations (Woodward and Gonzales, 1990; Gysling, et al., 1976) but not in others (Khatib et al., 1988). It also differs from an in vivo micodialysis study in which a 30% increase in DA was seen 40 min after a 2.5 g/kg ip injection (Imperato and Di Chiara, 1986). Differences between the two experimental designs may account for the discrepancy, for example: we start collecting our basal samples 45 min after probe insertion, whereas the other group waits 24 hours after their probe has been implanted, and secondly, we use a concentric probe design, whereas they use a transcerebral probe. Both post-implantation time and type of probe used have been shown to influence data obtained with microdialysis (Benveniste, 1989; Di Chiara, 1990). The main concern being that with shorter post-implantation intervals, neurotransmitter release may not be calcium-dependent. We have, however, seen a Ca^{++} dependency in our basal DA levels (data not shown).

Since our data shows that both NMDA and non-NMDA receptor subtypes may be involved in the NMDA-induced increase in the striatal dialysate DA, it seems reasonable to speculate that the lack of an ethanol inhibition on this effect may be due to a non-NMDA receptor component. It has previously been shown that although ethanol inhibits both NMDA-stimulated DA release in vitro (Woodward and Gonzales, 1990), and calcium-dependent, NMDA receptor stimulated cyclic GMP production (Hoffman et al., 1989a), it takes much higher concentrations to inhibit kainate-induced cyclic GMP production (Hoffman et al., 1989b). In hippocampal cell culture, Lovinger et al., (1989) also showed a much lower sensitivity of non-NMDA receptor-stimulated currents to ethanol inhibition than those produced by NMDA stimulation. An alternative suggestion could be that the NMDA-induced stimulation of dialysate DA may, in fact, not be a calcium-dependent process. It has been proposed that the inhibitory effect of ethanol on agonists acting at the NMDA receptor may be mediated via an action on agonist-stimulated calcium influx. A third possibility may be that the effect of NMDA to release striatal DA in vivo may involve processes or circuits not maintained in an in vitro preparation. Fourthly, ethanol does not inhibit NMDA-stimulated release of DA from striatal slices when glycine is added to the preparation (Woodward and Gonzales, 1990). Earlier in this discussion, we had pointed out that the inhibitory effect of DNQX on the NMDA-induced stimulation of dialysate DA may be due to its interaction at the glycine site of the NMDA receptor. We had also proposed that the lack of Mg^{++} inhibition on the NMDA-stimulated increase in extracellular DA may be due to the presence of sufficient amounts of glycine in the striatum in vivo. This rationale may also be used to explain the lack of ethanol inhibition on the NMDA-evoked increase in dialysate DA. Further work needs to be done to better understand the effects of ethanol on NMDA functions in living animals.

Acknowledgments

This work was supported by grants from NIAAA (AA08484 and AA00147). The excellent technical assistance by Stacie Westbrook is much appreciated.

REFERENCES

Benveniste, H., 1989, Brain microdialysis, J. Neurochem. 52:1667-1679.

Birch, P.J., Grossman, C.J. and Hayes, A.G., 1988, 6,7-Dinitro-quinoxaline-2,3-dione and 6-nitro-7-cyano-quinoxaline-2,3-dione antagonize responses to NMDA in the rat spinal cord via an action at the strychnine-insensitive glycine receptor, Eur. J. Pharmacol. 156:177-180.

Carter, C.J., L' Heureux, R., and Scatton, B., 1988, Differential control by N-methyl-D-aspartate and kainate of striatal dopamine release in vivo: a trans-striatal dialysis study. J. Neurochem, 51:462-468.

Clow, D.W., and Jhamandas K., 1989, Characterization of L-glutamate action on the release of endogenous dopamine from the rat caudate-putamen, J. Pharmacol. Exp. Ther. 248:722-728.

Di Chiara, G., 1990, In vivo brain dialysis of neurotransmitters, TiPS. 11:116-121.

Gonzales, R.A., 1990, NMDA receptors excite alcohol research, TiPS. 11:137-139.

Gothert, M. and Fink, K., 1989, Inhibition of N-methyl-D-aspartate (NMDA)- and L-glutamate-induced noradrenaline and acetylcholine release in the rat brain by ethanol, Naunyn-Schmied. Arch Pharmacol., 340:516-521.

Grant, K.A., Valverius, P., Hudspith, M., and Tabakoff, B., 1990, Ethanol withdrawal seizures and the NMDA receptor complex, Eur. J. Pharmacol. 176:289-296.

Gysling K., Bustos, G., Concha, I., and Martinez, G., 1975, Effect of ethanol on dopamine synthesis and release from rat corpus striatum, Biochem. Pharmacol. 25:157-162.

Hoffman, P.L., Rabe, C.S., Moses, F., and Tabakoff, B., 1989a, N-methyl-D-aspartate receptors and ethanol: inhibition of calcium flux and cyclic GMP production, J. Neurochem. 52:1937-1940.

Hoffman, P.L., Moses, F., and Tabakoff, B., 1989b, Selective inhibition by ethanol of glutamate-stimulated cyclic GMP production in primary cultures of cerebellar granule cells, Neuropharmacology 28:1239-1243.

Hoffman, P.L., Rabe, C.S., Grant K.A., Valverius, P., Hudspith, M., and Tabakoff, B., 1990, Ethanol and the NMDA receptor, Alcohol. 7:229-231.

Imperato, A., and Di Chiara G., 1986, Preferential stimulation of dopamine release in the nucleus accumbens of freely moving rats by ethanol, J. Pharmacol. Exp. Ther. 239:219-228.

Imperato, A., Honore, T., and Jensen, L.H., 1990, Dopamine release in the nucleus caudatus and in the nucleus accumbens is under glutamatergic control through non-NMDA receptors: a study in freely-moving rats, Brain Research 530:223-228.

Keefe, K.A., Zigmond, M.J., and Abercrombie, E.D., 1990, Excitatory amino acid receptor involvement in the regulation of striatal extracellular dopamine, Ann. N.Y. Acad. Sci. 604:614-615.

Khatib, S.A., Murphy, J.M. and McBride W.J., 1988, Biochemical evidence for activation of specific monoamine pathways by ethanol, Alcohol. 5:295-299.

Lovinger, D.M., White, G., and Weight, F.F., 1989, Ethanol inhibits NMDA-activated ion current in hippocampal neurons, Science 243:1721-1724.

Mayer, M.L. and Westbrook, G.L., 1987, The physiology of excitatory amino acids in the vertebrate central nervous system, Prog. Neurobiol. 28:197-276.

Moghaddam, B., Gruen, R.J., Roth, R.H., Bunney, B.S., and Adams, R.N., 1990, Effect of L-glutamate on the release of striatal dopamine: in vivo dialysis and electrochemical studies, Brain Research. 518:55-60.

Nowak, L., Bregestovski, P., Ascher, P., Herbet, A., Prochiantz, A., 1984, Magnesium gates glutamate-activated channels in mouse central neurons, Nature 309:261-263.

Patel, J., Zinkard, W.C., Klika, A.B., Mangano, T.J., Keith, R.A., and Salama, A.I., 1990, 6,7-Dintroquinoxaline-2,3-dione blocks the cytotoxicity of N-methyl-D-aspartate and kainate , but not quisqualate, in cortical cultures, J. Neurochem. 55:114-121.

Paxinos, G., and Watson, C., 1982, The Rat Brain in Stereotaxic coordinates, Academic Press, New York.

Wang, J., 1991, Presynaptic glutamate receptors modulate dopamine release from striatal synaptosomes, J. Neurochem. 57: 819-822.

Woodward J.J. and Gonzales, R.A. Ethanol inhibition of N-methyl-D-aspartate-stimulated endogenous dopamine release from rat striatal slices: reversal by glycine, J. Neurochem. 54:712-715.

ACTIONS OF ALCOHOLS AND OTHER SEDATIVE/HYPNOTIC COMPOUNDS ON CATION CHANNELS ASSOCIATED WITH GLUTAMATE AND 5-HT₃ RECEPTORS

David M. Lovinger[1] and Robert W. Peoples[2]

[1]Department of Molecular Physiology and Biophysics
Vanderbilt University Medical School
Nashville, TN 37232

[2]Laboratory of Molecular Physiology and Pharmacology
National Institute on Alcohol Abuse and Alcoholism
Rockville, MD 20852

INTRODUCTION

The neural actions of a number of sedative/hypnotic agents appear to be mediated via alterations in the function of neurotransmitter receptors of the ligand-gated ion channel class. For example, it has long been known that barbiturates and benzodiazepines potentiate responses mediated by $GABA_A$ receptors (Schmidt, 1963; Nicoll, 1972; Ransom and Barker, 1976; Choi et al., 1981; Study and Barker, 1981; Olsen, 1982; Gage and Robertson, 1985; Vicini et al., 1987) and that this action likely underlies the sedative and anesthetic actions of these compounds. Ethanol (EtOH) also has potent actions on a number of different ligand-gated channels. Intoxicating concentrations of ethanol have been shown to inhibit responses mediated by the NMDA-type glutamate receptor (Dildy and Leslie, 1989; Hoffman et al., 1989; Lima-Landman and Albuquerque, 1989; Lovinger et al., 1989; Göthert and Fink, 1989), while ethanol is less potent in inhibiting AMPA/kainate receptor-mediated responses (Hoffman et al., 1989; Lovinger et al., 1989; Lovinger et al., 1990). In contrast, ethanol potentiates the function of other ligand-gated ion channels including some types of $GABA_A$ receptors (Mehta and Ticku, 1988; Suzdak et al., 1986; Nakahiro et al., 1991; Wafford et al., 1990), 5-HT₃ receptors (Lovinger, 1991), strychnine-sensitive glycine receptors (Celentano et al., 1988) and nicotinic acetylcholine receptors (Okada, 1967; Miller et al., 1991).

We have investigated the actions of ethanol and other sedative/hypnotic agents on responses mediated by glutamate and 5-HT₃ receptors in a variety of neuronal preparations (Lovinger et al., 1989; Lovinger et al., 1990; Lovinger, 1991; Lovinger and White, 1991; Lovinger and Peoples, 1991). In addition to reviewing observations concerning ethanol's actions on these receptors we will contrast the actions of ethanol at the different receptor types. We will also compare the effects of ethanol to those of other sedative/hypnotic agents. Our findings will be discussed in light of recent observations concerning the molecular structure of the subunits which form the different ligand-gated channels.

Alcohol, Cell Membranes, and Signal Transduction in Brain
Edited by C. Alling *et al.*, Plenum Press, New York, 1993

GLUTAMATE RECEPTOR FUNCTION IS INHIBITED BY ETHANOL

Given the important role of glutamate as the primary excitatory neurotransmitter in the mammalian brain it is not surprising that much attention has been focussed on the actions of ethanol on glutamatergic transmission. It has now been widely demonstrated that ethanol inhibits the function of the ligand-gated ion channel type glutamate receptors (see Gonzalez, 1990; Samson and Harris, 1992; for review). The NMDA receptor appears to be especially sensitive to ethanol's actions with inhibition of responses to activation of this receptor occuring at ethanol concentrations from 5-100 mM. In general, the AMPA/kainate receptors are less sensitive to ethanol. However, preliminary reports suggest that brain AMPA/kainate receptors expressed in Xenopus oocytes may be more sensitive to ethanol than receptors examined in neuronal cultures or in brain slices (Dildy-Mayfield et al., 1991).

The mechanism of ethanol's action at the NMDA receptor appears to be a non-competitive inhibition of receptor-gated channel opening. EtOH does not alter the potency with which NMDA activates the receptor (Dildy-Mayfield and Leslie, 1991). In addition, EtOH does not alter the potency with which glycine activates the coagonist site on the receptor (Peoples and Weight, 1992). Observations from several laboratories indicate that high concentrations of glycine reduce the inhibitory effect of EtOH on NMDA receptor function (Woodward and Gonzalez, 1990; Rabe and Tabakoff, 1991; Dildy-Mayfield and Leslie, 1991). However, this may reflect an action distinct from the actions by which glycine potentiates receptor function. In addition, EtOH does not appear to interact with other modulatory sites on the NMDA receptor including the sites where divalent cations act to inhibit receptor function (Woodward and Gonzalez, 1990; Dildy-Mayfield and Leslie, 1991). The mechanism by which EtOH inhibits the function of AMPA receptors has not been characterized in detail. However, preliminary studies by Harris and colleagues suggest that responses activated by low concentrations of agonist may be more sensitive to EtOH than responses produced by high agonist concentrations (Dildy-Mayfield et al., 1991).

We have begun to examine the interaction between EtOH and the kinetics of NMDA receptor-mediated current in whole-cell patch-clamp recordings from rat cortical neurons in culture. Application of a low concentration of NMDA (15 μM) in the presence of a moderate concentration of glycine (200 nM) activates current which maintains a fairly constant amplitude during prolonged exposure to agonist (Fig. 1A). During simultaneous application of EtOH and NMDA, inhibition of the amplitude of current is of similar magnitude throughout the time course of application (Fig. 1A). Thus, EtOH does not induce any time-dependent component to responses activated by low NMDA concentrations. At higher NMDA concentrations (50 μM), current exhibits two components; a short-latency decaying component, and a longer latency "steady-state" component. The fast decaying component appears to correspond to fast desensitization of the receptor. The steady-state component likely reflects an equilibrium resulting from channel openings from both closed and desensitized channel states (Fig. 1B). Simultaneous application of EtOH and 50 μM NMDA activates current which is smaller in amplitude than that observed in NMDA alone, but does not differ substantially in its time course (Fig. 1B). In 5 neurons in which the effect of 50 mM EtOH on current activated by 50 μM NMDA was examined the rate of current decay did not differ from that observed in the presence of NMDA alone (decay time constant tau=0.62 ± 0.12 sec in the absence and 0.72 ± 0.22 sec in the presence of EtOH). We have observed similar results in the presence of a concentration of glycine (1 μM) which is nearly saturating for activation of the coagonist site. These observations suggest that EtOH does not alter the rate or extent of desensitization of the NMDA receptor under conditions where desensitization is minimal or quite pronounced.

To further examine the possibility that EtOH might have preferential effects on channel opening from desensitized or non-desensitized receptors we compared the actions of EtOH when applied at various times after the onset of application of 50 μM NMDA (Fig. 1D). We observed that the magnitude of EtOH inhibition of NMDA activated current was similar whether NMDA and EtOH were applied simultaneously or EtOH was applied at time points up to 20 sec after the onset of NMDA application. These observations indicate that EtOH does not preferentially inhibit non-desensitized or desensitized receptors. Thus, it is likely that channel openings from any non-conducting state are equally sensitive to EtOH. This could occur either through a reduction in channel open time or a reduction in the probability of channels opening from any state. Decreases in both open time and probability of opening during prolonged NMDA+alcohol application have been observed in single channel studies (Lima-Landman and Albuquerque, 1989; McLarnon et al., 1991). It remains to be determined which of these changes in channel behavior contributes more to the overall reduction in NMDA receptor-activated current observed in whole-cell recordings.

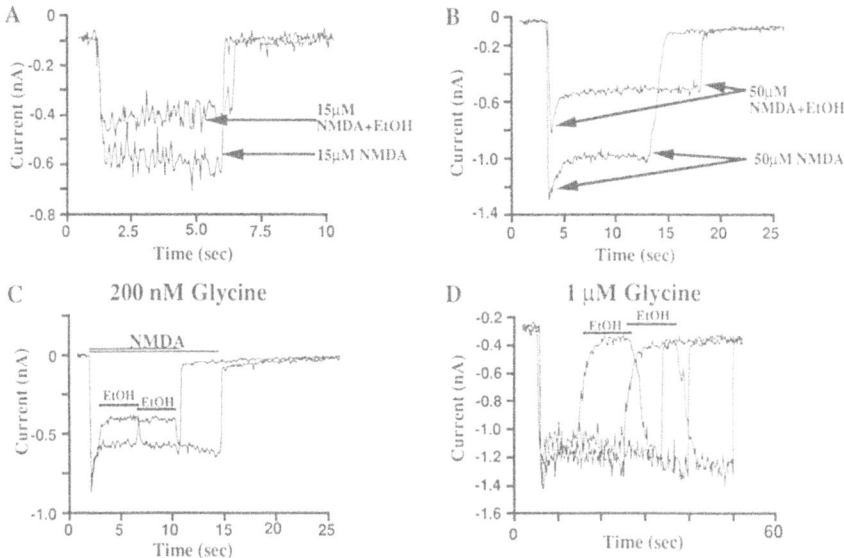

Figure 1. A) Non-desensitizing current activated by a low concentration of NMDA in the absence and presence of 50mM EtOH. B) Current activated by a higher NMDA concentration in the absence and presence of 50mM EtOH. Note the prominent desensitizing (fast decaying) component and the lack of effect of EtOH on the rate of decay. C) Inhibition of current activated by 50μM NMDA+200nM glycine when 50mM EtOH is applied beginning 2 or 5 sec after the onset of current. Note that EtOH produces a similar magnitude of inhibition regardless of the delay between current onset and EtOH application. D) Similar experiment to that shown in C only performed in the presence of 1μM glycine and 100mM EtOH was applied 10 or 20 sec after the onset of current.

Little is known about the interaction between EtOH and kinetics of AMPA receptors since most studies have examined ion current under conditions in which desensitizing components are not observed (Lovinger et al., 1989; Dildy-Mayfield et al., 1991). However, EtOH inhibits responses activated by the agonists kainate and quisqualate to a similar degree (Lovinger et al., 1989). Since kainate activates, but does not desensitize AMPA receptors, these findings suggest that EtOH does not inhibit receptor function by increasing the rate or extent of receptor desensitization.

ETHANOL POTENTIATES THE FUNCTION OF 5-HT$_3$ RECEPTORS AND ACTS PREFERENTIALLY ON NON-DESENSITIZED RECEPTORS

Intoxicating concentrations of ethanol increase the amplitude of ion current generated by activation of the 5-HT$_3$ receptor for the neurotransmitter serotonin. Potentiation of the function of this ligand-gated ion channel has been observed in neuroblastoma cells as well as in isolated neurons from adult rat nodose ganglion (Lovinger, 1991; Lovinger and White, 1991). Potentiation is observed in 70-80% of cells examined in both preparations suggesting some heterogeneity of alcohol sensitivity of receptors. We have previously reported that ethanol increases the potency of 5-HT for activating current and that ethanol increases the rate of current decay in the presence of agonist (Lovinger and White, 1991). This latter effect presumably reflects an increase in the rate of fast receptor desensitization. In more recent experiments, we have improved our method of drug superfusion to allow for faster application of 5-HT and EtOH. We were thus able to examine the time course of current in the absence and presence of EtOH with greater precision when a low concentration of 5-HT (1 μM) was used. During 5-HT applications lasting 10-20 sec EtOH increased peak current amplitude by 26±9% at 50 mM (n=5) and 61±17% at 100 mM (n=7). Fig. 2 shows an example of 5-HT-activated current before and during EtOH application. It can be seen from this figure that increases in both current amplitude and rate of current decay were observed.

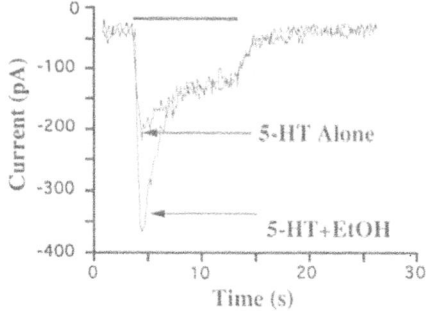

Figure 2. Ethanol potentiates a fast-decaying component of 5-HT$_3$ receptor-mediated current. Current activated by 1μM 5-HT in the absence and presence of 100mM EtOH. Note that potentiation is only observed during the first 5 sec of 5-HT+EtOH coapplication. The "steady-state" component of transmission is not potentiated. The line above the currents indicates the time during which 5-HT or 5-HT+EtOH was applied.

Current activated by 5-HT alone and 5-HT+EtOH exhibited single exponential decay kinetics during the first 5-10 sec of agonist application. The time constant of this fast decaying component averaged 2.47±0.26 sec when 5-HT was applied alone, and 1.37±0.16 sec in the presence of 5-HT+100 mM EtOH (n=7). A slower decaying component of current exhibited linear decay and thus could not be fit with exponential functions. Interestingly, this component of current was not potentiated in the presence of EtOH. The result of the increase in amplitude and rate of the fast-decaying component was that current measured 5-10 sec after the onset of 5-HT+EtOH was no longer potentiated relative to current activated by 5-HT alone at the same time point. Furthermore, application of 5-HT for 5 sec prior to application of EtOH completely eliminated potentiation.

These observations are best understood in light of what is known about the kinetic behavior of ligand-gated ion channels. Channels are able to interconvert between a number of functional states. In the absence of agonist, channels are non-conducting or "closed". When agonist is applied and binds to receptors channels rapidly "open" and begin to conduct ions. In the continous presence of agonist, channels close even though agonist is likely still bound to the receptors. Such agonist-bound but closed states are referred to as "desensitized" states. Our results suggest that EtOH increases the rate of the initial opening of channels when agonist binds to the receptor. In addition, EtOH appears to increase the rate at which receptors enter into desensitized states. Once receptors enter desensitized states, however, they appear to become insensitive to EtOH as indicated by the lack of potentiation seen 5 sec after the beginning of application of 5-HT+EtOH and the loss of

EtOH potentiation following 5-HT pretreatment. Thus, transitions from desensitized to open channel states appear to be EtOH-insensitive. It should be noted that the actions of EtOH differ from simply increasing the concentration of 5-HT since increased agonist concentrations increase the rate and extent of receptor desensitization whereas EtOH only increases rate.

ETHER HAS ACTIONS SIMILAR TO ETHANOL ON GLUTAMATE AND 5-HT$_3$ RECEPTORS

We have begun to determine the actions of other sedative/hypnotic compounds on ligand-gated ion channels. Our initial observations indicated that alkanols other than ethanol (e.g. methanol) can inhibit NMDA receptor-mediated current and potentiate 5-HT$_3$ receptor-mediated current (Lovinger et al., 1989; Lovinger and White, 1991). In addition, we have observed similar effects of trichloroethanol, an alcohol which is the active metabolite of the general anesthetic chloral hydrate (Peoples and Weight, 1991; Lovinger and Zhou, submitted). This alcohol acts on both receptor types with a potency greater than EtOH, and produces greater maximal potentiation of 5-HT$_3$ receptor function than EtOH. Diethyl ether also inhibits ion current mediated by glutamate receptors. Fig. 3 shows responses to NMDA before, during and after ether application. In addition, this figure shows the increase in inhibition with increasing ether concentrations. In general, the effects of ether appear similar to those of ethanol with NMDA receptor-mediated current being inhibited more potently than AMPA receptor-mediated current.

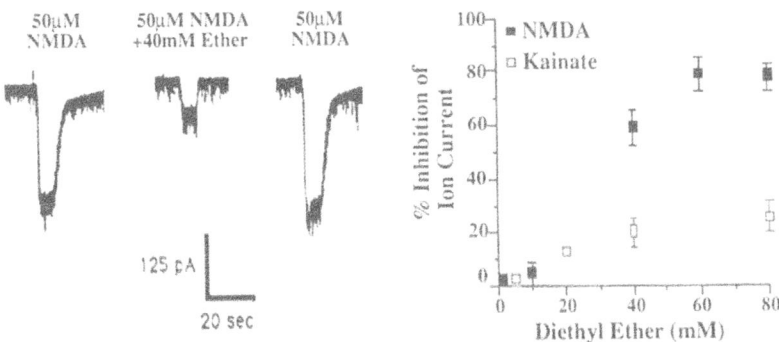

Figure 3. Left: Ion current activated by 50µM NMDA/200 nM glycine before, during and after ether application. Right: Inhibition of ion current activated by NMDA and kainate as a function of ether concentration. Number of neurons in NMDA experiments: 1mM ether=6, 10mM=3, 40mM=5, 60mM=3, 80mM=4. Number of neurons in kainate experiments: 5mM ether=4, 20mM=5, 40mM=6, 80mM=5.

We have also observed that ether potentiates the function of 5-HT$_3$ receptors at concentrations similar to those at which it inhibits NMDA receptor function (Zhou and Lovinger, in preparation). Ether appears to be more efficacious than ethanol; producing maximal potentiation of greater than 100% above baseline while EtOH potentiates current by a maximum of ~40-60% above baseline.

INHIBITION OF 5-HT$_3$ RECEPTOR FUNCTION BY BARBITURATES AND BENZODIAZEPINES

Our understanding of the mechanisms of action of benzodiazepines and barbiturates is more advanced than that of the mechanisms of the alcohols and anesthetics mentioned above. Benzodiazepines and barbiturates potentiate the function of GABA$_A$ receptors, and it is this action which is believed to underlie their sedative and anesthetic actions as mentioned above. These compounds have quite different actions on other ligand-gated ion channels. For example, barbiturates inhibit the function of AMPA-type glutamate receptors (Miljkovic and MacDonald, 1986; Peoples et al., 1990), but have little effect on NMDA receptors (Peoples et al., 1990). Benzodiazepines have little effect on these ligand-gated channels (Peoples et al., 1990). We examined the effects of pentobarbital and two benzodiazepines, midazolam and flurazepam, on 5-HT$_3$ receptor-mediated current in NCB-20 neuroblastoma cells. Interestingly, diazepam inhibited current by less than 20% even at a concentration of 10 µM. Thus, the potency for benzodiazepine inhibition may relate to the number of cyclic structures or halogen functional groups on the benzodiazepine molecule.

Concentration-dependent inhibition of current was also observed in the presence of the two benzodiazepines (Fig. 4A,B). Midazolam appeared to be slightly more potent than flurazepam in inhibiting current. Benzodiazepines did not appear to alter the decay rate of current in the presence of 5-HT as can be seen in Fig. 4A. Application of Na-pentobarbital (50-1000 µM) also inhibited current activated by 2 µM 5-HT, and the magnitude of inhibition increased with increasing concentrations of pentobarbital (Fig. 4C).

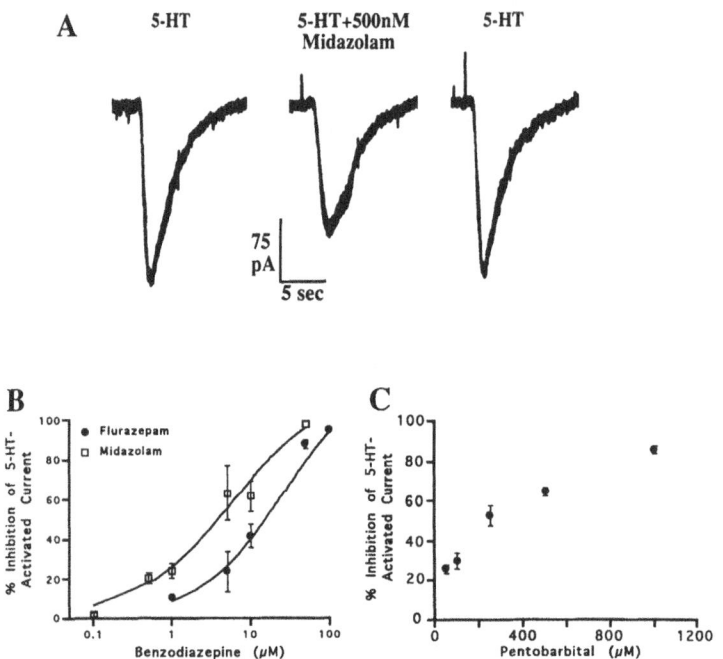

Figure 4. Inhibition of 5-HT$_3$ receptor-mediated ion current by barbiturates and benzodiazepines. A) Ion current activated in an NCB-20 neuroblastoma cell by application of 2µM 5-HT before, during and after application of midazolam as indicated. B) Graph showing percent inhibition of 5-HT-activated current as a function of midazolam and flurazepam concentration. Number of NCB-20 cells per concentration for midazolam: 0.1µM=6; 0.5=12; 1=12; 5=3; 10=4; 50=2, for flurazepam: 1µM=2; 5=6; 10=12; 50=4; 100=2. Lines are the best fit generated using the logistic equation previously described (Lovinger and White, 1991). The IC$_{50}$ for midazolam=5.8µM; for flurazepam IC50=26.2µM. C) Percent inhibition of 5-HT-activated ion current as a function of Na-pentobarbital concentration. Data are from 4 NCB-20 cells. Current in A was activated by 2µM 5-HT.

DISCUSSION

Differences in the interaction of Alcohols with the Kinetics of Glutamate and non-Glutamate-Gated Receptor/Channels

The interactions of ethanol with the kinetics of 5-HT$_3$ and NMDA receptors appear to differ. Inhibition of NMDA receptor-mediated current appears to occur regardless of the desensitization state of receptors. In contrast, 5-HT$_3$ receptors are most sensitive to EtOH when opening from a non-desensitized state. These differential effects of EtOH may reflect differences in the way that conformational changes produced by EtOH interact with different mechanisms by which neurotransmitters activate channel opening. A survey of the literature on the effects of alcohols and general anesthetics on ligand-gated ion channels indicates that the function of certain receptors is potentiated by these compounds while others are inhibited. Generally, the 5-HT$_3$, nicotinic ACh, GABA$_A$ and glycine receptors are potentiated by alcohols and anesthetics (Okada, 1967; Celentano et al., 1988; Lovinger, 1991; Nakahiro et al., 1991; Miller et al., 1991). In the case of the nicotinic receptor, most alcohols produce an increase in agonist potency similar to that reported for the 5-HT$_3$ receptor (Miller et al., 1991). However, some alcohols and halogenated hydrocarbon general anesthetics can also inhibit the function of nicotinic ACh receptors (c.f. Dilger and Brett, 1991; Murrell and Haydon, 1991). Glutamate receptor function appears to be inhibited by the alcohols and general anesthetics examined to date (Lovinger et al., 1989; Peoples et al., 1990; Peoples and Weight, 1991).

The observation that pentobarbital and benzodiazepines do not potentiate, but rather inhibit, 5-HT$_3$ receptor function is consistent with a profile of action of these compounds which differs from that of alcohols and ethers. One possibility is that the latter set of compounds act at similar sites on a given ligand-gated channel. Barbiturates and benzodiazepines might act on sites which are distinct from those affected by the other compounds. Furthermore, it appears that receptors for barbiturates are only present on certain subtypes of ligand-gated ion channels (Olsen, 1982; Miljkovic and MacDonald, 1986; Peoples et al., 1990). Potent effects of benzodiazepines have only been observed on GABA$_A$ receptors (Choi et al., 1981; Study and Barker, 1981), and the inhibition of 5-HT$_3$ receptor function reported at present represents a rather low potency effect of these compounds. Thus, benzodiazepines may be among the most specific sedative agents yet described with regard to their mechanism of action.

Differences in Subunit Structure of Glutamate and non-Glutamate Gated Receptor/Channels

Much is now known about the molecular makeup of the protein subunits which form the ligand-gated ion channels and a number of differences can be seen between the glutamate receptors and other receptor/channels. Most striking is the fact that the glutamate receptor subunits are larger than those which make up the other receptors (see Hollman et al., 1989; Keinänen et al., 1990; Barnard and Henley, 1990; Kutsuwada et al., 1992; for comparison of channel subunits). The bulk of the difference in size of the receptors can be accounted for by the large N-terminal region present on the glutamate receptor subunits (Hollman et al., 1989; Keinänen et al., 1990; Monyer et al., 1992; Kutsuwada et al., 1992). This N-terminal domain is considered to reside in the extracellular milieu when the receptor is inserted into the membrane. It may be involved in binding neurotransmitter and in initial gating of the channel. Another striking difference between glutamate receptors and other ligand-gated channels is the absence of the "cis-loop" in the N-terminal region of the glutamate receptors (Hollman et al., 1989; Keinänen et al, 1990; Monyer et al., 1992; Kutsuwada et al., 1992). Reduction at this site appears to inhibit the function of the nicotinic acetylcholine receptor (Aizenman et al., 1989), but its function at the 5-HT$_3$ receptor has not yet been explored. It will be interesting to examine the interaction between agents which potentiate the function of the 5-HT$_3$ receptor and the function of this cis-loop region of the receptor. It is possible that these differences in the N-terminal portion of glutamate vs. non-glutamate-activated receptor/channels are important for determining whether receptors are potentiated or inhibited by alcohols and general anesthetics.

Numerous other differences exist between glutamate receptors and the other ligand-gated ion channels whose molecular structures are known. The amino acid sequences of the membrane spanning domains differ between glutamate receptors and other receptors although certain residues like the proline in transmembrane domain I and the putative pore-lining residues in transmembrane domain II are conserved (Hollman et al., 1989; Barnard and Henley, 1991). The primary sequence of the membrane-spanning domains also differs between AMPA and NMDA receptors (Moriyoshi et al., 1991; Monyer et al., 1992; Kutsuwada et al., 1992) and between the 5-HT$_3$ receptor and GABA, glycine or acetylcholine receptors (Maricq et al., 1991). Given the relationship between the hydrophobicity of alcohols and the potency of their actions at ligand-gated ion channels, it is tempting to speculate that the membrane-spanning domains are targets for the actions of these compounds. Indeed, the differential sensitivity of ligand-gated ion channels to alcohols could result, in part, from differences in these regions.

Molecular Heterogeneity Within a Given Receptor Type

It is by now painfully obvious that each receptor of the ligand-gated ion channel superfamily can be formed from combinations of a variety of protein subunits. Indeed, within the last few years several glutamate receptor-forming subunits have been cloned including several NMDA receptor subunits (Keinänen et al., 1990; Barnard and Henley, 1991; Moriyoshi et al., 1991; Monyer et al., 1992; Kutsuwada et al., 1992). This molecular heterogeneity may underlie differences in drug sensitivity among the glutamate receptors. Indeed, molecular heterogeneity has been suggested to underlie differences in ethanol sensitivity of GABA receptors (Wafford et al., 1991). A similar explanation may eventually account for the differences in the ethanol sensitivity of 5-HT$_3$ receptor in different cells. However, other explanations deserve consideration. For example, it is possible that posttranslational modifications may alter ethanol sensitivity. On the other hand, the effects of ethanol may be sensitivite to small changes in channel state such that certain resting channel states may be more sensitive to ethanol than others. While we have no evidence of differential sensitivity of channel states for the glutamate receptors, the 5-HT$_3$ receptor appears to respond more readily to ethanol when opening from the rested state. This may provide a clue to explaining how channel states interact with the effects of ethanol and related compounds.

Ligand-Gated Ion Channels and the Neural Actions of Alcohol

Our concern with the mechanisms of alcohol actions on ligand-gated ion channels might be somewhat misplaced if we had no reason to believe that these receptors play some role in the neural effects of ethanol. However, evidence using a number of behavioral paradigms indicates involvement of NMDA and 5-HT$_3$ receptors in acute intoxication and consequences of chronic alcohol abuse (see Samson and Harris, 1992 for review). Recent studies indicate similarities between the actions of ethanol and those of other NMDA antagonists (Balster and Wessinger, 1983; Fidecka and Langwinski, 1989; Kulkarni et al., 1990; Wilson et al., 1990; Grant et al., 1991; Robledo et al., 1991). Greater than normal activation of NMDA receptors may also play a role in withdrawal hyperexcitability and seizures (Grant et al., 1990; Morrisett et al., 1990). Experiments with 5-HT$_3$ receptor antagonists suggest that these compounds can block recognition of intoxication (Grant and Barrett, 1991), reduce the ability of ethanol to stimulate increases in brain dopamine levels (Carboni et al., 1989; Wozniak et al., 1990), and decrease alcohol intake (Sellers et al., 1988; Fadda et al., 1991). Thus there is some reason to believe that by determining the nature of alcohol/receptor interactions we will learn more about the molecular basis of intoxication and alcohol toxicity.

ACKNOWLEDGEMENTS

This work was supported in part by grant AA08986 from NIAAA and a fellowship to D.M.L. from the Alcoholic Beverage Medical Research Foundation.

REFERENCES

Aizenman, E., Lipton, S.A. and Loring, R.H., 1989, Selective modulation of NMDA responses by reduction and oxidation, Neuron 2:1257.

Balster, R.L. and Wessinger, W.D., 1983, Central nervous system depressant effects of phencyclidine, in: Phencyclidine and Related Arylcyclohexylamines: Present and Future Applications, J.M. Kamenka, E.F. Domino and P. Geneste, eds., NPP Books, Ann Arbor, MI, USA, pp. 291-309.

Barnard, E.A. and Henley, J.M., 1990, The non-NMDA receptors: types, protein structure and molecular biology, Trends Pharmacol. Sci. 11:500.

Carboni, E., Acquas, E., Frau, R. and Di Chiara, G., 1989, Differential inhibitory effects of a 5-HT3 antagonist on drug-induced stimulation of dopamine release, Eur. J. Pharmacol. 164:515.

Celentano, J.J., Gibbs, T.T. and Farb, D.H., 1988, Ethanol potentiates GABA- and glycine-induced chloride currents in chick spinal cord neurons, Brain Res. 455:377.

Choi, D.W., Farb, D.H. and Fishbach, G.D., 1981, Chlordiazepoxide selectively potentiates GABA conductance of spinal cord and sensory neurons in cell culture, J. Neurophysiol. 45:621.

Dildy, J.E. and Leslie, S.R., 1989, Ethanol inhibits NMDA-induced increases in intracellular Ca++ in dissociated brain cells, Brain Res. 499:383.

Dildy-Mayfield, J.E. and Leslie, S.W., 1991, Mechanism of inhibition of N-methyl-D-aspartate-stimulated increases in free intracellular calcium by ethanol, J. Neurochem. 56:1536.

Dildy-Mayfield, J.E., Sikela, J.M. and Harris, R.A., 1991, Evidence for kainate receptor subtypes based on differential ethanol sensitivity of brain and GluR3 receptors expressed in Xenopus oocytes, Soc. Neurosci. Abstr. 17:1167.

Dilger, J.P. and Brett, R.S., 1991, Actions of volatile anesthetics and alcohols on cholinergic receptor channels, in: Molecular and Cellular Mechanisms of Alcohol and Anesthetics, Annals of the New York Academy of Sciences Vol. 625, E. Rubin, K.W. Miller and S.H. Roth, eds., N.Y. Acad. Sci., New York, NY, pp. 616-627.

Fadda, F., Garau, B., Marchei, F., Colombo, G. and Gessa, G.L., 1991, MDL 72222, a selective 5-HT3 receptor antagonist, suppresses voluntary ethanol consumption in alcohol-preferring rats, Alcohol and Alcoholism 26:110.

Fidecka, S. and Langwinski, R., 1989, Interaction between ketamine and EtOH in rats and mice, Pol. J. Pharmacol. Pharm. 41:32.

Gage, P.W. and Robertson, B., 1985, Prolongation of inhibitory postsynaptic currents by pentobarbitone, halothane and ketamine in CA1 pyramidal cells in rat hippocampus, Br. J. Pharmacol. 85:675.

Grant, K.A., Valverius, P., Hudspith, M. and Tabakoff, B., 1990, Ethanol withdrawal seizures and the NMDA receptor complex, Eur. J. Pharmacol. 176:289.

Grant, K.A., Knisely, J.S., Tabakoff, B., Barrett, J.E. and Balster, R.L., 1991, Ethanol-like discriminative stimulus effects of non-competitive N-methyl-D-aspartate antagonists, Behav. Pharmacol. 2:95.

Grant, K.A. and Barrett, J.E., 1991, Blockade of the discriminative stimulus effect of ethanol with 5-HT3 receptor antagonists, Psychopharmacol. 104:451.

Göthert, M. and Fink, K., 1989, Inhibition of N-methyl-D-aspartate (NMDA)- and L-glutamate-induced noradrenaline and acetylcholine release in the rat brain by ethanol, Nauyn-Schmiedeberg's Arch. Pharmacol. 340:516.

Gonzalez, R.A., 1990, NMDA receptors excite alcohol research, Trends Pharm. Sci. 11:137.

Hollman, M., O'Shea-Greenfield, A., Rogers, S.W. and Heinemann, S., 1989, Cloning by functional expression of a member of the glutamate receptor family, Nature 342:643.

Hoffman, P.L., Rabe, C.S., Moses, F. and Tabakoff, B., 1989, N-methyl-D-aspartate receptors and ethanol: inhibition of calcium flux and cyclic GMP production, J. Neurochem. 52:1937.

Keinänen, K., Wisden, W., Sommer, B., Werner, P., Herb, A., Verdoorn, T.A., Sakmann, B. and Seeburg, P.H., 1990, A family of AMPA-selective glutamate receptors, Science 249:556.

Kulkarni, S.K., Mehta, A.K. and Ticku, M.K., 1990, Comparison of anticonvulsant effect of ethanol against NMDA-, kainic acid-, and picrotoxin-induced convulsions in rats, Life Sci. 46:481.

Kutsuwada, T., Kashiwabuchi, N., Mori, H., Sakimura, K., Kushiya, E., Araki, K., Meguro, H., Masaki, H., Kumanishi, T., Arakawa, M. and Mishina, M., 1992, Molecular diversity of the NMDA receptor channel, Nature 358:36.

Lima-Landman, M.T.R. and Albuquerque, E.X., 1989, Ethanol potentiates and blocks NMDA-activated single-channel currents in rat hippocampal pyramidal cells, FEBS Lett. 247:61.

Lovinger, D.M., White, G. and Weight, F.F., 1989, Ethanol inhibits NMDA-activated ion current in hippocampal neurons, Science 243:1721.

Lovinger, D.M., White, G. and Weight, F.F., 1990, NMDA receptor-mediated synaptic excitation selectively inhibited by ethanol in hippocampal slice from adult rat, J. Neurosci. 10:1372.

Lovinger, D.M., 1991, Ethanol (EtOH) potentiates ion current mediated by 5-HT_3 receptors in NCB-20 cells, Alcoholism Clin. Exper. Res. Abstr. 14:313.

Lovinger, D.M. and White, G., 1991, Ethanol potentiation of 5-hydroxytryptamine$_3$ receptor-mediated ion current in neuroblastoma cells and isolated adult mammalian neurons, Mol. Pharmacol. 40:263.

Lovinger, D.M. and Peoples, R.W., 1991, Ethanol (EtOH) Potentiation of 5-HT_3 receptor-mediated ion current: Comparison to actions of other sedative/hypnotic agents, Alcohol Clin. Exper. Res. 15:325.

McLarnon, J., Sawyer, D. and Baimbridge, K., 1991, Actions of intermediate chain-length n-alkanols on single channel NMDA currents in rat hippocampal neurons, in: Molecular and Cellular Mechanisms of Alcohol and Anesthetics, Annals N.Y. Acad. Sci. Vol. 625, E. Rubin, K.W. Miller, and S.H. Roth, eds., N.Y. Acad. Sci., New York, N.Y., pp. 283-286.

Maricq, A.V., Peterson, A.S., Brake, A.J., Myers, R.M. and Julius, D., 1991, Primary structure and functional expression of the 5HT_3 receptor, a serotonin-gated ion channel, Science 243:432.

Mehta, A.K. and Ticku, M.K., 1988, Ethanol potentiation of GABAergic transmission in cultured spinal cord neurons involves gamma-aminobutyric acidA-gated chloride channels, J. Pharmacol. Exp. Therap. 246:558.

Miljkovic, Z. and MacDonald, J.F., 1986, Voltage-dependent block of excitatory amino acid currents by pentobarbital, Brain Res. 376:396.

Miller, K.W., Wood, S.C., Forman, S.A., Bugge, B., Hill, W.A.G. and Abadji, V., 1991, The nicotinic acetylcholine receptor in its membrane environment, in: Molecular and Cellular Mechanisms of Alcohol and Anesthetics, Annals N.Y. Acad. Sci. Vol. 625, E. Rubin, K.W. Miller, and S.H. Roth, eds., N.Y. Acad. Sci., New York, N.Y., pp. 600-615.

Monyer, H., Sprengel, R., Schoepfer, R., Herb, A., Higuchi, M., Lomeli, H., Burnashev, N., Sakmann, B. and Seeburg, P.H., 1992, Heteromeric NMDA receptors: Molecular and functional distinction of subtypes, Science 256:1217.

Moriyoshi, K., Masayuki, M., Ishii, T., Shigemoto, R., Mizuno, N. and Nakanishi, S., 1991, Molecular cloning and characterization of the rat NMDA receptor, Nature 354:31.

Morrisett, R.A., Rezvani, A.H., Overstreet, D., Janowsky, D.S., Wilson, W.A., Swartzwelder, H.S., 1990, MK-801 potently inhibits alcohol withdrawal seizures in rats, European J. Pharmacol. 176:103.

Murrell, R.D. and Haydon, D.A., 1991, Actions of n-alcohols on nicotinic receptor ion channels in cultured rat muscle cells, in: Molecular and Cellular Mechanisms of Alcohol and Anesthetics, Annals of the New York Academy of Sciences, Vol. 625, E. Rubin, K.W. Miller, and S.H. Roth, eds., N.Y. Acad. Sci., New York, NY, pp. 365-374.

Nakahiro, M., Arakawa, O. and Narahashi, T., 1991, Modulation of gamma-aminobutyric acid receptor-channel complex by alcohols, J. Pharmacol. Exper. Therap. 259:235.

Nicoll, R.A., 1972, The effects of anaesthetics on synaptic excitation and inhibition in the olfactory bulb, J. Physiol. 223:803.

Okada, K., 1967, Effects of alcohols and acetone on the neuromuscular junction of frog, Japanese J. Physiol., 17:245.

Olsen, R.W., 1982, Drug interactions at the GABA receptor ionophore complex, Ann. Rev. Pharm. Toxicol. 22:245.

Peoples, R.W., Lovinger, D.M. and Weight, F.F., 1990, Inhibition of excitatory amino acid currents by general anesthetic agents, Soc. Neurosci., Abst. 16:1017.

Peoples, R.W. and Weight, F.F., 1991, Modulation of amino acid activated ion currents by trichloroethanol, IBRO World Congress of Neurosci. Abstr. 3:63.

Peoples, R.W. and Weight, F.F., 1992, Ethanol inhibition of N-methyl-D-aspartate-activated ion current in rat hippocampal neurons is not competitive with glycine, Brain Res. 571:342.

Rabe, C.S. and Tabakoff, B., 1991, Glycine site-directed agonists reverse the actions of ethanol at the N-methyl-D-aspartate receptor, Mol. Pharm. 38:753.

Ransom, B.R. and Barker, J.L., 1976, Pentobarbital selectively enhances GABA-mediated postsynaptic inhibition in tissue cultured mouse spinal neurons, Brain Res. 114:530.

Robledo, P., Kaneko, W. and Ehlers, C.L., 1991, Combined effects of ethanol and MK-801 on locomotor activity in the rat, Pharmacol. Biochem. Behav. 39:513.

Samson, H.H. and Harris, R.A., 1992, Neurobiology of alcohol abuse, Trends Pharmacol. Sci. 13:206.

Schmidt, R.F., 1963, Pharmacological studies on the primary afferent depolarization of the toad spinal cord, Pflügers Archiv. 277:325.

Sellers, E.M., Kaplan, H.L., Lawrin, M.O., Somer, C.A., Naranjo, C.A. and Frecker, R.C., 1988, The 5-HT_3 antagonist GR38032F decreases alcohol consumption in rats, Soc. Neurosci. Abstr. 14:41.

Suzdak, P.D., Schwartz, R.D., Skolnick, P. and Paul, S.M., 1986, Ethanol stimulates gamma-aminobutyric acid receptor-mediated Cl- transport in rat brain synaptoneurosomes, Proc. Natl Acad. Sci. U.S.A. 83:4071.

Study, R.E. and Barker, J.L., 1981, Diazepam and (-)-pentobarbital: Fluctuation analysis reveals different mechanisms for potentiation of gamma-aminobutyric acid responses in cultured central neurons, Proc. Nat'l Acad. Sci. U.S.A. 78:7180.

Vicini, S., Mienville, J.-M. and Costa, E., 1987, Actions of benzodiazepine and beta-carboline derivatives on gamma-aminobutyric acid-activated chloride channels recorded from membrane patches of neonatal rat cortical neurons in culture, J. Pharm. Exper. Therap. 243:1195.

Wafford, K.A., Burnett, D.M., Dunwiddie, T.V. and Harris, R.A., 1990, Genetic differences in the ethanol sensitivity of $GABA_A$ receptors expressed in Xenopus oocytes, Science 249:291.

Wafford, K.A., Burnett, D.M., Leidenheimer, N.J., Burt, D.R., Wang, J.B., Kofuji, P., Dunwiddie, T.V., Harris, R.A. and Sikela, J.M., 1991, Ethanol sensitivity of the $GABA_A$ receptor expressed in Xenopus oocytes requires 8 amino acids contained in the y2L subunit, Neuron 7:27.

Wilson, W.R., Bosy, T.Z. and Ruth, J.A., 1990, NMDA agonists and antagonists alter the hypnotic response to ethanol in LS and SS mice, Alcohol 7:389.

Woodward, J.J. and Gonzalez, R.A., 1990, EtOH inhibition of N-methyl-D-aspartate-stimulated endogenous dopamine release from rat striatal slices: reversal by glycine, J. Neurochem. 54:712.

Wozniak, K.M., Pert, A. and Linnoila, M., 1990, Antagonism of 5-HT_3 receptors attenuates the effects of ethanol on extracellular dopamine, Eur. J. Pharmacol. 187:287.

MECHANISMS THAT MEDIATE ETHANOL-INDUCED INCREASES IN DIHYDROPYRIDINE-SENSITIVE CALCIUM CHANNELS

Robert O. Messing

Gallo Clinic & Research Center
Department of Neurology
University of California
San Francisco, CA 94110
U.S.A.

INTRODUCTION

Chronic alcoholics can develop a hyperexcitable withdrawal syndrome when they abruptly discontinue drinking ethanol. Features of this syndrome include tremor, confusional states, convulsions, delirium, and autonomic dysfunction (Charness et al., 1989). Some manifestations of alcohol withdrawal may result from ethanol-induced changes in the function of voltage-dependent calcium channels, since treatment of rodents made dependent on alcohol with calcium channel antagonists reduces tremors, seizures and mortality during alcohol depravation (Bone et al., 1989, Little et al., 1986, Littleton et al., 1990).

We first reported that ethanol regulates the function of dihydropyridine-sensitive calcium channels. Using the neural cell line PC12, we found that ethanol acutely inhibits voltage-dependent ^{45}Ca uptake through dihydropyridine-sensitive (L-type) channels, and exposure to ethanol for 2-6 days causes a reciprocal increase in uptake measured in cells deprived of ethanol for 15-25 minutes (Messing et al., 1986). This enhanced capacity of cells for calcium influx appears to be due to an increased number of binding sites for dihydropyridines, suggesting that prolonged ethanol exposure increases the density of functional L channels. Similar increases in dihydropyridine binding sites have been reported by others in PC12 cells (Skattebol and Rabin, 1987) and in brain membranes from alcohol-dependent rats (Dolin et al., 1987). This paper reviews our findings regarding mechanisms by which ethanol increases expression of L channels.

ACUTE INHBITION OF L CHANNEL FUNCTION AS A STIMULUS FOR UP-REGULATION OF CHANNEL NUMBER

We first considered whether acute inhibition of L channel function by ethanol triggered a compensatory increase in channel number. We reasoned that if acute inhibition is the trigger for such an adaptive response, then prolonged exposure to other drugs that inhibit L channels should also increase L channel density. However, exposure of PC12 cells for 6 days to the L channel blockers nifedipine (4 nM), verapamil (900 nM), or diltiazem (1900 nM), did not alter calcium uptake or dihydropyridine binding measured in the absence of these antagonists (Marks et al., 1989). Thus, acute inhibition of calcium influx is not the cause of increased expression of L channels following chronic ethanol exposure.

INHIBITION OF PROTEIN KINASE C PREVENTS ETHANOL-INDUCED INCREASES IN L CHANNELS

Increases in calcium channel number have been noted following activation of protein kinase C (PKC) in *Aplysia* bag cell neurons (Strong et al., 1987) and chick skeletal myocytes (Navarro, 1987). In PC12 cells, we found that ethanol-induced increases in calcium uptake are prevented by the kinase inhibitors sphingosine and polymixin B (Fig. 1). In addition, sphingosine prevents ethanol-induced increases in the number of dihydropyridine binding sites (Messing et al., 1990). The effect of sphingosine is reversed by 10 nM phorbol 12,13 dibutyrate (PDBu; Fig. 1), which activates PKC, suggesting that PKC mediates up-regulation of L channels by ethanol. PDBu alone, however, did not increase L channel function (Fig. 1), indicating that additional mechanisms are involved in this effect of ethanol.

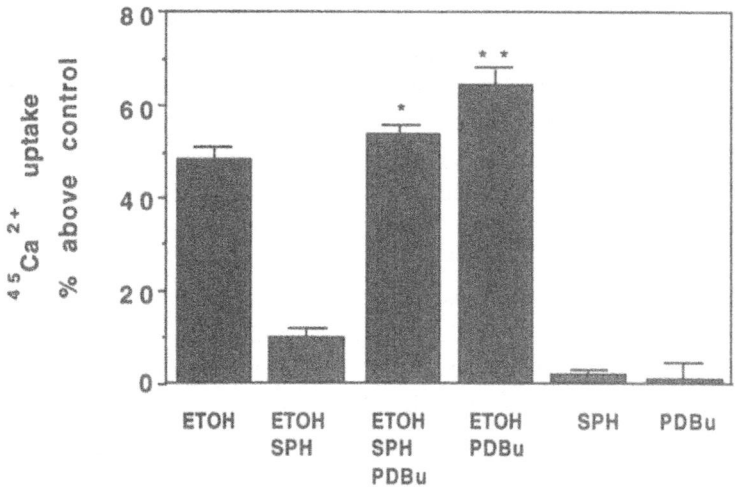

Fig. 1. K+-stimulated calcium uptake in PC12 cells cultured for 4 days in 200 mM ethanol (ETOH), 10 μM Sphingosine (SPH), or 10 nM phorbol 12, 13-dibutyrate (PDBu). * $p < 0.004$ compared with ETOH + SPH; **$p < 0.025$ compared with ETOH. Reprinted with permission from Messing, R.O., Sneade, A.B. and Savidge, B., 1990, *J. Neurochem.* 55:1383.

ETHANOL REGULATES PROTEIN KINASE C BY INCREASING LEVELS OF TWO ISOZYMES

Since PKC inhibitors block up-regulation of calcium channels by ethanol, we investigated whether prolonged exposure to ethanol regulates PKC. Using the peptide KRTLRR, which is selectively phosphorylated by PKC (Heasley and Johnson, 1989), we found that PKC-mediated phosphorylation was increased by $32 \pm 4\%$ in cells treated with 100 mM ethanol for 6 days (Messing et al., 1991). Acute treatment with 100 mM ethanol for 10 minutes did not alter phosphorylation. On the other hand, treatment for 10 minutes with 1 μM phorbol 12-myristate, 13-acetate (PMA), which potently activates PKC, increased PKC-mediated phosphorylation by $177 \pm 22\%$. The combination of ethanol (100 mM for 6 days) and PMA (1 μM for 10 min) was additive ($205 \pm 25\%$ above untreated control; $p < 0.01$ compared to PMA alone).

Since the effects of ethanol and PMA were additive and followed different time courses, it is likely that these agents stimulate PKC by different mechanisms. Phorbol esters, like diacylglycerols, activate PKC by binding to the enzyme and enhancing its affinity for calcium and phospholipid (Nishizuka, 1986). However, we found that ethanol did not alter the EC_{50} for calcium, phosphatidylserine, or diacylglycerol (DG), nor did ethanol change the K_m or V_{max} for histone phosphorylation (Messing et al., 1991). We also found that DG levels are not altered by treatment with 100 or 200 mM ethanol for 6 days or 200 mM for 2 min (Messing et al., 1991). Moreover, levels of free intracellular calcium in PC12 cells are not altered by treatment with 100 mM ethanol (Rabe and Weight, 1988). Therefore, ethanol does not appear to activate PKC by increasing levels of cofactors or by enhancing the affinity of the enzyme for cofactors or substrate.

Another mechanism by which ethanol might activate PKC is by increasing levels of PKC isozymes. To investigate this possibility, we measured PKC activity in detergent-solubilized homogenates of PC12 cells under conditions where activity varies with enzyme concentration. We found that treatment with 25-200 mM ethanol caused a concentration-dependent increase in PKC activity that was maximal ($34 \pm 5\%$ above control) after exposure to 100 mM ethanol. This effect was time-dependent and was maximal after 6 days in cells exposed to 200 mM ethanol. This was associated with an 82% increase in the maximal number of high affinity ($K_d = 11 \pm 2$ nM) binding sites for [^{3}H]PDBu, indicating that the increase in PKC activity was due to an increase in the level of PKC. PKC is a family of isozymes and to date 9 isoforms (α, βI, βII, γ, δ, ϵ, ϵ', ζ, η) have been described (Bacher et al., 1991, Bell and Burns, 1991). Using isozyme-specific antibodies, we found immunoreactivity to α, β, δ, ϵ, and ζ in our clone of PC12 cells. As shown in Fig. 2, treatment with 100 mM ethanol for 6 days increased δ and ϵ immunoreactivity, but did not alter α, β, or ζ immunoreactivity.

These findings suggest that chronic exposure to ethanol increases levels of δ and ϵPKC, and thereby increases PKC-mediated phosphorylation. Similar results have been reported by Obeid et al. (Obeid et al., 1990) in HL-60 cells where treatment with 1,25 dihydroxyvitamin D_3 increases expression of α and βPKC and stimulates PKC-mediated phosphorylation without altering DG levels. Physiologic activation of PKC is generally thought to result from phospholipase-induced generation of diacylglycerol. Our results and those of Obeid et al. suggest that in some tissues, increased expression of PKC isozymes is another mechanism by which certain drugs and hormones regulate PKC activity.

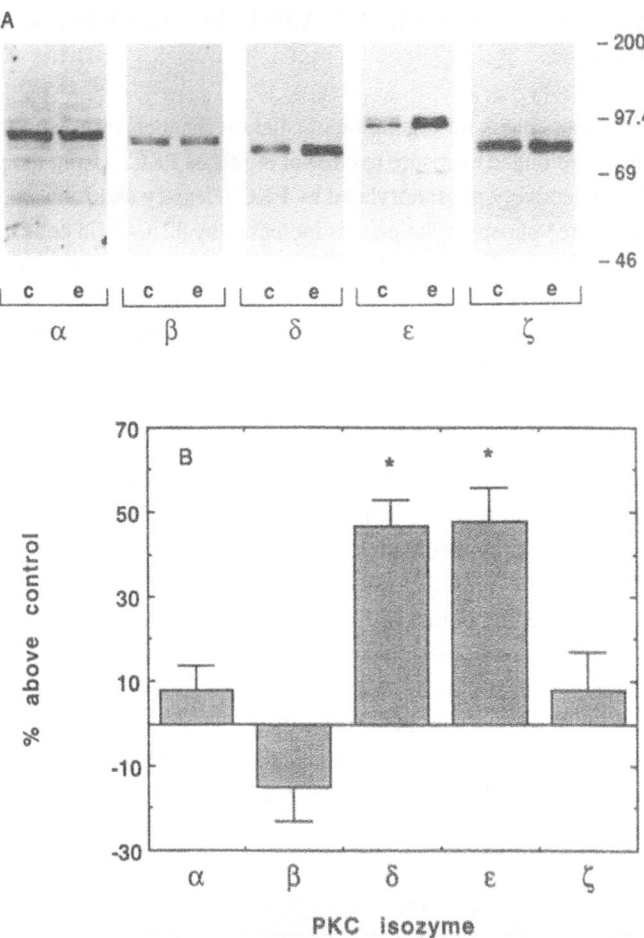

Fig. 2. PKC immunoreactivity in ethanol-treated cells. *A*. Western blots from cells cultured with (e) or without (c) 100 mM ethanol for 6 days. *B*. Mean ± SE values from 3-7 experiments. * p < 0.001 for δ (n=6) and ε (n=7). p > 0.19 for other conditions. Reproduced with permission from Messing, R.O., Petersen, P.J. and Henrich, C.J., 1991, *J. Biol. Chem.* 266:23428.

cAMP DOES NOT MEDIATE ETHANOL'S EFFECT ON CALCIUM CHANNELS

Since treatment with 10 nM PDBu for several days failed to increase L channel activity (Fig. 1), additional mechanisms besides activation of PKC must be involved in up-regulation of calcium channels by ethanol. In hippocampal neurons and heart muscle, L channel function is enhanced by cAMP (Gray and Johnston, 1987, Reuter et al., 1986). Acute ethanol exposure increases cAMP levels and after several days of ethanol exposure, cells generate larger increases in cAMP upon rechallenge with ethanol (Rabin, 1990). However, we found that treatment with dibutyryl cAMP or 8-bromoadenosine cAMP did not increase K+-stimulated calcium uptake (Table 1). Furthermore, the combination of 8BrcAMP and 3 nM PMA markedly decreased rather than increased uptake. This suggests that cAMP does not mediate increases in L channel number following ethanol exposure. Further studies are required to identify the other mechanisms that, together with PKC activation, increase L channel density.

Table 1. Calcium uptake after treatment with cAMP analogues.

Treatment	Duration	^{45}Ca uptake, % of control
Dibutyryl cAMP, 1 mM	4 days	101 ± 7
8-Bromoadenosine cAMP, 1 mM	4 days	86 ± 7
8-Bromoadenosine cAMP 1 mM, PMA, 3 nM	6 days	45 ± 3

ACKNOWLEDGEMENTS

This work was supported by grants from the National Institue on Alcohol Abuse and Alcoholism, the National Institute of Neurological Disorders and Stroke, the Alcoholic Beverage Medical Research Foundation and by Basil O'Connor Starter Scholar Research Award no. 5-696 from the March of Dimes Birth Defects Foundation.

REFERENCES

1. Bacher, N., Zisman, Y., Berent, E. and Livneh, E., 1991, Isolation and characterization of PKC-L, a new member of the protein kinase C-related gene family specifically expressed in lung, skin, and heart, *Mol Cell Biol.* 11:126.
2. Bell, R.M. and Burns, D.J., 1991, Lipid activation of protein kinase C, *J Biol Chem.* 266:4661.
3. Bone, G.H., Majchrowicz, E., Martin, P.R., Linnoila, M. and Nutt, D.J., 1989, A comparison of calcium antagonists and diazepam in reducing ethanol withdrawal tremors, *Psychopharmacology.* 99:386.
4. Charness, M.E., Simon, R.P. and Greenberg, D.A., 1989, Ethanol and the nervous system, *N Engl J Med.* 321:442.
5. Dolin, S., Little, H., Hudspith, M., Pagonis, C. and Littleton, J., 1987, Increased dihydropyridine-sensitive calcium channels in rat brain may underlie ethanol physical dependence, *Neuropharmacology.* 26:275.
6. Gray, R. and Johnston, D., 1987, Noradrenaline and B-adrenoceptor agonists increase activity of voltage-dependent calcium channels in hippocampal neurons, *Nature* 327:620.
7. Heasley, L.E. and Johnson, G.L., 1989, Regulation of protein kinase C by nerve growth factor, epidermal growth factor, and phorbol esters in PC12 pheochromocytoma cells, *J Biol Chem.* 264:8646.
8. Little, H.J., Dolin, S.J. and Halsey, M.J., 1986, Calcium channel antagonists decrease the ethanol withdrawal syndrome, *Life Sci.* 39:2059.

9. Littleton, J.M., Little, H.J. and Whittington, M.A., 1990, Effects of dihydropyridine calcium channel antagonists in ethanol withdrawal; doses required, stereospecificity and actions of Bay K 8644, *Psychopharmacology*. 100:387.

10. Marks, S.S., Watson, D.L., Carpenter, C.L., Messing, R.O. and Greenberg, D.A., 1989, Comparative effects of chronic exposure to ethanol and calcium channel antagonists on calcium channel antagonist receptors in cultured neural (PC12) cells, *J Neurochem*. 53:168.

11. Messing, R.O., Carpenter, C.L. and Greenberg, D.A., 1986, Ethanol regulates calcium channels in clonal neural cells, *Proc Natl Acad Sci USA*. 83:6213.

12. Messing, R.O., Petersen, P.J. and Henrich, C.J., 1991, Chronic ethanol exposure increases levels of protein kinase C δ and ε and protein kinase C-mediated phosphorylation in cultured neural cells, *J Biol Chem*. 266:23428.

13. Messing, R.O., Sneade, A.B. and Savidge, B., 1990, Protein kinase C participates in up-regulation of dihydropyridine-sensitive calcium channels by ethanol, *J Neurochem*. 55:1383.

14. Navarro, J., 1987, Modulation of [3H]dihydropyridine receptors by activation of protein kinase C in chick muscle cells, *J Biol Chem*. 262:4649.

15. Nishizuka, Y., 1986, Studies and perspectives of protein kinase C, *Science*. 233:305.

16. Obeid, L.M., Okazaki, T., Karolak, L.A. and Hannun, Y.A., 1990, Transcriptional regulation of protein kinase C by 1,25-dihydroxyvitamin D3 in HL-60 cells, *J Biol Chem*. 265:2370.

17. Rabe, C.S. and Weight, F.F., 1988, Effects of ethanol on neurotransmitter release and intracellular free calcium in PC12 cells, *J Pharmacol Exp Ther*. 244:417.

18. Rabin, R.A., 1990, Chronic ethanol exposure of PC 12 cells alters adenylate cyclase activity and intracellular cyclic AMP content, *J Pharmacol Exp Ther*. 252:1021.

19. Reuter, H., Kokubun, S. and Prod'hom, B., 1986, Properties and modulation of cardiac calcium channels, *J Exp Biol*. 124:191.

20. Skattebol, A. and Rabin, R., 1987, Effects of ethanol on $^{45}Ca^{2+}$ uptake in synaptosomes and in PC12 cells, *Biochem Pharmacol*. 36:2227.

21. Strong, J.A., Fox, A.P., Tsien, R.W. and Kaczmarek, L.K., 1987, Stimulation of protein kinase C recruits covert calcium channels in Aplysia bag cell neurons, *Nature*. 325:714.

ALCOHOL: EFFECTS ON GABA$_A$ RECEPTOR FUNCTION AND GENE EXPRESSION

Molina C. Mhatre and Maharaj K. Ticku

Department of Pharmacology
The University of Texas Health Science Center
7703 Floyd Curl Drive
San Antonio, TX 78284-7764, U.S.A.

The GABA synapse is a site of action for a variety of centrally acting convulsant, anticonvulsant, depressant, and anxiolytic drugs (Barnard, 1988; Olsen and Tobin, 1990; Ticku, 1991). These drugs bind to one of several allosterically linked sites on the GABA$_A$ receptor complex and either enhance or inhibit GABAergic transmission. GABA$_A$ receptor is a member of the ligand-gated ion channel family of receptors (Schofield et al., 1987). There are numerous reports which indicate that ethanol affects the receptors belonging to this family. Several lines of behavioral, electrophysiological and biochemical functional evidences implicate an involvement of the GABA-benzodiazepine receptor system in the action of ethanol in the CNS (reviewed in Ticku, 1990; 1991; Buck and Harris, 1991). Ethanol potentiates GABA-induced ^{36}Cl-flux in cultured embryonic spinal cord neurons (Mehta and Ticku, 1988, 1989), rat cerebral cortical synaptoneurosomes (Suzdak et al., 1986a) and microsacs (Allan and Harris, 1986, 1987). Chronic exposure to ethanol results in tolerance and physical dependence. The mechanism of these actions of ethanol, its tolerance, physical dependence and withdrawal have not been clearly established. Chronic exposure to ethanol might be leading to cellular adaptive changes in the central nervous system, which result in craving due to the requirement of ethanol for normal functioning, which is a physical dependence. This adaptive change in the excitability of neurons due to withdrawal of ethanol administration causes anxiety and seizures. However, which cellular changes lead to the development of tolerance, physical dependence, and withdrawal symptoms are not clear. Ethanol has a pharmacological profile very similar to benzodiazepines and barbiturates, and a partial inverse agonist of the benzodiazepine receptor, Ro15-4513 (ethyl-8-azido-5,6-dihydro-5-methyl-6-oxo-4H-imidazo[1,5-α]-[1,4]benzodiazepine-3-carboxyate) was reported to be a selective antagonist of the effects of pharmacologically relevant concentration of ethanol on GABA-gated chloride influx (Suzdak et al., 1986). Furthermore, behavioral studies have also demonstrated the ability of Ro15-4513 to antagonize some of the behavioral effects following acute administration of ethanol in animals (Suzdak et al., 1986; Ticku and Kulkarni, 1988). It is a known fact that ethanol withdrawal symptoms can be treated with benzodiazepines which potentiate GABAergic transmission. Also, the severity of withdrawal symptoms get enhanced by treatments which reduce GABAergic transmission e.g. picrotoxin, as well as benzodiazepine inverse agonists. Thus, it is reasonable to assume that the molecular and cellular mechanisms underlying some of the behavioral effects of ethanol, including the development of tolerance and dependence, could be explained by understanding its unique interaction with the $GABA_A$-benzodiazepine receptor complex.

We and others have previously reported that chronic ethanol treatment does not alter the binding parameters of benzodiazepine agonists to the central type of receptors (Rastogi et al., 1986). However, since Ro15-4513 is a partial inverse agonist and it reverses ethanol's behavioral and biochemical effects and its binding was shown to be different from that of diazepam and Ro15-1788 (Sieghart et al., 1987), we investigated the effect of chronic ethanol treatment on the binding of [^{3}H]Ro15-4513 in brain regions of the rat as well as primary cultured neurons (Mhatre et al., 1988; Mhatre and Ticku, 1989). Our results demonstrate that chronic ethanol treatment produces an upregulation of binding sites for Ro15-4513 in cerebral cortex and cerebellum of the rat and embryonic spinal cord neurons in primary culture. Similar treatment did not alter the binding of benzodiazepine agonists and antagonists to their receptor sites (Mhatre and Ticku, 1989). The ability of chronic ethanol treatment to selectively enhance [^{3}H]Ro15-4513 binding further supports the role of GABAergic system in the actions of ethanol and the potential role of Ro15-4513 binding sites in alcohol dependence. There are reports that chronic ethanol treatment increases the efficacy of benzodiazepine inverse agonists e.g. DMCM (methyl-6,7-dimethoxyl-4-ethyl –β-carboline-3-carboxylate) and Ro15-4513.

Furthermore, chronic ethanol administration has also been found to decrease the efficacy of GABA-induced ^{36}Cl-influx in synaptoneurosomes (Morrow et al., 1988) and cultured spinal cord neurons (Mhatre and Ticku, unpublished data). The maximal response to GABA was significantly lowered in the chronically ethanol treated primary cultures of spinal cord neurons compared to the untreated neurons. Also, Buck and Harris (1990a) in their study reported that chronic ethanol exposure (50 mM) of Xenopus oocytes microinjected with mouse brain mRNA and treated chronically with ethanol reduces the amplitude of current activated by muscimol (Buck and Harris, 1990a). Thus, chronic ethanol induces subsensitivity of the $GABA_A$ receptor, which could be related to the phenomenon of ethanol dependence, tolerance as well as ethanol withdrawal; and may be a consequence of a change in the $GABA_A$ receptor gene transcription or altered post translational events in the expression of receptor.

The $GABA_A$-benzodiazepine receptor is a complex protein consisting of several homologous membrane spanning glycoprotein subunits (α, β, γ, δ) (Schofield et al., 1987; Levitan et al., 1988a, b; Pritchett et al., 1989). The available evidence suggests that this receptor is a hetero oligomer of Mr=220-400 KDa, formed by the assembly of presumably five polypeptides with unknown stoichiometry (Olsen and Tobin, 1990). Numerous studies have indicated structural as well as functional heterogeneity of the $GABA_A$ receptor in different regions of the brain; which is related to differential expression of the α variants within the CNS (Wisden et al., 1988, 1989). These α-subunit mRNAs when expressed in combination with the β subunit mRNA exhibit different pharmacological properties (Levitan et al., 1988b). In order to investigate a molecular basis for chronic ethanol induced changes in the $GABA_A$ receptor, we evaluated steady state levels of mRNA for $α_1$, $α_2$, $α_3$, $α_5$ and $α_6$ subunits of the receptor. Chronic ethanol was administered by intragastric intubation method. The $GABA_A$ receptor α subunit mRNA levels were measured in the poly (A+) RNA fraction purified from samples of total RNA individually isolated from cerebral cortex and cerebellum of ethanol treated, ethanol withdrawn (24 and 36 hrs) and control rats. Following chronic ethanol treatment, the mRNA levels for the $α_1$ (4.2 and 3.8 Kb), the $α_2$ (6 and 3 Kb), and the $α_5$ (2.8 Kb; sequence in Ymer et al., 1989) subunits were found to be decreased (40%) in cerebral cortex and these changes were found to persist following 24 hr ethanol withdrawal, and returned to control values at 36 hr withdrawal (Mhatre and Ticku, 1992). The α_3 subunit mRNA level remained unchanged. Similarly, the same treatment reduced the α_1 subunit mRNA level in cerebellum (Ticku and Mhatre 1991; Ticku et al., 1992a; Mhatre and Ticku, 1992). There was no change in the poly (A)+ RNA levels, which rules out a generalized change in the mRNA turnover. Montpied et al. (1991b) have also observed a similar reduction in the α_1 and the α_2 subunit mRNA levels in the cerebral cortex, following chronic ethanol treatment by vapor inhalation method. This downregulation of mRNA levels of the α-subunits could be associated with the subsensitivity of GABAergic transmission which might lead to the tolerance. However, this reduction in mRNA levels for the receptor is not reflected in the changes in the number of binding sites but it is correlated with the responsiveness of the receptor. It could be hypothesized that the differences in the coupling mechanism of these sites rather than the number of sites might be the cause of functional disturbance induced by chronic treatment, as supported by the findings of Harris and Allan (1989).

Ro15-4513 is reported to be a selective antagonist of the sedative and intoxicating effects of ethanol (Suzdak et al., 1986a), and is a partial inverse agonist for the benzodiazepine receptor. Recently, Luddens and coworkers (1990) have discovered a new α subunit variant (α_6) present in cerebellar granular cells. The recombinant receptors composed of the α_6, β_2 and γ_2 subunits bind with high affinity to the GABA agonist [^{3}H]muscimol and [^{3}H]Ro15-4513 but not the other benzodiazepines. We observed a marked increase in the α_6 subunit mRNA level in the cerebellum, following chronic ethanol treatment (Figure 1) which is consistent with an increase in the number of binding sites and photolabeling of [^{3}H]Ro15-4513 to the P_{55} (~56 K) band in the cerebellum following chronic ethanol treatment and withdrawal (Mhatre et al., 1988; Ticku and Mhatre, 1991; Mhatre and Ticku, 1992). The increase in the α_6 mRNA might be related to the increased labelling of P_{55} and partially responsible for the increased binding in cerebellum. Photoaffinity studies indicated a significant increase in P_{41}, P_{50} protein components in the cerebral cortex and P_{50} and P_{56} components in the cerebellum following ethanol treatment (Mhatre and Ticku, 1992). Thus, it is feasible that the increased binding observed in P_{50} and P_{41} components in the cerebral cortex and P_{50} component in the cerebellum may be due to increased levels of some other untested subunit polypeptide (possibly α_4 subunit mRNA). Though the α_6 subunit is not expressed in cerebral cortex, diazepam insensitive [^{3}H]Ro15-4513 binding sites are present in this region (Turner et al., 1991; Mhatre and Ticku, unpublished observation). Behavioral studies also support these data as the chronic ethanol treatment was found to increase the receptor sensitivity to Ro15-4513 selectively (Mehta and Ticku, 1989). Acute exposure to ethanol was also found to enhance the ability of benzodiazepine inverse agonist (DMCM) in reduction of muscimol activated ^{36}Cl- uptake in the membrane isolated from mouse cerebral cortex (Buck and Harris, 1990b). From our data, it appears that the interaction between ethanol and Ro15-4513 is unique and mediated by a change in the subunit expression which might be leading to functionally different receptors. However, it is important to note that since the α_6 subunit is unique to cerebellum and other GABA antagonists can reverse many of the ethanol's effects, other $GABA_A$ isoreceptors are also important in the actions of ethanol. After longer ethanol treatment, a change in the expression of the receptor subunits might lead to a decrease in sensitivity in GABAergic transmission, causing symptoms of withdrawal and dependence. This might explain both the development of tolerance and the phenomenon of kindling and seizures produced during withdrawal.

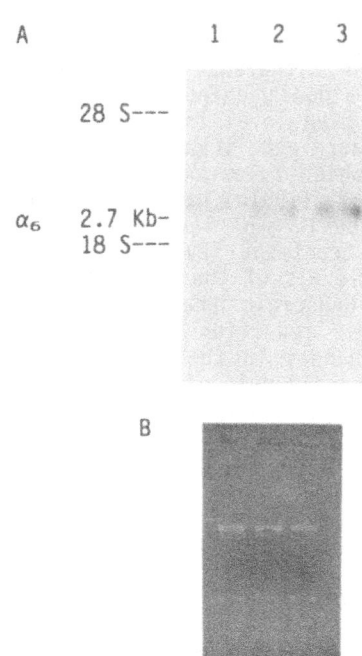

Figure 1.   Effect of chronic ethanol treatment and withdrawal on the GABA$_A$ receptor α_6 subunit mRNA level in rat cerebellum. Experiments were performed as described elsewhere (Mhatre and Ticku, 1992). The lanes represent control (1), ethanol treated (2), and ethanol withdrawn (3). A photograph of the ethidium bromide stained gel prior to transfer, to demonstrate normalized loading of the samples is presented below the autoradiograph (B).

Ethanol and other drugs reacting at the GABA$_A$ receptor sites such as the benzodiazepines and barbiturates share a number of pharmacological properties like the development of cross tolerance and cross-dependence following chronic administration of these drugs (Boisse, H.N. and M. Okamoto, 1980; Le, A.D. et al., 1986). Benzodiazepines and barbiturates bind to the GABA$_A$ receptors and enhance GABAergic neurotransmission. It is interesting that both chronic benzodiazepine and ethanol treatments produce a decrease in the GABA$_A$ receptor gene expression. Chronic benzodiazepine treatment has also been reported to produce downregulation of the α_1 subunit mRNA of the GABA$_A$ receptor (Miller et al., 1988; Heninger et al., 1989). We and other workers found marked reductions in the levels of the GABA$_A$ receptor α subunit mRNAs in primary cultured neurons exposed to GABA (Ticku et al., 1992; Montpied et al., 1991a). These reports support the hypothesis that downregulation of the α transcripts following chronic ethanol, benzodiazepine or GABA treatment could be a cellular adaptive response to potentiation of chloride conductance due to these drugs. Interestingly, though barbiturates also enhance chloride ion conductance, chronic administration of pentobarbital fails to alter the GABA$_A$ receptor α_1 subunit mRNA level (Morrow et al., 1990). Thus it appears that subunit composition required for modulation by particular drug could be a critical factor in this phenomenon. When expression of brain mRNA or cDNA in Xenopus oocytes was used to determine what subunits of the GABA$_A$ receptor are required for modulation by barbiturates, antisense oligonucleotides to the α_1, β_1, γ_1 or $\gamma_2S + 2L$; γ_2L or γ_3 subunits did not alter GABA action or enhancement by pentobarbital. Probably, determination of other subunit mRNA levels following chronic pentobarbital administration will provide an answer for this.

Recently, the γ_2 subunit was found to be particularly important for enhancement of GABA action by benzodiazepines (Pritchett et al., 1989; Sigel et al., 1990). Wafford and coworkers (1991) have shown that enhancement of GABA action by ethanol is dependent upon 8 amino acids contained in the intracellular loop of the γ_2L subunit. In their study, using the complex mixture of brain mRNA, hybridization of only the γ_2L subunit with complementary oligonucleotide was sufficient to block ethanol action. However, the resistance of the GABA_A receptor of short sleep mice to ethanol was not found to be due to a lack of γ_2L subunit (Buck et al., 1991).

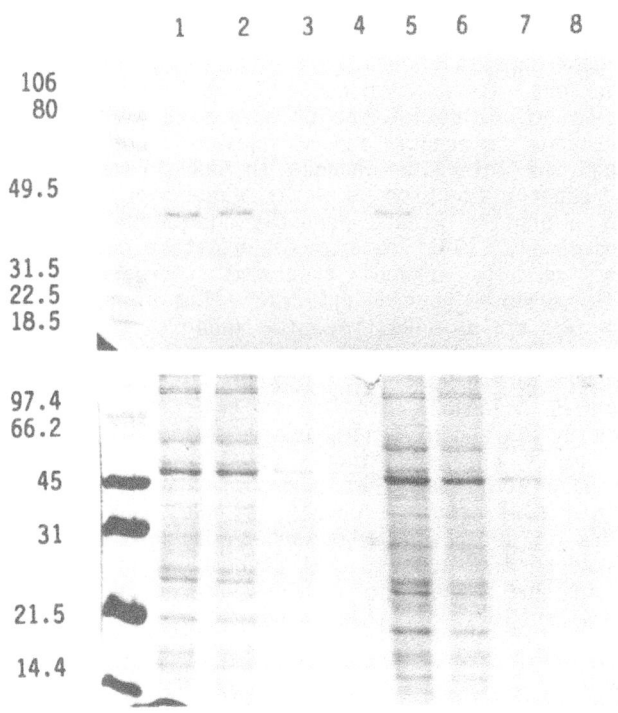

Figure 2. Quantitative immunoblot analysis of the α_1 subunit polypeptide in cerebellar membrane of vehicle and ethanol treated rats. Decreasing amounts of cerebellar membrane were loaded in the different lanes as follows: lanes 1 and 5 = 12 μg, lanes 2 and 6 = 6 μg, lanes 3 and 7 = 3 μg, lanes 4 and 8 = 1.5 μg. The lanes represent vehicle (1 – 4) and ethanol (5 – 8) treated groups. The α_1 subunit band is seen at 51 KDa.

To assess whether there was any change in the concentration of the α subunit proteins encoded by these subunit mRNAs following chronic ethanol treatment, we investigated the expression of α_1, α_2 and α_3 subunits of GABA_A receptor by immunoblotting using polyclonal antibodies raised against α_1, α_2 and α_3 subunit polypeptides (Zezula et al., 1991).

Comparison between the control and ethanol treated rats revealed 61% reduction in the α_1 subunit (51 KDa), 47% reduction in the α_2 subunit (53 KDa) and 30% reduction in α_3 subunit polypeptide (59 KDa) and 58% reduction in the lower band (54 KDa) of α_3 subunit in the cerebral cortex and 46% reduction in the α_1 subunit (51 KDa) in the cerebellum (Fig. 2; Mhatre et al., 1992). The α_2 and α_3 subunits were not detectable in the cerebellum. In conclusion, these data indicate that chronic ethanol treatment alters the expression of GABA$_A$ receptor subunit proteins which could underlie changes in GABA$_A$ receptor sensitivity.

Thus the change in the function of the GABA$_A$ receptor may be associated with an alteration in the α-subunit mRNA as well as protein levels, which could involve both transcriptional and post-transcriptional elements. The increased binding of [^{3}H]Ro15-4513 in the cerebellum is partly due to increased level of α_6 subunit mRNA. Also, the available data indicate the possible existence of another subunit(s) which could be important in understanding the mechanism of chronic ethanol action. The changes observed in mRNA steady state levels of a subpopulation of the GABA$_A$ receptors could be the consequence of the altered regulation of the rate of transcription or processing of the transcripts or changes in mRNA stability. Thus, chronic ethanol treatment and its withdrawal alters GABA$_A$ receptor gene expression which could be related to cellular adaptation to the functional disturbance caused by ethanol.

These results do not define the primary site of ethanol action, but suggest that ethanol tolerance and withdrawal effect could be partly explained through the molecular changes in GABA$_A$ receptor. How ethanol enhances GABA$_A$ receptor function is still a question. Whether it alters the activity of a protein kinase probably linked to phosphorylation of the γ_2L (Wafford et al., 1991) remains to be determined.

In summary, chronic ethanol treatment increased the binding of [^{3}H]Ro15-4513 in cerebral cortex and cerebellum of rats and produced a reduction in the α_1, α_2, α_5 GABA$_A$ receptor subunits mRNAs in the cerebral cortex and α_1 subunit mRNA in the cerebellum, and an increase in α_6 subunit mRNA in the cerebellum. An increase in the α_6 subunit mRNA which selectively encodes Ro15-4513 binding in the cerebellum is consistent with an increase in the photolabeling of [^{3}H]Ro15-4513 to the 55 KDa band in the cerebellum of chronic ethanol treated rats. Using polyclonal antibodies, we have confirmed that chronic ethanol treatment decreased the α_1 subunit (51 KDa), the α_2 subunit (53 KDa) and the α_3 subunit (59 KDa) polypeptides in cerebral cortex and α_1 subunit polypeptide in the cerebellum. These results suggest that chronic ethanol treatment produces a decrease in the expression of GABA$_A$ receptor subunit which may underlie a molecular basis for ethanol tolerance.

ACKNOWLEDGEMENTS

This work was supported by NIAAA grant (AA04090).

REFERENCES

Allan, A.M. and Harris, R.A., 1986, Gamma-aminobutyric acid and alcohol actions: neurochemical studies of long sleep and short sleep mice, Life Sci. 39, 2005-2015.

Allan, A.M. and Harris, R.A., 1987, Acute and chronic ethanol treatments alter GABA receptor-operated chloride channels, Pharmacol. Biochem. Behav. 27, 665-670.

Barnard, E.A., 1988, The structure of the GABA/benzodiazepine receptor complex with its gated ion channel, in: GABA and Benzodiazepine Receptors, Vol. II, R.P. Squires, ed., CRC Press, Boca Raton, FL.

Buck, K.J. and Harris, R.A., 1990a, Chronic ethanol exposure of Xenopus oocytes expressing mouse brain mRNA reduces GABA receptor-activated current and benzodiazepine receptor ligand modulation, Neurosci. Abstrs. 16, 260.

Buck, K.J. and Harris, R.A., 1990b, Benzodiazepine agonist and inverse agonist actions on $GABA_A$ receptor-operated chloride channels. I. Acute effects of ethanol, J. Pharmacol. Exp. Ther. 253, 706-712.

Buck, K.J. and Harris, R.A., 1991, Neuroadaptive responses to chronic ethanol. Alcohol: Clin. Exp. Res. 15, 460-470.

Buck, K.J., Hahner, L., Sikela, J. and Harris, R.A., 1991, Chronic ethanol treatment alters brain levels of γ-aminobutyric $acid_A$ receptor subunit mRNAs: relationship to genetic differences in ethanol withdrawal seizure severity, J. Neurochem. 57, 1452-1455.

Harris, R.A. and Allan, A.M., 1989, Genetic differences in coupling of benzodiazepine receptors to chloride channels, Brain Res. 490, 26-32.

Heninger, C., Saito, N., Tallman, J.F., Duman, R.S., Garrett, K.M., Vitele, M.P. and Gallager, D.W., 1989, Effects of continuous diazepam administration on $GABA_A$ subunit mRNA in rat brain, Neurosci. Abstrs. 15, 831.

Levitan, E.S., Blair, L.A.C., Dionne, V.E. and Barnard, E.A., 1988a, Biophysical and pharmacological properties of cloned $GABA_A$ receptor subunits expressed in Xenopus oocytes, Neuron 1, 773-781.

Levitan, E.S., Schofield, P.R., Burt, D.R., Rhee, L.M., Wisden, W., Konler, M., Fujita, N., Rodriguez, H.F., Stephenson, F.A., Darlison, M.A., Barnard, E.A. and Seeburg, P.H., 1988b, Structural and functional basis for $GABA_A$ receptor heterogeneity, Nature 335, 76-79.

Ludden, H.D.B., Pritchett, M., Kohler, I., Killish, K., Keinanen, H., Monyer, H., Sprengel, R., and Seeburg, P.H., 1990, Cerebellar $GABA_A$ receptor selective for a behavioral alcohol antagonist, Nature (London) 346, 648-665.

Mehta, A.K. and Ticku, M.K. 1988, Ethanol potentiation of GABAergic transmission in cultured spinal cord neurons involves γ-aminobutyric $acid_A$-gated chloride channels, J. Pharmacol. Exp. Ther. 296, 558-564.

Mehta, A.K. and Ticku, M.K. 1989, Chronic ethanol treatment alters the behavioral effects of Ro15-4513, a partially negative ligand for benzodiazepine binding sites, Brain Research 489, 93-100.

Mhatre, M., Mehta, A.K. and Ticku, M.K., 1988, Chronic ethanol administration increases the binding of the benzodiazepine inverse agonist and alcohol antagonist [^{3}H]Ro15-4513 in rat brain, Eur. J. Pharmacol. 153, 141-145.

Mhatre, M. and Ticku, M.K., 1989, Chronic ethanol treatment selectively increases the binding of inverse agonists for benzodiazepine binding sites in cultured spinal cord neurons, J. Pharmacol. Exp. Ther. 251, 164-168.

Mhatre, M.C. and Ticku, M.K., 1992, Chronic ethanol administration alters $GABA_A$ receptor gene expression, Mol. Pharmacol., 49, 415-422.

Mhatre, M.C , Pena, G. and Ticku, M.K., 1992, Antibodies specific for $GABA_A$ receptor α subunits reveal that chronic alcohol treatment downregulates alpha subunit expression in rat brain regions, Submitted.

Miller, L.G., Smith, W.R. and Claycomb, W.D., 1988, Acute benzodiazepine administration alters $GABA_A$ receptor gene expression, Neurosci. Abstrs., 167.

Montpied, P., Ginns, E.I., Martin, B.M., Roca, D., Farb, D.H. and Paul, S.M., 1991a, γ-Aminobutyric acid (GABA) induces a receptor-mediated reduction in $GABA_A$ receptor α subunit messenger RNAs in embryonic chick neurons in culture, J. Biol. Chem. 266, 6011-6014.

Montpied, P., Morrow, A.L., Karanian, J.W., Ginns, E.I., Martin, B.M. and

Paul, S.M., 1991b, Prolonged ethanol inhalation decreases γ-aminobutyric acidA receptor α subunit mRNAs in the rat cerebral cortex, Mol. Pharmacol. 39, 157–163.

Morrow, A.L., Suzdak, P.D., Karanian, J.W. and Paul, S.M., 1988, Chronic ethanol administration alters γ-aminobutyric acid, pentobarbital and ethanol-mediated ^{36}Cl-uptake in cerebral cortical synaptoneurosomes, J. Pharmacol. Exp. Ther. 246, 158–164.

Morrow, A.L., Montpied, P., Lingford-Huges, A. and Paul, S.M., 1990, Chronic ethanol and pentobarbital administration in the rat: effects on GABAA receptor function and expression in brain, Alcohol, 7, 237–244.

Olsen, R.W. and Tobin, A.J. 1990, Molecular biology of GABAA receptors, FASEB J. 4, 1469–1480.

Pritchett, D.B., Sontheimer, H., Shivers, B.D., Ymer, S., Kettenmann, H., Schofield, P.A. and Seeburg, P., 1989, Importance of a novel GABAA receptor subunit for benzodiazepine pharmacology, Nature 338, 582–585.

Rastogi, S.K., Thyagarajan, R., Clothier, J. and Ticku, M.K., 1986, Effect of chronic treatment of ethanol on benzodiazepine and picrotoxin sites on the GABA receptor complex in regions of the brain of the rat, Neuropharmacol. 25, 1179–1184.

Schofield, P.R., Darlison, M.G., Fujita, N., Burt, D.R., Stephenson, F.A., Rodriguez, H., Rhee, L.M., Ramachandran, J., Reale, V., Glencorse, T.A., Seeburg, P.H. and Barnard, E.A., 1987, Sequence: functional expression of the GABAA receptor shows a ligand-gated receptor super-family, Nature (London) 328, 221–227.

Sieghart, W., Eichinger, A., Richards, J.G. and Mohler, H., 1987, Photoaffinity labeling of benzodiazepine receptor proteins with the partial inverse agonist [^{3}H]Ro15-4513: a biochemical and autoradiography study, J. Neurochem. 48, 46–52.

Sieghart, W., 1988, Multiple benzodiazepine binding sites, in: GABA and Benzodiazepine Receptors, Vol. II, R.F. Squires, ed.

Suzdak, P.D., Schwartz, R.D., Skolnick, P. and Paul, S.M., 1986a, Ethanol stimulates γ-aminobutyric acid receptor-mediated chloride transport in rat brain synaptoneurosomes, Proc. Natl. Acad. Sci. U.S.A. 83, 4071–4075.

Suzdak, P., Glowa, J.R., Crawley, J.N., Schwartz, R.D., Skolnick, P. and Paul, S.M., 1986b, A selective imidazodiazepine antagonist of ethanol in the rat, Science 234, 1243.

Ticku, M.K., 1990, Alcohol and GABA-benzodiazepine receptor function, Ann. Med. 22, 241–246.

Ticku, M.K., 1991, Drug modulation of GABAAergic transmission, in: Seminars in the Neurosci. 3, 211–218.

Ticku, M.K., Lowrimore, P. and Lehoullier, P., 1986, Ethanol enhances GABA-induced ^{36}Cl-influx in primary spinal cord cultured neurons, Brain Res. Bull. 17, 123–126.

Ticku, M.K. and Kulkarni, S.K., 1988, Molecular interactions of ethanol with GABAergic system and potential of Ro15-4513 as an ethanol antagonist, Pharm. Biochem. Behav. 30, 501–510.

Ticku, M.K. and Mhatre, M.C., 1991, Chronic ethanol administration induces changes in GABAA receptor gene expression, Neurosci. Abstracts 17, 263(#109.9).

Ticku, M.K., Mhatre, M. and Mehta, A.K., 1992a, Modulation of GABAergic transmission by ethanol, in: GABAergic Synaptic Transmission, Vol. 47, G. Biggio, A. Concas and E. Costa, ed., Raven Press, New York, pp 255–268.

Ticku, M.K., Mhatre, M.C. and Mehta, A.K., 1992b, Chronic GABA treatment downregulates GABAA receptor complex and α mRNA subunits in mammalian cortical neurons. Neurosci. Abstr., in press.

Turner, D.M., Sapp, E.W. and Olsen, R.W. (1991), The benzodiazepine/

alcohol antagonist Ro15-4513: binding to a GABA$_A$ receptor subtype that is insensitive to diazepam, J. Pharmacol. Exp. Ther., 257, 1236-1242.

Wafford, K.A., Burnett, D.M., Leidenheimer, N.J., Burt, D.R., Wang, J.B., Kofuyi, P., Dunwiddie, T.V., Harris, R.A. and Siketa, J.M., 1991, Ethanol sensitivity of the GABA$_A$ receptor expressed in Xenopus oocytes requires 8 amino acids contained in the γ_2L subunit, Neuron, 7, 27-31.

Wisden, W., Morris, B.J., Darlison, M.G., Hunt, S.P. and Barnard, E.A., 1988, Distinct GABA$_A$ receptor α subunit mRNAs show differential patterns of expression in bovine brain, Neuron 1, 937-947.

Wisden, W., Morris, B.J., Darlison, M.G., Hunt, S.P. and Barnard, E.A., 1989, Localization of GABA$_A$ receptor α-subunit mRNAs in relation to receptor subtypes, Mol. Brain Res. 5, 305-310.

Ymer, S., Draughn, A., Kohler, M., Schofield, P.R. and Seeburg, P.H., 1989, Sequence and expression of a novel GABA$_A$ receptor α subunit. Fedn. Eur. Biochem. Soc. Lett. 258, 119-122.

Zezula, J., Fuchs, K. and Sieghart, W., 1991, Separation of α_1, α_2 and α_3 subunits of the GABA$_A$-benzodiazepine receptor complex by immunoaffinity chromatography. Brain Res. 563, 325-329.

CONTROL OF ALCOHOLISM BY TREATMENT WITH SKV, A HERBAL DRUG MIXTURE FROM INDIA

V. Nachiappan, K.R. Shanmugasundaram*, and S.I. Mufti

Department of Pharmacology and Toxicology and The Arizona Cancer Center, University of Arizona, Tucson, AZ 85721 and
*Madras University, India

ABSTRACT

Male Wistar rats were used in this study and were given free access to 25% ethanol in drinking water. The rats thus exposed to alcohol developed addiction in 90 days. At 120 days, a group of rats was given a daily dose of 5.0 ml SKV, a herbal drug mixture that is often used to stop craving and induce voluntary reduction of alcohol consumption. Both control and SKV-treated rats were fed a modified chow diet ad libitum and the experiment was continued for 240 days. SKV administration caused a slow but progressive drop in ethanol consumption. With reduced voluntary intake of ethanol, the rats exhibited normal weight gain and these rats developed no withdrawal symptoms with blood alcohol levels reduced from 180 mg/dl to 63 mg/dl after the drug treatment. An increase in (Na^+-K^+) ATPase and Mg^{2+}-ATPase was observed in all regions except Pons-medulla with alcohol consumption. A decrease was recorded in Ca^{2+}-ATPase in all regions except Pons-medulla and cerebrum. Partial reversal of these changes was observed with ethanol restriction. In summary, changes in brain enzyme activity were observed due to ethanol toxicity and SKV treatment led to a restriction of voluntary ethanol ingestion and consequent correction of some of the abnormalities.

INTRODUCTION

Alcohol abuse and dependence is a major health problem. A community based study in the U.S. carried out from 1981 to 1983 using Diagnostic and Statistical Manual (DSM) criteria by the American Psychiatric Association (1987) indicated that approximately 13% of the adult population has had alcohol abuse and dependence at some time in their life. The proportion of population that is alcohol dependent is even larger in some other countries. With a problem affecting such a proportion of population, therefore, it is important to devise strategies that would help reduce alcohol dependence in the afflicted individuals.

According to the DSM III R criteria alcohol dependence is characterized by excessive use of alcohol and the loss of ability to reduce or control it. It is imperative, therefore, that due emphasis be given to devices that could help reduce craving for alcohol and becoming dependent on it. Using SKV, a brew of herbal drugs in India presents such an attempt. This drug mixture was formulated by Shanmugasundaram and Shanmugasundaram (1986). Previous animal studies have shown that SKV use leads to a slow but progressive voluntary reduction of ethanol consumption and lowers blood ethanol levels that could be well tolerated by rats without ensuing withdrawal symptoms (Prabakaran et al., 1988). This report examines effects of this drug on ionic transport mechanisms in the brain that may play a significant role in alcohol dependence.

MATERIALS AND METHODS

Preparation of SKV

Preparation of SKV is given in detail in Shanmugasundaram and Shanmugasundaram (1986). But briefly one gram amounts of *Piper nigrum* Linn. (Piperaceae) seeds, *Piper longum* Linn. (Piperaceae) seeds, *Santalum album* Linn. (Santalaceae) heartwood, *Pterocarpus santalinus* Linn. (Papilonaceae) heartwood, *Nardostachys jatamansi* DC. (Valerianaceae) roots, *Symplocos racemosa* Roxb. (Styraceae) bark, *Andropogen muricatus* Rets. (Gramineae) roots, *Elettaria carcomum* Maton (Scitaminacea) seeds, *Berberis aristata* DC. (Berberideae) root/bark/stem, *Plumbago zeylonica* Linn. (Plumbaginaceae) roots and *Cyprus rotundus* Linn (Cyperaceae) tubers, were soaked in 1 l of water containing 300 g cane sugar, 40 g of raisins and 5 g of *Woodfordia floribunda* Salisb. (Lythraceae) flowers and allowed to ferment for 30 days. At the end of this time the mixture was stirred, strained through cheese cloth and allowed to age for 3 months. The clear brown liquor on analysis contained 1-2 g ethanol, 3-4 g reducing sugar, less than 20 mg total lipids, 25-30 mg nitrogen, 13-16 mg phosphorous, 0.8-1.0 mg iron, 1.0-1.5 mg copper and 0.9-0.95 m Equiv per 100 ml of final product. The pH of SKV was around 5.5.

Animals and Their Treatment

Male rats of wistar strain weighing between 100-120 g were given a semisolid diet containing the purina chow and wheat flour in the ratio of 1:2 (w/w). This diet lowered protein from 21% to 14% and increased carbohydrate content from 55% to 68%. SKV prepared as above was given to a group of rats at a dose of 5 ml/day mixed with a small amount of diet and fed in the mornings. The animals were divided into the following groups.

Group I rats were given the mixed diet and water ad libitum and served as a control group. Part of this group was terminated at 120 days and another part at 240 days. The data obtained from the two parts was pooled, since it did not show a significant difference.

Group II rats were given the same diet as in group I for the first 120 days and from 121-240 days SKV was added to the diet. This group was to assess the toxicity of SKV, if any at 240 days.

Group III rats were given the same modified diet and ethanol (25% v/v) in drinking water ad libitum for 120 days.

Group IV rats were given the same treatment as in group III and terminated at the end of 240 days.

Group V rats were given the same diet as in group III and ethanol continued until 240th day. SKV was given from 121-240 days along with the diet.

Animals were terminated at the end of the experimental period, their cranial cavity was cut open and brain was separated into different regions namely cerebrum, cerebellum, thalamus-hypothalamus, midbrain and pons-medulla as described by Glowinsky and Iversen (1966). Different regions were homogenised in 0.01 M Tris-buffer pH 7.4 and (Na$^+$-K$^+$) ATPase, Mg^{2+} ATPase and Ca^{2+} ATPase were assayed according to Evans (1969) with some modifications as described below.

Assay for Total ATPase

Aliquots of homogenised tissue samples were incubated at 37°C in 0.1 M Tris pH 7.4 containing 0.5 mM ATP (substrate), 5 mM each of sodium chloride, potassium chloride and magnesium chloride. The reaction was arrested by addition of 10% trichloracetic acid and the solution was centrifuged at 3000 rpm for 10 min. Thus the phosphorous liberated stays in the supernatant to which ammonium molybdate and aminonaphthol sulfonic acid (ANSA) are added. The phosphomolybdenum blue color formed was measured at 660 nm in a spectrophotometer as described by Fiske and Subbarow (1925).

Assay for Mg^{2+}-ATPase

For Mg^{2+}-ATPase an assay similar to the above was used excepting replacement of potassium chloride by oubain (5 μM) to inhibit the (Na^+-K^+) ATPase activity.

Assay for (Na^+-K^+) ATPase

Activity of (Na^+-K^+) ATPase was measured by determining the difference between total ATPase activity and Mg^{2+}-ATPase activity.

Assay for Ca^{2+} ATPase Activity

For Ca^{2+}-ATPase assay, calcium chloride (5 mM) was added as the activator and the remainder of the procedure used was the same as in the above.

All enzyme activities were expressed as μmoles of phosphorous liberated/mg protein/hour.

RESULTS

Table 1 represents the distribution of (Na^+-K^+) ATPase in different regions of the brain. Both the control diet (Group I) and the control diet with SKV (Group II) had the same enzyme activity in their respective brain regions indicating that SKV alone did not alter the enzyme levels. When rats consumed alcohol for 120 days (Group III) there was 31%, 68% and 84% increased enzyme activity in cerebellum, midbrain and cerebrum respectively. Continuing alcohol for another 120 days (Group IV) enhanced the enzyme activity to 55% in cerebellum and to a 2-fold increase in mid-brain and cerebrum. SKV treatment partially reversed alterations made by alcohol ingestion. Pons-medulla and thalamus-hypothalamus showed no changes with ethanol consumption.

Table 2 depicts the pattern of Mg^{2+}-ATPase in different brain regions. As shown in Table 1 the enzyme concentration was unchanged in all the brain regions by the presence of SKV in the control diet (Group II). Alcohol administered for 120 days (Group III) had no alterations in the enzyme levels in Pons-medulla, mid-brain and cerebrum, whereas cerebellum and thalamus-hypothalamus showed a 45% increase. Enzyme activity in Pons-medulla remained the same even after prolonged ethanol treatment for 240 days (Group IV), but there was a 20%, 40% and 65% increase in mid-brain, cerebrum and thalamus-hypothalamus respectively. Cerebellum showed a 2-fold increase in the same group. All the regions showed a significant reversal of their enzyme activities towards normal levels with SKV therapy (Group V).

Table 3 illustrates the levels of Ca^{2+}-ATPase in different brain regions. Unlike (Na^+-K^+) ATPase and Mg^{2+}-ATPase, where an increase was observed with ethanol consumption, Ca^{2+}-ATPase showed the opposite effect, i.e. a decrease was noted with alcohol ingestion; cerebrum and cerebellum showed 12% and 14% decrease in Group III rats and there was approximately a 20% decrease in the enzyme activities in cerebellum, cerebrum and thalamus-hypothalamus in Group IV animals. With SKV therapy the enzyme activity was partially reversed in thalamus-hypothalamus and completely reversed in cerebellum. SKV showed no alterations in Ca^{2+}-ATPase in cerebrum. When SKV was administered with the control diet (Group II), Ca^{2+}-ATPase like other enzymes showed no change in enzyme activity in any of the brain regions.

Table 1. Distribution of (Na$^+$, K$^+$)-ATPase in the different regions of the brain in control and experimental animals. Values are expressed in μmoles of Pi liberated/mg protein/hr (mean ± S.D. for six animals in each group).

Regions	Group I	Group II	Group III	Group IV	Group V
Pons-medulla	6.44 ± 0.71	6.60 ± 0.76	6.81 ± 0.77	6.73 ± 0.71	6.30 ± 0.65
Cerebellum	8.61 ± 0.82	8.83 ± 0.91	11.28 ± 1.35**	13.37 ± 1.42***	10.58 ± 1.32[@@]
Mid-brain	7.50 ± 0.81	7.48 ± 0.64	12.62 ± 1.21***	15.61 ± 1.85***	9.38 ± 1.05***[@@@]
Cerebrum	8.86 ± 0.75	9.04 ± 0.86	16.31 ± 1.75***	20.22 ± 2.12***	10.96 ± 1.85***[@@@]
Thalamus-hypothalamus	7.31 ± 0.81	7.28 ± 0.76	7.91 ± 0.65	8.56 ± 0.82*	7.71 ± 0.75

For statistical evaluation of significant variations, Group I compared with Groups II, III and IV (*). Group V is compared with Group IV (@) and Group III (*). Statistically significant alterations are expressed as *p < 0.05; **p < 0.01; ***p < 0.001; @ p < 0.05; @@ p < 0.01; @@@ p < 0.001.

Table 2. Distribution of Mg^{2+}-ATPase in the different regions of the brain in control and experimental animals. Values are expressed in µmoles of Pi liberated/mg protein/hr (mean ± S.D. for six animals in each group).

Regions	Group I	Group II	Group III	Group IV	Group V
Pons-medulla	8.31 ± 0.91	8.62 ± 0.88	8.91 ± 0.85	8.99 ± 0.92	8.21 ± 0.85
Cerebellum	9.65 ± 0.86	9.58 ± 1.05	14.07 ± 1.52***	18.28 ± 1.76***	11.07 ± 1.07**@@@
Mid-brain	9.63 ± 1.06	9.17 ± 0.98	10.83 ± 1.09	11.65 ± 1.25*	9.83 ± 1.07@
Cerebrum	10.42 ± 1.09	10.65 ± 1.42	12.25 ± 1.42	14.65 ± 1.61***	11.09 ± 1.25@@@
Thalamus-hypothalamus	8.90 ± 0.79	9.06 ± 0.95	12.94 ± 1.41***	14.72 ± 1.51***	10.85 ± 1.25*@@@

For statistical evaluation of significant variations, Group I compared with Groups II, III and IV (*). Group V is compared with Group IV (@) and Group III (*). Statistically significant alterations are expressed as *p < 0.05; **p < 0.01; ***p < 0.001; @ p < 0.05; @@ p < 0.01; @@@ p < 0.001.

Table 3. Distribution of Ca^{2+}-ATPase in the different regions of the brain in control and experimental animals. Values are expressed in µmoles of Pi liberated/mg protein/hr (mean ± S.D. for six animals in each group).

Regions	Group I	Group II	Group III	Group IV	Group V
Pons-medulla	6.81 ± 0.58	6.85 ± 0.69	7.01 ± 0.69	7.24 ± 0.71	6.98 ± 0.73
Cerebellum	8.42 ± 0.71	8.56 ± 0.81	7.26 ± 0.73*	6.85 ± 0.69**	8.10 ± 0.84@@
Mid-brain	7.21 ± 0.91	7.28 ± 0.65	7.56 ± 0.75	7.84 ± 0.81	7.45 ± 0.69
Cerebrum	8.65 ± 0.66	8.72 ± 0.78	7.65 ± 0.77*	7.21 ± 0.74**	7.85 ± 0.81
Thalamus–hypothalamus	7.85 ± 0.83	7.74 ± 0.81	6.25 ± 0.58	6.05 ± 0.58**	6.85 ± 0.61@

For statistical evaluation of significant variations, Group I compared with Groups II, III and IV (*). Group V is compared with Group IV (@) and Group III (*). Statistically significant alterations are expressed as *p < 0.05; **p < 0.01; ***p < 0.001; @ p < 0.05; @@ p < 0.01; @@@ p < 0.001.

DISCUSSION

Ethanol due to its amphipathic characteristics, is able to easily penetrate biological membranes and affects the functions of most organs in the body (Wing et al, 1984). The intoxicating effects of ethanol are a result of the influence of ethanol on the central nervous system (Tabakoff et al., 1987).

The generation and maintenance of ionic gradients is a prerequisite for conduction and transmission of information by neurons (Liljequist and Tabakoff, 1986). Major function of (Na^+-K^+) ATPase is to maintain and restore proper ionic gradient across cell membranes, thus controlling neuronal excitability as well as many membrane transport functions (Foster et al., 1989). Chronic ethanol consumption alters the (Na^+-K^+) ATPase activity in the synaptosomal plasma membrane (Shiohara et al., 1984). Specific regions within the neuronal membranes show selective sensitivity to ethanol (Chin and Goldstein, 1981) suggesting that various membrane bound proteins, including enzymes are dependent on their immediately surrounding lipids for optimal activity and differ in their sensitivity to ethanol (Farias et al., 1975).

Chronic exposure to alcohol changes the membrane composition that becomes less easily reversible (Sun and Sun, 1985). The results presented here show that ethanol ingestion increased (Na^+-K^+) ATPase activity in cerebrum, cerebellum, mid-brain to different degrees (Group III and Group IV in Table 1) but had no effect on the enzyme activity in Pons-medulla. SKV drug partially reversed the enzyme levels towards normal levels. These results are in conformity with the observations of Rao et al. (1985) who found significant increase in the activity of (Na^+-K^+) ATPase in cerebral cortex, cerebellum following short term administration of ethanol. Rangaraj and Kalant (1984) found that chronic ethanol administration followed by 24 hour withdrawal increased the basal (Na^+-K^+) ATPase activity to different degrees in cerebral cortex, cerebellum but not in Pons-medulla, hypothalamus and thalamus-midbrain.

Our results also showed that Mg^{2+}-ATPase activity increased in all the regions except Pons-medulla after alcohol treatment for 240 days and SKV treatment reversed the enzyme activity in all the regions to near normal levels (Table 2). A similar increase was observed in the synaptosomal fraction after chronic ethanol treatment by Shiohara et al. (1984).

Ca^{2+}-ATPase regulates calcium homeostasis by actively extruding excess intracellular calcium into the extra cellular mileu (VanBreeman et al., 1978). The results presented here indicate that Ca^{2+}-ATPase activity is decreased in cerebellum, cerebrum and thalamus-hypothalamus to varying degrees with ethanol ingestion (Group IV) which was partially reversed by SKV therapy. Pons-medulla and mid-brain remained unaltered with any of the treatments (Table 3). Ca^{2+}-ATPase decreased in amygdala after chronic ethanol administration (Keane and Leonard, 1986). Ca^{2+}-ATPase may sequester Ca^{2+} ions into intraterminal mitochondria which then act as Ca^{2+} storage vesicles (Michell, 1975).

When calcium ions accumulate inside the cell they potentially activate destructive enzymes that result in serious cellular injury or death (Schanne et al., 1979). Chronic effects of ethanol on synaptosomal calcium buffering mechanisms suggest that ethanol may act through membrane lipids to inhibit calcium stimulated ATP hydrolysis (Ross et al., 1988) resulting in calcium accumulation (Ross, 1977).

Chronic ethanol ingestion leads to development of tolerance and physical dependence in animals and man. Alcohol dependence is associated with biological adaptation which is expressed through changes in cell and organ functions (Sun and Sun, 1985). Alterations in the membrane structure and lipid composition are responsible for the central nervous system dysfunction suffered by alcoholics. However, ethanol tolerance does not affect all regions of the brain uniformly. A single mechanism, such as cell membrane rigidity, is insufficient to explain all the phenomena of ethanol tolerance and its regional differences (Rangaraj and Kalant, 1984). Alcohol effect on the plasma membrane depends on protein modifications (Ledig et al., 1985). Displacement of water by ethanol in the membrane would alter the receptor-protein conformations associated with ion channel functions (Klemm, 1990).

Results from our previous study have shown that the alterations brought by alcohol in the liver, kidney and pancreas were reversed by SKV (Prabakaran et al., 1988) and the blood alcohol levels were reduced from 180 mg/dl (Group IV) to 63 mg/dl with SKV therapy (Group V). Results presented here reveal that ethanol ingestion alters the membrane bound ATPases in various regions of brain to different degrees and the changes are completely or partially reversed by SKV.

Therapeutic drugs are used to control alcohol addiction and antagonise ethanol intoxication. Nitrenidipine, a calcium channel antagonist, when administered for a certain length of time is reported to prevent withdrawal syndrome with mice exposed to alcohol for a long time (Whittington et al., 1991). Craving for alcohol is mediated through the binding of opiate receptors by tetrahydroisoquinolines obtained by the interaction of acetaldehyde with catecholamines in the brain (Olson et al., 1984). When Naloxone and Naltrexone (opiate antagonists) are given to rats that had an access to alcohol they reduced their alcohol consumption (Sinclair, 1990). Similarly, SKV appears to break alcohol dependence in the rats, permit lower blood ethanol levels and reduce drinking while reversing the pathological changes. It might be mediated through the binding to the opiate receptors. Further work is warranted.

ACKNOWLEDGMENTS

The research described here was supported by the Council of Scientific and Industrial Research and Univeristy Grants Commission, New Delhi, India. It was also facilitated by help and discussions from NIH grant No. CA51088 to Dr. Mufti.

REFERENCES

American Psychiatric Association, 1987, Diagnostic and Statistical Manual, in: "Mental Disorders", American Psychiatric Association, Washington.

Chin, J.H., and Goldstein, D.B., 1981, Membrane-disordering action of ethanol: Variation with membrane cholesterol content and depth of the spin label probe, Molec. Pharmacol. 19:425.

Evans, D.J., Jr., 1969, Membrane adenosine triphosphatase of E. Coli activation by Ca^{2+} ions and inhibition by monovalent cations, J. Bacteriol. 100:914.

Farias, R.N., Bloj, B., Morero, R.D., Sineriz, F., Trucco, R.E., 1975, Regulation of allosteric membrane bound enzymes through changes in membrane lipid composition, Biochim. Biophys. Acta, 415:231.

Fiske, C.H., and Subbarow, Y., 1925, The colorimetric determination of phosphorous, J. Biol. Chem., 66:375.

Foster, D.M., Huber, M.D., Klemm, W.R., 1989, Ethanol may stimulate or inhibit (Na^+-K^+) ATPase, depending upon Na^+ and K^+ concentrations, Alcohol, 6:437.

Glowinski, J., and Iversen, L.L., 1966, Regional studies on catecholamine in the rat brain. I. The disposition of $^3[H]$ DOPA in various regions of the brain, J. Neurochem., 13:655.

Goldstein, D.B., and Pal, N., 1971, Alcohol dependence produced in mice by inhalation of ethanol: Grading withdrawal reaction, Sicence, 172:288.

Keane, B., and Leonard, B.E., 1985, The effect of ethanol administration on the activities of synaptosomal Ca^{2+}-ATPases in rat brain (612th meeting, London) in: "Biochem. Soc. Trans.", B. Moreland, ed., The Royal Society, London.

Klemm, W.R., 1990, Dehydration: A new alcohol theory, Alcohol, 7:49.

Ledig, M., Kopp, P., and Mandel, P., 1985, Effect of ethanol on adenosine triphosphatase and enolase activities in rat brain and in cultured nerve cells, Neurochem. Res., 9:1311.

Liljequist, S., and Tabakoff, B., 1986, Neurochemical effects of alcohol, in: "Alcohol and Disease", Morgan, ed, (Churchill Livingstone) Edinburgh.

Michell, R.H., 1975, Inositol phospholipids and cell surface receptor function, Biochim. Biophys. Acta, 415:81.

Olson, G.A., Olson, R.D., and Kastin, A.J., 1984, Endogenous opiates:1983, Peptides, 5:975.

Prabakaran, K., Ramanujam, K.S., Vasanthi, N., Shanmugasundaram, K.R., and Shanmugasundaram, E.R.B., 1988, Control of addiction by SKV therapy - its action on water, food intake, brain function and cell membrane composition, Pharmacol. Res. Commun. 20:99.

Rangaraj, N., and Kalant, H., 1984, Effect of ethanol tolerance on norepinephrine-ethanol inhibition of (Na+-K+) adenosine triphosphatase in various regions of rat brain, J. Pharmac. Exp. Ther., 231:416.

Rao, P.A., Kumari, C.L., and Sadasivu, B., 1985, Acute and short term effects of ethanol on membrane enzymes in rat brain, Neurochem. Res., 10:1577.

Ross, D.H., 1977, "Actions of Alcohol Toxication and Withdrawal", Plenum Publishing Corporation, New York.

Ross, D.H., Garrett, K.M., and Cardenas, H.L., 1985, The effects of lubrol wx on brain membrane Ca^{2+}/Mg^{2+}-ATPase and ATP dependent Ca^{2+} uptake activity following acute and chronic ethanol, Neurochem. Res., 10:283.

Schanne, F.A.X., Kane, A.B., Young, E.E., and Farber, J.L., 1979, Calcium dependence of toxic cell death: A final common pathway, Science, 206:700.

Shanmugasundaram, E.R.B., and Shanmugasundaram, K.R., 1986, An Indian herbal formula (SKV) for controlling voluntary ethanol intake in rats with chronic alcoholism, J. Ethno. Pharmacol., 17:171.

Shiohara, E., Tsukada, M., Chiba, S., Yamazaki, H., and Suchiro, N., 1984, Effect of prolonged ethanol consumption on (sodium and potassium). ATPase activity of rat brain membranes, Arukoru. Kenkyu to Yakubutso. Izon., 19:302.

Sinclair, J.D., 1990, Drugs to decrease alcohol drinking, Ann. Med., 22:357.

Sun, G.Y. and Sun, A.Y., 1985, Ethanol and membrane lipids, Alcohol. Clin. Exp. Res., 9:164.

Tabakoff, B., Hoffman, P.L., and Liljequist, S., 1987, Effects of ethanol on the activity of brain enzymes, Enzymes 37:70.

Van Breeman, C., Aaronson, P., Loutzenhiser, R., and Meisheri, K.D., 1980, Ca^{2+}-movements in smooth muscle, Chest (Suppl. 1), 78:157.

Whittington, M.A., Dolin, S.J., Patch, T.L., Siarey, R.J., Butterworth, A.R., and Little, H.J., 1991, Chronic dihydropyridine treatment can reverse the behavioral consequences of and prevent adaptations to, chronic ethanol treatment, 103:1669.

Wing, D.R., Harvey, D.J., Belcher, S.J., and Paton, W.D.M., 1984, Changes in membrane lipid content after chronic ethanol administration with respect to fatty acyl composition and phospholipid type, Biochem. Pharmacol., 33:1625.

NUCLEOSIDE TRANSPORT AND ETHANOL-INDUCED HETEROLOGOUS DESENSITIZATION

Adrienne Gordon[1,2,3], Mohan K. Sapru[1,3], Sharon W. Krauss[3], and Ivan Diamond[1,2,3]

Department of Neurology[1]
Department of Pharmacology[2]
The Gallo Center[3]
University of California, San Francisco General Hospital
San Francisco, CA 94110

Chronic ethanol exposure causes heterologous desensitization of cAMP production by receptors coupled to $G\alpha_s$, the GTP-binding subunit of G_s (Mochly-Rosen et al., 1988). This occurs in several cultured cell lines as well as in brains from rats chronically exposed to ethanol (Gordon et al., 1986; Nagy et al., 1989; Hoffman and Tabakoff, 1977). Moreover, freshly isolated circulating lymphocytes from alcoholics also show heterologous desensitization when compared to cells from non-drinking controls (Diamond et al., 1987), suggesting that alterations in cAMP production may be important in the pathophysiology of alcoholism. Our studies on the mechanism underlying the decrease in cAMP production indicate that ethanol inhibition of nucleoside transport is a requisite first step in ethanol-induced heterologous desensization. Inhibition of adenosine transport by ethanol results in an extracellular accumulation of adenosine. This adenosine binds to adenosine A_2 receptors causing an increase in intracellular cAMP levels which leads to a cascade of events resulting in heterologous desensitization of cAMP signal transduction (Nagy et al., 1989; Nagy et al., 1990).

ADENOSINE MEDIATES HETEROLOGOUS DESENSITIZATION

When NG108-15 cells, a neuronal x glioma hybrid cell line, are exposed chronically to ethanol, both adenosine receptor- and PGE1 receptor-stimulated cAMP production are reduced from 25-40% compared with control cells (Fig. 1). Heterologous desensitization in other systems is due to continued exposure to agonist, suggesting that ethanol caused the extracellular accumulation of an agonist for a receptor coupled to $G\alpha_S$. Since adenosine has been implicated in mediating some of the effects of ethanol in the central nervous system (Proctor and Dunwiddie, 1984; Dar et al., 1983), we examined the possibility that heterologous desensitization was due to an ethanol-induced extracellular accumulation of adenosine. If adenosine were responsible for chronic ethanol-induced heterologous desensitization, then coincubation of ethanol with adenosine deaminase should prevent heterologous desensitization. Figure 1 shows that heterologous desensitization did not

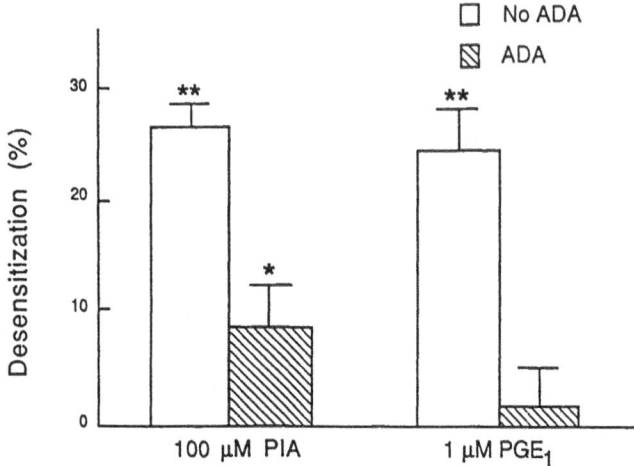

Figure 1. Chronic effects of ethanol on desensitization of receptor-stimulated cAMP production in NG108-15 cells. Cells were maintained for 48 hr with or without 100 mM ethanol in the presence or absence of 1 unit/ml ADA. Cell were washed and incubation with 100 μM PIA or 1 μM PGE₁ for 30 min in the absence of ethanol. cAMP levels were measured by radioimmunoassay. Desensitization is expressed as the percentage decrease in ethanol-treated cells, compared with cells never exposed to ethanol; cells treated with ethanol and ADA are compared with cells treated with ADA alone. Asterisks indicate a significant difference from cells not treated with ethanol (*p < 0.025, **p < 0.001, Students t test). Used with permission, Mol Pharmacol.

occur in NG108-15 cells when adenosine deaminase was present during the ethanol exposure (Nagy et al., 1989). In addition, mutant cells lacking adenosine transporter also should not show heterologous desensitization after chronic exposure to ethanol since adenosine is not released from cells. Indeed, we found that S49 lymphoma mutant cells lacking adenosine transport did not show heterologous desensitization after chronic exposure to ethanol (Nagy et al., 1989). These results support our hypothesis that extracellular adenosine accumulation is required for ethanol-induced heterologous desensitization.

The experiments described above suggest that ethanol causes an extracellular accumulation of adenosine. Such accumulation could either be due to ethanol inhibition of adenosine uptake, stimulation of adenosine efflux, or inhibition of adenosine metabolism leading to increased efflux of adenosine from the cell. We have found that ethanol inhibits adenosine uptake into S49 wild type cells (Fig. 2) (Nagy et al., 1990), NG108-15 cells (Sapru et al., unpublished observations), and human lymphocytes (Table 4 and Krauss et al., manuscript in preparation). In S49 cells, uptake of adenosine is nearly linear for 60 seconds and ethanol inhibits uptake ~35% at all time points (Fig. 2A). Inhibition of adenosine uptake increased with increasing ethanol concentration (Fig. 2B) and was nearly maximal at 200 mM ethanol (K_i = 74 ± 8 mM). An Eadie-Hofstee plot (Fig. 3) showed that acute ethanol was a noncompetitive inhibitor of adenosine uptake. Ethanol decreased the V_{max} for adenosine uptake by 50% but caused only a small change in affinity. Ethanol also inhibited the uptake of uridine (Fig. 2B) and formycin (Krauss et al., unpublished observations), a non-metabolizable nucleoside, suggesting that the target of ethanol is the nucleoside transporter rather than nucleoside metabolism.

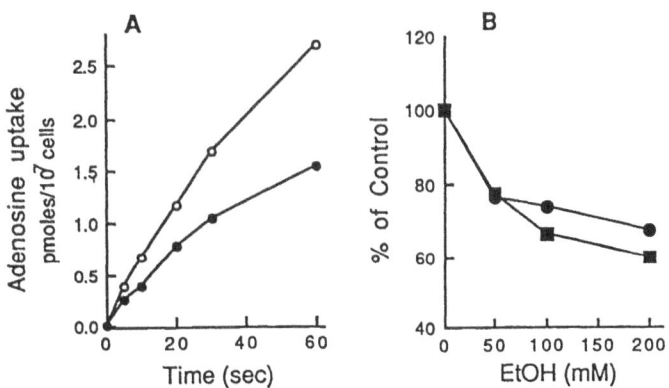

Figure 2. Acute effects of ethanol on adenosine uptake. "A", time course of [^{3}H]adenosine uptake in S49 cells in the presence and absence of ethanol. After preincubation for 4 min in 0 or 200 mM ethanol, uptake of 0.3 μM [3H]adenosine was measured in the absence (open circles) and presence (filled circles) of 200 mM ethanol by rapid centrifugation through oid. Nonspecific uptake was determined in the presence of 10 μM dipyridamole and subtracted from total uptake to determine specific uptake. Nonspecific uptake averaged 0.03 pmol/10^7 cells at 10 sec. Values are from a representative experiment done in triplicate, replicated 6 times. Mean values for ethanol-treated cells from all 6 experiments are significantly lower than controls (p < 0.02). "B", uptake of nucleosides, leucine, and deoxyglucose in S49 cells in the presence of varying concentrations of ethanol. Uptake of 0.3 μM [^{3}H]adenosine (filled circles) and 0.3 μM [^{3}H]uridine (filled squares) was measured at 90 sec in the presence of 50-200 mM ethanol and compared to cells incubated in the absence of ethanol. Points represent mean values from 3 to 6 experiments done in triplicate. For greater clarity, standard errors (from 6 to 12% of mean value) are not shown. Used with permission, J Biol Chem.

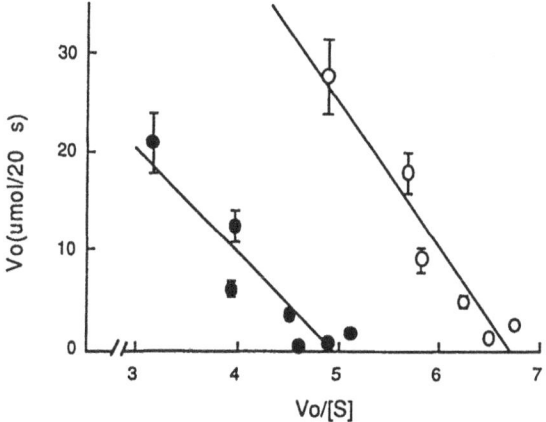

Figure 3. Eadie-Hofstee analysis of ethanol inhibition of adenosine uptake. S49 cells were preincubated with 0 (open circles) or 200 mM ethanol (filled circles) for 3 min and [^{3}H] adenosine uptake measured in the presence of 0.1-6.4 μM adenosine at 20 sec V_{max} for control cells was 292.2 ± 62.2 μMol of adenosine/min/10^7 cells and 154.2 ± 26.3 for ethanol-treated cells. K_m values were 14.7 ± 3.3 μM in control cells and 10.4 ± 2.0 μM in ethanol-treated cells. K_m values were 14.7 ± 3.3 μM in control cells and 10.4 ± 2.0 μM in ethanol-treated cells. Values represent mean ± S.E. from 6 experiments, each carried out in triplicate. S.E. is not shown when smaller than the symbols in the fig. Used with permission, J Biol Chem.

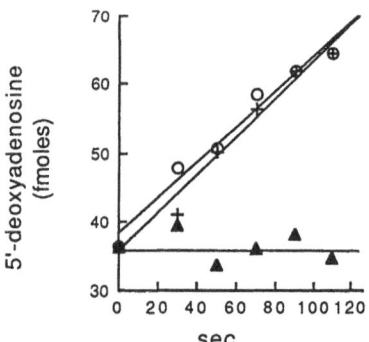

Figure 4. Time course of [^{3}H]5'-deoxyadenosine efflux. CCRF-CEM cells were incubated in the presence (open circles) and absence of ethanol (crosses). Efflux of pre-loaded [^{3}H]5"-deoxyadenosine was measured after 30-110-s exposure to 200 mM ethanol at 12°C. Nonspecific efflux was determined in the presence of 10 μM dipyridamole (filled triangles). Ethanol inhibition of [^{3}H]-deoxyadenosine influx under these conditions (data not shown) was similar to results at 22°C. Values are from a representative experiment done in triplicate and replicated 3 times. Used with permission, J Biol Chem.

We also determined that ethanol does not affect efflux of nucleosides (Nagy et al., 1990). We measured efflux of 5'-deoxyadenosine in CCRF-CEM cells, a human lymphoblastoid cell line that does not metabolize this nucleoside and found that 200 mM ethanol has no effect on efflux from 0-20 seconds (Fig. 4). Taken together, our results suggest that ethanol specifically inhibits uptake of nucleosides via nucleoside transporters.

CHRONIC EFFECTS OF ETHANOL ON ADENOSINE TRANSPORT

Chronic exposure of cells to ethanol causes many adaptive changes. Heterologous desensitization occurs after a 48 hour exposure to 200 mM ethanol (Fig. 1). This desensitization is accompanied by a corresponding decrease in mRNA, protein, and function of $G\alpha_S$ (Mochly-Rosen et al., 1988). In addition, after chronic exposure to ethanol, adenosine transport is no longer inhibited by ethanol (Nagy et al., 1990). Based on our studies in model cell culture systems, we proposed that

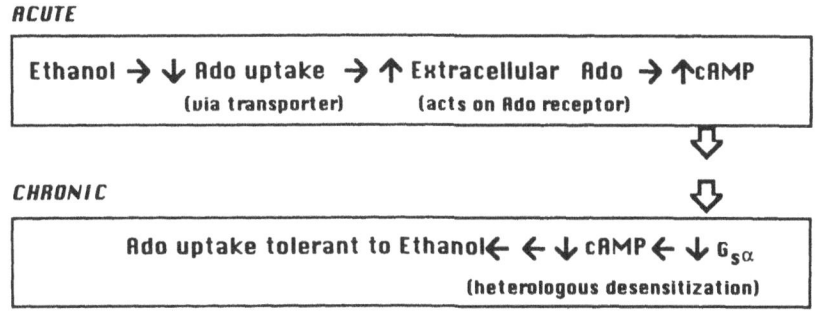

Figure 5

ADENOSINE RECEPTOR ANTAGONISTS PREVENT ADAPTATION TO CHRONIC ETHANOL

Using the above model, we predicted that adenosine receptor antagonists should prevent the adaptive changes due to chronic ethanol exposure. Table 1 shows that when NG108-15 cells are incubated with 200 mM ethanol for 48 hours, adenosine receptor-dependent cAMP production is reduced by 38% and PGE_1 receptor-dependent cAMP production by 29%. However, when the cells are incubated with ethanol in the presence of the adenosine receptor antagonist, BW A1434U, heterologous desensitization does not occur. Both adenosine receptor and PGE_1 receptor-stimulated cAMP levels are essentially the same as in control cells not exposed to ethanol.

Table 1. Adenosine antagonist BW A1434U prevents chronic ethanol-induced heterologous desensitization

| | cAMP (% Control) | |
| | PIA Stimulation | PGE1 |
Stimulation		
Chronic EtOH	62 ± 13	71 ± 10
BW A1434U	121 ± 25	94 ± 9
Chronic EtOH + BW A1434U	105 ± 4.2	95 ± 11

NG108-15 cells were grown for 48 h with or without 200 mM ethanol in the presence or absence of 10 μM BW A1434U. The cells were then stimulated with 100 μM PIA or 1 μM PGE1 and cAMP levels measured by radioimmunoassay. Values represent mean ± S.D. of a representative experiment performed in triplicate.

In addition to its effects on cAMP production, A1434U prevents adaptation of the adenosine transport system to ethanol. When naive NG108-15 cells are exposed to ethanol for 60 seconds, adenosine uptake is inhibited by 33.8% (Table 2). After exposure to 200 mM ethanol for 48 hours, there is no longer inhibition of adenosine uptake on rechallenge with ethanol. However, when A1434U is present during the ethanol exposure, adenosine uptake is inhibited 29.8% by ethanol, similar to inhibition in control cells (Table 2). These results support our hypothesis that the chronic effects of ethanol are mediated by adenosine through activation of adenosine receptors.

Table 2. Effect of adenosine antagonist BW A1434U on chronic ethanol-induced tolerance of adenosine uptake

Chronic Treatment	Inhibition of Adenosine Uptake (% Inhibition)
---	33.8 ± 0.7
EtOH	2.25 ± 0.04
BW A1434U	31.4 ± 0.56
EtOH + BW A1434U	29.8 ± 0.5

NG108-15 cell were incubated for 48 h with or without 200 mM ethanol (EtOH) or with 10 μM BW A1434U in the presence or absence of 200 mM ethanol. The cells were washed and uptake of 0.3 μM [^{3}H]adenosine measured at 60 sec in the presence or absence (control) of 200 mM acute ethanol. Values represent mean ± S.D. of representative experiment experiment performed in triplicate.

APPLICATION TO ALCOHOLISM

We have shown that freshly isolated lymphocytes from alcoholics show heterologous desensitization similar to that found in cultured cell lines. Both PGE$_1$ receptor- and adenosine receptor-stimulated cAMP accumulation are reduced by ~75% in cells from alcoholics as compared to non-drinking controls ((Diamond et al., 1987) and Nagy, Diamond and Gordon, unpublished observations). These results suggest that the adenosine transporter may also be a primary target for ethanol in human cells and that the heterologous desensitization we observe in freshly isolated circulating cells from alcoholics may be due to ethanol inhibition of adenosine uptake in vivo. Therefore, we have investigated which types of nucleoside transporters are present on human lymphocytes and which of those present are inhibited by ethanol.

Multiple nucleoside transport systems mediate passage of adenosine across cell membranes. There are two classes of sodium-independent facilitative membrane transporters which are inhibited by dipyridamol. One type has high affinity binding sites for the inhibitor, nitrobenzylmercatopurine (NBMPR), ("NBMPR-sensitive" transporters) and the other type lacks NBMPR binding sites and is relatively NBMPR-insensitive ("NBMPR-resistant" transporters) (reviewed in Jarvis, 1988; Plagemann et al., 1988; Paterson and Cass, 1986). There are also two sodium-dependent concentrative nucleoside transporters that are relatively insensitive to both dipyridamole and NBMPR.

Table 3. NBMPR-sensitive transporters predominate in human lymphocytes and model cell lines.

Cell Type	NBMPR-Sensitive Uptake
Freshly Isolated Lymphocytes	94%
Cultured Human Lymphocytes	75%
S49	93%
L1210	78%
NG108-15	82%

Lymphocytes were cultured in L2 media as described previously (Nagy et al., 1988). Uptake of 5 mM formycin was measured for 12 seconds (human lymphocytes and L1210) or 45 seconds (S49) in the presence or absence of 1 μM NBMPR. Uptake of 0.3 μM adenosine was measured after 60 seconds in NG108-15 cells. After uptake, cells were collected under oil and processed as described previously (3). All assays were within the linear range with 9-21 determinations having no more than 10% error and non-specific uptake (in the presence of 10 μM dipyridamole) subtracted. Data for human lymphocytes was from 5 non-alcoholic subjects.

Table 3 shows that NBMPR inhibits 75% of nucleoside uptake in cultured human lymphocytes and 94% of uptake in freshly isolated circulating human lymphocytes. This is similar to other cell types such as L1210, a mouse leukemic cell line, NG108-15 cells, and S49 cells. In cultured human lymphocytes, there was no difference in nucleoside uptake in the presence or absence of sodium (Krauss, Diamond and Gordon, manuscript in preparation), indicating that concentrative transporters are not present in human lymphocytes. Thus, human lymphocytes appear to possess only NBMPR-sensitive and NBMPR-resistant facilitative nucleoside transporters.

We next determined which of these two types of nucleoside transporters is sensitive to ethanol. Table 4 shows that ethanol inhibited nucleoside uptake by 41% in cultured human lymphocytes. When NBMPR is present, ethanol has no effect on the residual transport activity, suggesting that only nucleoside uptake by the NBMPR-sensitive transporter is inhibited by ethanol. To confirm this observation, we investigated the effects of ethanol on nucleoside uptake in the N1S1 hepatoma cell line, which contains mainly NBMPR-resistant nucleoside transporters. After NBMPR-treatment, 77% of nucleoside transport activity remains and the remaining uptake is not inhibited by ethanol (Table 4). Taken together, our results suggest that only NBMPR-sensitive nucleoside transporters are sensitive to ethanol and that, in vivo, ethanol acting on these transporters may be responsible for the heterologous desensitization observed in circulating lymphocytes isolated from alcoholics.

Table 4. Effect of acute ethanol on formycin uptake in human lymphocytes.

Cells	Uptake Relative to Untreated Cells		
	+ EtOH	+NBMPR	NBMPR + EtOH
Cultured Human Lymphocytes	59%	25%	21%
N1S1	84%	77%	76%

Cells were grown and assayed as described in the legend of Table 2. Cells were preincubated for 4 minutes with buffer ("untreated"), 200 mM EtOH, 1 μM NBMPR, or NBMPR + EtOH. Results are presented as the average of 6-24 determinations with no more than 10% errors and non-specific uptake (in the presence of 10 μM dipyridamole) subtracted. Data for cultured human lymphocytes was from 2 non-alcoholic subjects.

ACKNOWLEDGMENT

This work was supported in part by grants from the NIAAA and the March of Dimes Birth Defects Foundation.

REFERENCES

M. S. Dar, S. J. Mustafa, and W. R. Wooles, (1983), Possible role of adenosine in the CNS effects of ethanol, *Life Sci* 33:1363-1374.

I. Diamond, B. Wrubel, W. Estrin, and A. Gordon, (1987), Basal and adenosine receptor-stimulated levels of cAMP are reduced in lymphocytes from alcoholic patients, *Proc Natl Acad Sci USA* 85:6973-6976.

A. S. Gordon, K. Collier, and I. Diamond, (1986), Ethanol regulation of adenosine receptor- stimulated cAMP levels in a clonal neural cell line: An in vitro model of cellular tolerance to ethanol, *Proc Natl Acad Sci* 83:2105-2108.

P. L. Hoffman and B. Tabakoff, (1977), Alterations in dopamine receptor sensitivity by chronic ethanol treatment, *Nature* 268:551-553.

S. M. Jarvis, (1988), Adenosine receptors, In: "Adenosine Receptors", D. M. F. Cooper and C. Londos, eds., Alan R. Liss, Inc., New York, (1988).

D. Mochly-Rosen, F.-H. Chang, L. Cheever, M. Kim, I. Diamond, and A. S. Gordon, (1988), Chronic ethanol causes heterologous desensitization of receptors by reducing as messenger RNA, *Nature* 333:848-850.

L. E. Nagy, I. Diamond, D. J. Casso, C. Franklin, and A. S. Gordon, (1990), Ethanol increases extracellular adenosine by inhibiting adenosine uptake via the nucleoside transporter, *J Biol Chem* 265:1946-1951.

L. E. Nagy, I. Diamond, K. Collier, L. Lopez, B. Ullman, and A. S. Gordon, (1989), Adenosine is required for ethanol-induced heterologous desensitization, *Mol Pharmacol* 36:744-748.

L. E. Nagy, I. Diamond, and A. Gordon, (1988), Cultured lymphocytes from alcoholic subjects have altered cAMP signal transduction, *Proc Natl Acad Sci USA* 85:6973-6976.

A. R. P. Paterson and C. E. Cass, Transport of nucleoside drugs in animal cells, in "Membrane Transport of Antineoplastic Agents", I. D. Goldman, ed., Pergamon Press, Oxford, (1986).

P. G. W. Plagemann, R. M. Wohlhueter, and C. Woffendin, (1988), Nucleoside and nucleobase transport in animal cells. Biochim Biophys Acta 947:405-443.

W. R. Proctor and T. V. Dunwiddie, (1984), Behavioral sensitivity to purinergic drugs parallels ethanol sensitivity in selectively bred mice, *Science* 224:519-521.

EFFECTS OF ACUTE AND CHRONIC ETHANOL ADMINISTRATION ON THE POLY-PHOSPHOINOSITIDE SIGNALING ACTIVITY IN BRAIN

Grace Y. Sun, Jian-ping Zhang and Tai-An Lin

Biochemistry Department, University of Missouri
Columbia, MO 65212 U.S.A.

INTRODUCTION

Ethanol ingestion is known to exert diverse physiological effects on many body organs including the central nervous system (CNS). While excess acute ethanol intake can result in sedation, chronic ingestion is associated with development of tolerance and physical dependence and in turn, this may lead to the manifestation of withdrawal hyperexcitability. The biochemical mechanisms underlying the effects of ethanol are not well understood although alterations of many cell surface processes including membrane transport enzymes, receptors and ion channel activities have been implicated (see review by Deitrich et al., 1989). Some of these effects may be due partly to ethanol's ability to disorder membrane lipids, resulting in an alteration of the intricate relationship between proteins and lipids within the membrane. However, it is clear that ethanol does not act globally on the membrane lipids, rather, changes are attributed to the effects of ethanol on specific types of lipids present in different membrane domains (Wood and Schroeder, 1988). In order to better understand the physiological manifestations related to acute and chronic ethanol ingestion, it is important first to identify the ethanol-sensitive biochemical mechanisms in brain and then followed by attempts to understand how adaptation to the changes is developed.

Despite some discrepancies due mainly to the methods used for alcohol adminstration, procedures for analysis, age and species of the animal model, many studies have successfully related changes in membrane phospholipids and their fatty acids to the effects of chronic ethanol administration (see review by Sun and Sun, 1985). It appears that acidic phospholipids bearing negative charges are particularly sensitive to the effects of ethanol (Sun et al., 1987ab; 1989). This notion is supported by studies from our laboratory indicating that chronic ethanol administration resulted in an increase in the levels of acidic phospholipids such as phosphatidylserine (PS) and phosphatidylinositol (PI) in brain synaptic membranes (Sun and Sun, 1983; Sun et al., 1984). Studies in

Rubin's laboratory have also recognized the importance of acidic phospholipids in conferring membrane tolerance to ethanol (Taraschi et al., 1986; Hoek and Taraschi, 1988). Interestingly, results from these studies further showed that the action of ethanol on these acidic phospholipids varies depending on the type of tissue and subcellular membrane used for analysis.

Acidic phospholipids such as phosphoinositides (PI, PIP and PIP_2) are important in conferring negative charges to membrane, especially the inner monolayer of the plasma membrane. PIP_2 also participates directly in the signal transduction cascade by serving as substrate for the receptor-mediated phospholipase C reaction (Fig 1). Stimulation of this signaling pathway results in the release of inositol 1,4,5-trisphosphates ($Ins(1,4,5)P_3$) and diacylglycerols (DG), which are second messengers for mobilization of intracellular calcium stores and activation of protein kinase C, respectively (Berridge, 1987). In brain, a number of neurotransmitters and hormones are known to transmit signals through activation of receptors that are coupled to this signaling pathway (Fowler and Tiger, 1991). In turn, activation of this signaling pathway is associated with regulation of cellular events including protein phosphorylation, induction of early response gene expression, regulation of ion channel activity and secretion. The effect of ethanol on the poly-PI signaling system has not been examined in detail although previous studies have suggested possible interation with the guanine nucleotide binding protein (Hoffman and Tabakoff, 1990). In order to better understand the effect of ethanol on the poly-PI signaling system in brain and to identify possible site(s) in this pathway that shows special sensitivity to the effect of ethanol, it is necessary to develop suitable procedures for studying this event in brain in vivo.

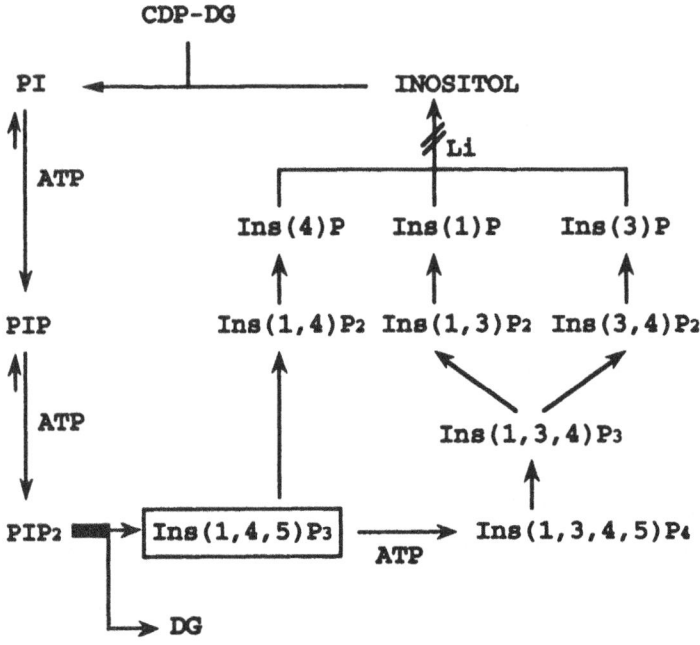

Fig. 1. The signal transduction pathway involving poly-PI.

ETHANOL EFFECTS ON POLY-PI SIGNALING ACTIVITY IN BRAIN

Studies with cultured cells and brain slices

The cultured cell system has been a useful model for studying the molecular action of ethanol on receptors involved in the poly-PI signaling pathway (Alling et al., 1989). Initial studies with primary astrocytes have indicated changes in agonist-stimulation of the poly-PI signaling pathway on exposure of these cells to ethanol (Ritchie et al., 1988). Simonsson et al. (1989a) showed that short-term exposure of primary astrocytes to ethanol could potentiate the serotonin-stimulated inositol lipid metabolism. On the other hand, long term exposure of neuroblastoma-glioma hybrid cells (NG108-15) to ethanol resulted in a decrease in the response to bradykinin-stimulated poly-PI hydrolysis (Simonsson et al., 1989b). Similarly, exposure of N1E-115 neuroblastoma cells to ethanol also resulted in a decrease in the response to stimulation by neurotensin (Smith, 1990). Among cells that were derived from peripheral organs, hepatocytes seem to show special sensitivity to the effects of ethanol (Hoek et al., 1987). In fact, addition of ethanol alone to these cells could result in stimulation of poly-PI hydrolysis and increase in cytosolic calcium (Hoek et al., 1987; Rubin and Hoek, 1989; Hoek et al., 1992). The differences in ethanol sensitivity towards agonist-induced hydrolysis of poly-PI among different cell systems is an intriguing phenomenon and definitely requires further investigation.

Brain slice preparations have been used to study the effects of ethanol on the poly-PI signaling pathway in brain. In most studies, relatively high concentrations of ethanol in vitro were needed to elicit an inhibitory effect on the agonist-mediated poly-PI response (Gonzales et al., 1986; Hoffman et al., 1986). However, it is important to recognize that some agonists are more sensitve than others and differences in response exist among different brain regions and animal species. When brain slices were prepared from rats after chronic ethanol treatment and subsequently stimulated by neurotransmitter agonists such as carbachol and norepinephrine, little changes in basal or stimulated poly-PI signaling activity were observed between the alcohol-treated group and controls (Gonzales and Crews, 1988). On the other hand, ethanol-induced changes in the carbachol stimulated poly-PI hydrolysis were found in C57BL mice that were administered ethanol chronically (Hoffman et al. 1986). The increase in potency (decrease in EC_{50}) for carbachol to stimulate PIP_2 hydrolysis in the chronic ethanol mice was in agreement with the increase in number of cholinergic receptors reported previously (Tabakoff et al., 1979; Rabin et al., 1980; Smith, 1983). Studies by Smith et al. (1986) further observed the development of tolerance to the cholinergic signaling pathway with respect to chronic ethanol administration. Although these studies illustrated the diverse effects of ethanol on the poly-PI signaling pathway, there are several limitations with regard to using brain slice preparations to assess these changes. This pertains specially to ability to label the slices with [^{3}H]inositol and to maintain the slices under conditions for maximal response to agonists.

In vivo procedures to assess ethanol effects on poly-PI turnover in brain

Due to the complex cellular make-up, few attempts have been made previously to elucidate the effects of ethanol on the poly-PI signaling activity in brain in vivo. Allison and Cicero (1980) were the first to observe the effect of acute ethanol in depressing the levels of inositol 1-phosphate in brain, although at the time of their study a metabolic relationship between inositol phosphate and the poly-PI signaling pathways had not been recognized. Thus, a major objective of this chapter is to describe the <u>in vivo</u> procedures developed in our laboratory to further explore the effects of ethanol on the poly-PI signaling activity in brain <u>in vivo</u>.

Poly-PI metabolism in brain labeled with [^{32}P]Pi. One of the original methods for assessing quantitatively agonist-stimulation of poly-PI hydrolysis in cultured cell systems is to labeled the cells with [^{32}P]Pi. With cells labeled with this precursor, stimulation by agonists that are coupled to the poly-PI signaling pathway would result in a transient decrease in [^{32}P]PIP$_2$. Based on these findings, we explored the possibility of labeling the poly-PI in brain by injecting intracerebrally ^{32}Pi or [^{32}P]ATP into rodent brain (Sun and Lin, 1989). Initial studies indicated that both types of label were effectively incorporated into the phosphoinositides (PI, PIP and PIP$_2$) as well as other phospholipids. Besides, a high proportion of radioactivity is incorporated into PIP and PIP$_2$ during the initial 2 to 4 hr after injection. Thus, a predictable labeling pattern among the poly-PI and other phospholipids can be observed among different brain regions and subcellular fractions with respect to time after injection of the label. We further observed that although the myelin membrane contained high proportions of labeled poly-PI, this labeled pool is resistant to hydrolysis due to ischemia (Sun et al., 1990). On the contrary, labeled poly-PI in the synaptosomal fraction is highly susceptible to post mortem hydrolysis as well as other factors.

Initially, experiments were designed to examine the effects of acute ethanol administration on poly-PI metabolism in rat brain after labeling with ^{32}P[Pi] (Chandrasekhar et al., 1988a). Results indicated an increase in labeled PIP and PIP$_2$ in these brain samples after administering acute doses of ethanol (2-6 g/kg) by gavage. Since the increase was marked by a corresponding decrease in levels of diacylglycerol (DG) as determined by the HPTLC-GC method (Sun, 1988), it is concluded that the increase in labeled poly-PI is due to inhibition of receptor-mediated poly-PI hydrolysis. Previous study by Erickson and Graham (1973) had demonstrated the ability of acute ethanol to suppress acetylcholine release in brain. Thus, the ethanol-induced decrease in poly-PI turnover may be directly related to the decrease in receptor activity.

Experiments using the [^{32}P]Pi labeling procedure to examine the effects of chronic ethanol administration on poly-PI metabolism in rat brain did not show obvious changes (Chandrasekhar et al., 1988b). One of the problems with this type of experiment is the difficulty in injecting a constant amount of label into the brain of adult rats. Consequently, an experiment was carried out using the C57Bl mice (Sun et al., 1993). This strain of mice has been used previously for study of their sedative response to ethanol with age (Wood et al. 1982). In the present study, C57BL/NNIA mice at 12 and 26 months of age were administered ethanol 3 g/kg twice daily by gavage for three weeks. At the end of the feeding period, animals were injected intracerebrally with [^{32}P]ATP and sacrified 4 hr later. The brain homogenates were subjected to a subcellular fractionation procedure and the phospholipid labeling patterns among the fractions were analyzed. As shown in Fig 2a, there was an increase in labeling of poly-PI in the synaptosomal fractions from the 12 month-old ethanol group as compared to controls. Interestingly, synaptosomes isolated from 26 month-old mice showed instead a reduced poly-PI labeling in the ethanol group as compared to controls (Fig 2b). When the phospholipid labeling patterns from these two age groups (controls) were compared, the levels of labeled poly-PI were higher in the older age group than in the younger age group. The increase in poly-PI labeling with age is in agreement with the results observed previously (Sun and Lin, 1990). Therefore, it can be concluded from this study that poly-PI metabolism in brain is affected by both chronic ethanol administration and age.

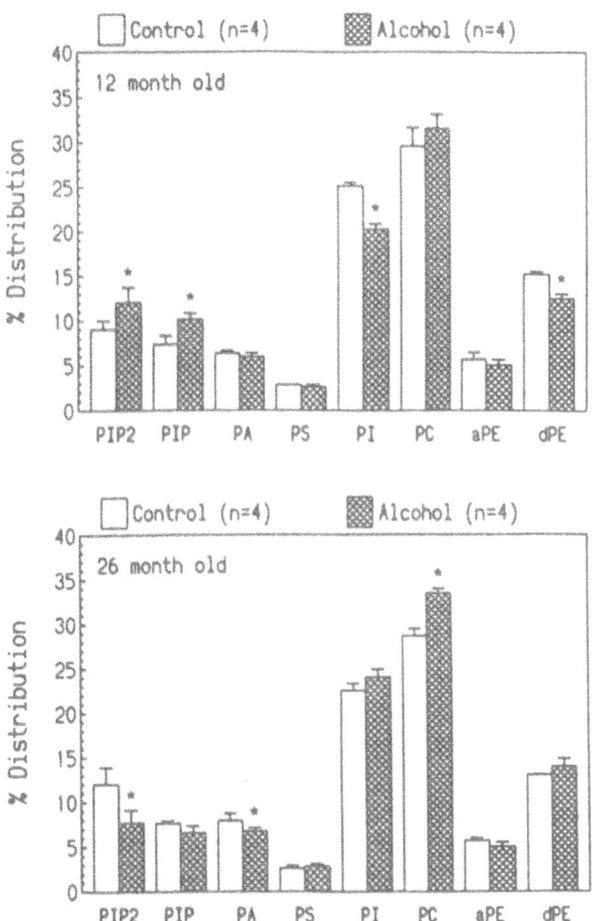

Fig 2. Percent distribution of radioactivity among phospholipids in synaptosomal fractions of control and alcoholic mice at 12 (2a) and 26 (2b) months of age. C57BL/6NNIA mice at the respective ages were administered an alcohol or an isocaloric liquid diet with sucrose by gavage for 3 weeks. Animals were injected intracerebrally with [^{32}P]ATP (10 μCi) 4 hr prior to killing and dissection of brain cortex. Data represent percent distribution (mean ± SD, n=4) of radioactivity among phospholipids in the synaptosomal fraction. Statistical evaluation based on analysis of variance indicated differences comparing the ethanol group with controls, $^*p < 0.05$. Abbreviations: PIP$_2$, phosphatidylinositol 4,5-bisphosphate; PIP, phosphatidylinositol 4-phosphate; PA, phosphatidic acid; PS, phosphatidylserine; PI, phosphatidylinositol; PC, phosphatidylcholine; aPE, phosphatidylethanolamine (alkenylacyl form) or ethanolamine plasmalogen; dPE, phosphatidylethanolamine (diacyl form). (Data abstracted from Sun et al., 1993).

Using a similar labeling procedure, an experiment was carried out to examine the effect of chronic ethanol on poly-PI metabolism in mice that were pair-fed a nutritionally complete liquid diet (Sustacal, chocolate flavor, Evansville, IN) containing either ethanol (5% v/v) or sucrose for 2 months. Results from this method of ethanol administration indicated also a similar increase in labeled poly-PI in the ethanol group as compared to the pair-fed controls (Sun et al., 1992).

In conclusion, these studies indicate that _in vivo_ labeling with either [^{32}P]Pi or [^{32}P]ATP can be used to assess changes in poly-PI metabolism in brain due to acute and chronic ethanol administration. There were no obvious differences in the phospholipid labeling patterns with either precursor, indicating that [^{32}P]ATP is rapidly hydrolyzed by ecto-ATPase and the labeled phosphate group is transported into the cells. Using this labeling protocol, an increase in poly-PI labeling was observed due to acute as well as chronic ethanol administration. Although it is likely that these increases are mediated by different mechanisms, more studies are needed to further examine the mechanism underlying the changes.

Metabolism of inositol phosphates in brain labeled with [^{3}H]inositol. There is convincing evidence that poly-PI in brain undergo rapid turnover and the inositol metabolites are maintained in a dynamic equilibrium (Fig 1). It is possible that this equilibrium is affected by agents such as alcohol and drugs. Since previous studies with [^{32}P]Pi labeling were limited to observing changes in the inositol phospholipids, we explore the possibility of using [^{3}H]inositol to label both inositol phospholipids and the water soluble inositol phosphates in brain. However, studies involving this label procursor require the use of inhibitors to dissociate the cycle. Earlier studies by Sherman et al. (1985) had indicated the effectiveness of systemic injection of lithium to inhibit the inositol monophosphatase activity in brain. Under this condition, the increase in levels of inositol monophosphates can be used to represent the amount of signals transduced due to agonist stimulation of the poly-PI signaling pathway. When rats were injected intracerebrally with [^{3}H]inositol and subsequently with lithium, there was a large increase in labeled inositol monophosphates (IP$_1$) in brain (Sun et al., 1992). Analysis of the inositol monophosphate isomers by ion chromatography further indicated a 40-fold increase in the level of Ins(1)P and a 20-fold increase in Ins(4)P in brain after lithium administration (Sun et al., 1992). This labeling procedure was further used to assess the amount of signals that are due to activation of the muscarinic cholinergic receptors by injecting mice with atropine (50 mg/kg), the cholinergic antagonist, prior to lithium adminstration. Atropine treatment resulted in a large decrease (>50%) in labeled IP$_1$, suggesting that the release of acetylcholine constituted a major portion of the signal transduced through poly-PI hydrolysis (Lin et al., 1993).

An experiment involving labeling brain with [^{3}H]inositol was used to examine the effect of ethanol on poly-PI turnover. In this experiment, mice were first injected intracerebrally with [^{3}H]inositol for 16 hr and subsequently injected i.p. with lithium (6-8 meq/kg body wt) 4 hr prior to sacrifice. A suboptimal dose of lithium was chosen in order to avoid adverse effects due to combined administration of both drugs. In a recent study by Atack et al. (1992), a similar dose of lithium·was used to elicit changes in the pilocarpine stimulation of inositol 1-phosphate in mouse brain. As shown in Fig 3, acute administration of ethanol (2 to 6 g/kg) by gavage resulted in a dose- and time-dependent decrease in labeled IP$_1$ in the ethanol group as compared to controls (Lin et al., 1993).

Furthermore, we observed obvious differences in the labeling of inositol metabolites between the cerebrum and cerebellum. The higher proportion of labeled IP$_1$ in the cerebrum is in agreement with the higher proportion of cholinergic receptors present in this part of the brain.

The in vivo experimental protocol involving injections of [H]inositol and lithium was used to examine the effect of chronic ethanol administration to mice pair-fed a liquid diet for 2 months. In most instances, an increase in labeled IP$_1$ can be found in the cerebrum after chronic ethanol administration (Sun et al., 1991). Nevertheless, variances in results were observed among different experiments probably due to multiple steps for labeling the brain with [³H]inositol and administering lithium.

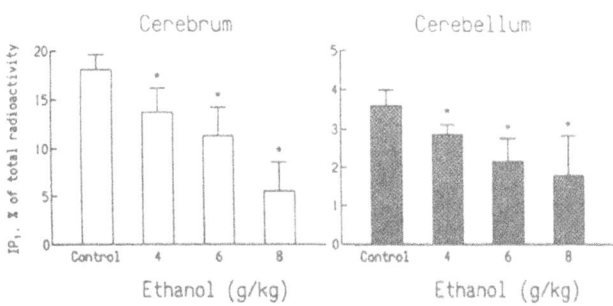

Fig 3 The levels of labeled IP$_1$ in mouse cerebrum and cerebellum with respect to acute ethanol administration. C57BL/6J mice were injected intracerebrally with [³H]inositol (10 µCi) for 20 hr and then i.p. with lithium (6 meq/kg body wt) for 4 hr prior to killing. Animals were given the respective dose of ethanol by gavage 2 hr after lithium injection. Data represent the percent of total activity and are mean ± SD from 5 animals in each group. Analysis of data by unpaired Student's t-test indicated significant differences, p < 0.05, comparing the ethanol groups with controls. (Data taken from Lin et al., 1993 with permission.)

Several experiments were carried out using the [³H]inositol labeling procedure to test whether mice administered chronically with an ethanol diet develop tolerance to the acute ethanol-induced inhibition of poly-PI turnover. In this study, chronic ethanol mice and pair-fed controls were first injected intracerebrally with [³H]inositol and then followed by i.p. injection of lithium (8 meq/kg) for 4 hr. Subsequently, both chronic ethanol and control groups were challenged with an acute dose of ethanol (5 or 6 g/kg) and analyses of blood ethanol concentration and labeled IP$_1$ in cerebrum took place at 3.5 and 7.5 hr after ethanol administration. Results of analysis of blood ethanol concentration indicated no difference between the chronic ethanol and control groups at 3.5 hr after acute administration of 5 g/kg. However, at 7.5 hr, blood ethanol was completely eliminated in the chronic ethanol group whereas a small amount was found remaining in the pair-fed control group (Table 1). When labeled IP$_1$ was compared between the chronic ethanol and pair-fed control groups prior to and after acute ethanol administration, the chronic ethanol mice showed a higher level of labeled IP$_1$ as compared to the pair-fed control mice that were administered the same acute dose of ethanol (Table 1). These results seem to suggest that chronic ethanol administration leads to development of tolerance to the inhibitory action of acute ethanol on the poly-PI signaling pathway.

Table 1. Poly-PI turnover in brain: Effects of acute ethanol on chronic ethanol and pair-fed control mice.

Acute ethanol	Time	IP/lipid, %		Blood ethanol, mg%	
		Pair-fed control	Chronic ethanol	Pair-fed control	Chronic ethanol
-	-	22.3±1.2 (n = 6)	N.D.	N.D.	N.D.
5g/kg	3.5hr	15.0±3.0[a] (n = 4)	23.1±2.5[b] (n = 3)	198±55 (n = 5)	217±70 (n =5)
5g/kg	7.5hr	18.5±2.1[a] (n = 4)	23.6±2.2[b] (n = 4)	37±12 (n = 4)	5±6[b] (n = 5)
5g/kg	14hr	16.2±2.6[a] (n = 4)	17.2±3.2 (n = 4)	3±1 (n = 5)	2±2 (n = 5)

C57Bl/6J mice were given either a liquid diet containing ethanol (3 to 4 g/kg) or pair-fed with the same diet with sucrose by gavage for 3 weeks. After the chronic feeding, mice were injected with [^{3}H]inositol for 20 hr and subsequently lithium (8 meq/kg) for 4 hr prior to killing. At 20 hr after [^{3}H]inositol injection, each group was administered either an acute dose of ethanol (5 g/kg) or the liquid diet for the time indicated. Results are percent radioactivity of total inositol phosphate (IP) eluted from the Dowex column against that in the organic phase. Blood ethanol concentrations (expressed as mg%) were determined using the alcohol dehydrogenase kit from Sigma Chem. Co. (St. Louis, MO). Data for pair-fed control mice that were not administered acute ethanol were derived from both pair-fed (n=3) and chow fed (n=3) animals which showed no difference in the level of labeled IP$_1$. Results are mean ± SD from the number of samples indicated in paratheses. Statistical evaluation based on analysis of variance indicated significant differences, p < 0.05, [a]comparing pair-fed control given acute ethanol with no ethanol and [b]comparing chronic ethanol with pair-fed controls that were given acute ethanol. N.D., not determined.

ETHANOL EFFECTS ON INS(1,4,5)P$_3$ BINDING AND METABOLISM

Radioreceptor assay of levels of Ins(1,4,5)P$_3$

In order to further relate the ethanol-induced changes in labeled IP$_1$ with other inositol metabolites within the poly-PI cycle, a radioreceptor binding assay similar to that described by Bredt et al. (1989) was use to determine the levels of Ins(1,4,5)P$_3$ in brain with respect to ethanol treatment. This assay procedure has the advantange that frozen brain tissue can be used directly for determination of Ins(1,4,5)P₃ but care has to be taken in the assay because this compound undergoes rapid changes during tissue processing. In agreement with the study with [³H]inositol, results of this study show a decrease in the levels of Ins(1,4,5)P₃ in both cerebrum and cerebellum after acute ethanol administration (6g/kg) (Fig 4).

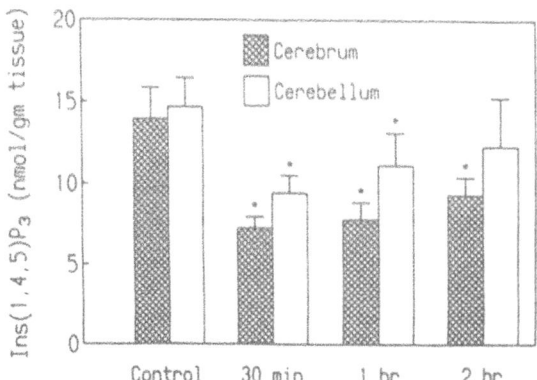

Fig 4 The levels of Ins(1,4,5)P₃ in cerebrum and cerebellum of mouse brain at various times after acute ethanol administration (6 gm/kg) by gavage. Ins(1,4,5)P₃ was determined using the radioreceptor assay as described by Bredt et al. (1989). Data are mean ± SD from 4 animals in each group. *Analysis of the data by unpaired Student's t-test indicated significant differences, $p < 0.05$, comparing the ethanol groups with controls. (Reproduction from Lin et al. 1993 with permission.)

The radioreceptor method was used also to test whether chronic ethanol mice develop tolerance to the acute ethanol-induced inhibition of Ins(1,4,5)P₃ release. In this experiment, mice were given the same feeding paradigm as described above (feeding with liquid diet with and without ethanol for 3 wk). At the end of the experiment, both groups of mice were administered either an acute dose of ethanol (6 gm/kg) or the control liquid diet. At 30 min after ethanol administration, mice were sacrificed and the heads were quickly frozen in liquid nitrogen. Brain tissue was dissected for determination of Ins(1,4,5)P₃ and trunk blood was collected for assay of blood ethanol concentrations. As indicated in Fig 5, there was no obvious difference in the level of Ins(1,4,5)P₃ in brain between the controls and chronic ethanol group. Nevertheless, the chronic ethanol group given an acute dose of ethanol showed a significantly higher ($p < 0.05$) level of Ins(1,4,5)P₃ as compared to the naive group given the same amount of acute ethanol. Again, results here are in agreement with previous study with [³H]inositol labeling in supporting the notion of tolerance development to the acute ethanol-induced inhibition of poly-PI signaling acitivity after chronic ethanol administration.

213

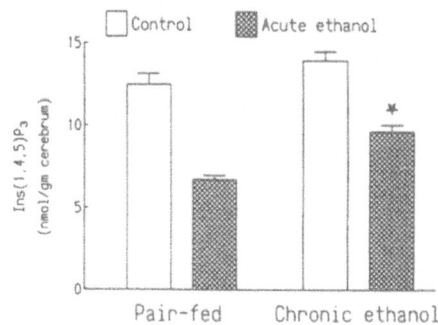

Fig 5 Effect of chronic ethanol administration on the response to inhibition of Ins(1,4,5)P₃ levels in mouse cerebrum due to acute ethanol administration. C57BL/6J mice were administered a liquid diet by gavage containing either ethanol or isocaloric amount of sucrose for 3 wk. At the end of the feeding period, both controls and chronic ethanol mice were administered either the liquid diet (control) or an acute dose of ethanol (6 gm/kg) and mice were sacrificed 30 min later. Blood alcohol levels for the pair-fed and chronic ethanol groups at 30 min after the acute dose were 354 ± 88 and 445 ± 42 (mean $\pm$ SD, n= 5) mg/dL, respectively. *Statistical evaluation based on analysis of variance and Bonferroni post-test indicated significant differences between chronic ethanol group given acute ethanol versus pair-fed controls, $p < 0.05$.

Effects of chronic ethanol on Ins(1,4,5)P₃ 3-kinase and 5-phosphatase in brain regions

Besides interacting with its intracellular receptor, Ins(1,4,5)P₃ is metabolized by the Ins(1,4,5)P₃ 3-kinase and 5-phosphatase. These enzymes are responsible for converting the second messenger to Ins(4)P and Ins(1)P, respectively, and finally to free inositol (Fig 1). Interestingly, these two enzymes differ greatly in their regional distribution in brain. The Ins(1,4,5)P₃ 3-kinase which converts Ins(1,4,5)P₃ to Ins(1,3,4,5)P₄ is of special interest because it shows many properties indicative of a regulatory enzyme, including activation by Ca²⁺/calmodulin kinase (Takazawa et al., 1990) and limited proteolysis by calpain (Lee et al., 1990). Furthermore, the 3-kinase but not the 5-phosphatase is sensitive to cerebral ischemic insult (Lin et al., 1992). An experiment was carried out to examine the effects of chronic ethanol administration on the 3-kinase and 5-phosphatase activities in different mouse brain regions. As shown in Fig 5, despite the large differences in enzyme activity among different brain regions, no obvious differences were found between the ethanol group and controls. Therefore, it is reasonable to conclude that development of tolerance to acute ethanol-induced inhibition of the poly-PI activity after chronic ethanol administration is probably not due to an adaptive change by these two enzymes.

CONCLUSION

Several in vivo protocols were used to examine the effects of acute and chronic ethanol administration on the poly-PI signaling acitivity in the rodent brain. It is recognized that each method has its own advantages and limitations. Studies with [³²P]Pi injection is an important procedure to relate metabolism of inositol phospholipids to that of other phospholipids. Since the labeling of the PIP and PIP₂ with respect to other phospholipids is highly predictable with time after injection of label, it is possible to

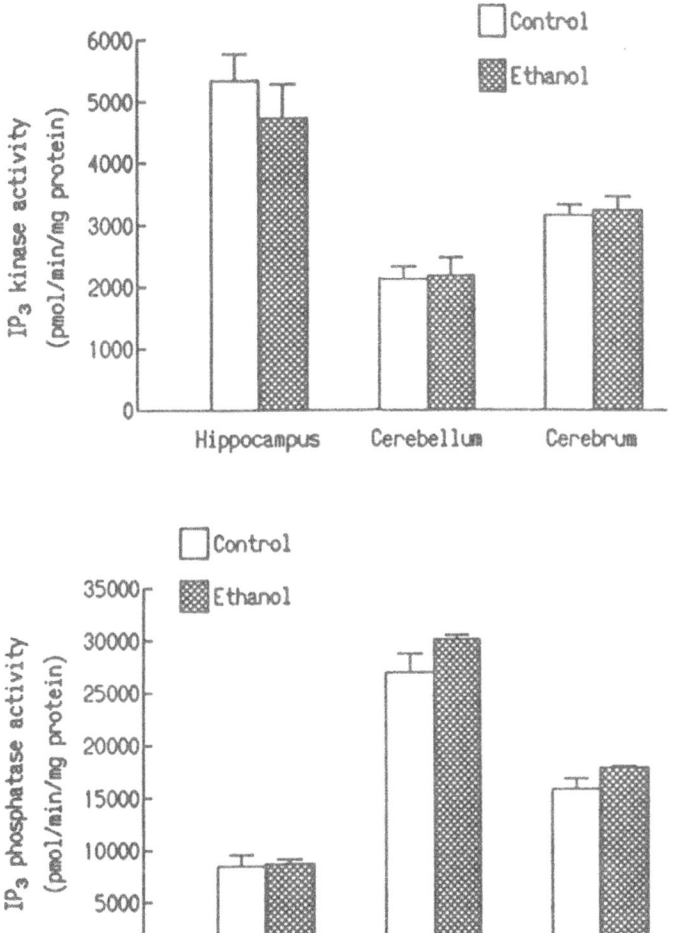

Fig 6 Activity of Ins(1,4,5)P_3 3-kinase and 5-phosphatase in mouse brain regions and with respect to chronic ethanol administration. Mice were fed the control and ethanol liquid diet for 3 wk as described in Table 1. After the feeding period, brains were removed and brain regions were dissected on ice and stored frozen at -70°C until time for analysis (less than 1 wk). Assay of the Ins(1,4,5)P_3 3-kinase and 5-phosphatase activity was according to the procedure described by Lin et al. (1992).

comparison of the labeling patterns among different treatment groups may give information on how ethanol affects the poly-PI metabolism in brain. In most instances, changes in labeling are confined mainly in the synaptosomal fraction. This observation supports the notion that ethanol alters the activity of a metabolically poly-PI pool linked to the signaling pathway.

Labeling brain with [^{3}H]inositol has the advantage that the labeled precursor is incorporated into both inositol phospholipids and inositol phosphates. Therefore, a more complete picture of the poly-PI cycling activity can be related to ethanol administration. However, these studies necessitate the use of lithium to block the cycle. Under this condition, the amount of signals transduced through the poly-PI pathway can be depicted by the accumulation of labeled IP$_1$ within a fix time period after lithium injection. By administering specific receptor antagonist to the mice prior to lithium administration, it was found that over 50% of the signals generated in the cerebrum is associated with the cholinergic muscarinic receptor type. Although this protocol is feasible to examine the effect of ethanol on poly-PI turnover in brain, several methodological limitations in the procedure have resulted in large variances in data obtained. Obviously, an improvement on the methodology is important in obtaining useful data from this protocol.

Based on results of studies using both labeling procedures and the receptor binding assay for mass determination of Ins(1,4,5)P$_3$, it is possible to associate acute ethanol sedation with an inhibition of the poly-PI signaling activity in brain. Since a large portion of the signals can be attributed to activation of the muscarinic cholinergic receptors, the decrease in poly-PI turnover reflects the decrease in the release of acetylcholine. Although chronic ethanol administration did not cause obvious changes in the poly-PI turnover in brain, there is evidence that animals administered chronic ethanol developed tolerance to the inhibitory effect of acute ethanol on poly-PI turnover. Obviously, more studies are needed to further delineate the mechanism underlying the adaptive changes due to chronic ethanol and to ascertain whether tolerance development is related to alteration of the enzymes involved in this signaling pathway.

Acknowledgments

This research project was supported in part by DHHS grants RO1 AA 06661 (GYS) and T32 AA 07458 (TAL) from NIAAA. Thanks are due to Dr. Patricia Wixom for critical reading of the manuscript.

REFERENCES

Alling C., Hansson E. and Simonsson P., 1989, Cell cultures as a model for alcohol research. In "Molecular Mechanisms of Alcohol: Neurobiology and Metabolism" Sun et al., eds., Humana Press, Clifton, NJ. p193-206.

Allison J.H. and Cicero T.J., 1980, Alcohol acutely depresses myo-inositol 1-phosphate levels in the male rat cerebral cortex. *J. Pharmacol. Exp. Therap.*, 213:24-27.

Atack J.R., Cook S.M., Watt A.P. and Ragan C.I., 1992, Measurement of lithium-induced changes in mouse inositol(1)phosphate levels in vivo. *J. Neurochem.*, 59:1946-1954.

Berridge, M.J., 1987, Inositol trisphosphate and diacylglycerol: Two interacting second messengers. *Ann. Rev. Biochem.*, 56:159-193.

Bredt D.S., Mourey R.J. and Snyder S.H., 1989, A simple, sensitive, and specific radioreceptor assay for inositol 1,4,5-trisphosphate in biological tissues. *Biochem. Biophys. Res. Comm.*, 159:976-982.

Chandrasekhar, R., Lin, T-N, and Sun, G.Y., 1988a, Labeling of rat brain phospholipids by $[^{32}P]$-ATP: effects of chronic ethanol administration, in "Biomedical and Social Aspects of Alcohol and Alcoholism," K. Kuriyama, A. Tanaker and H. Ishic, eds., Excerpta Medica, International Congress Series 805, pp. 295-298.

Chandrasekhar, R., Huang, H-M. and Sun, G.Y., 1988b, Alterations in rat brain polyphosphoinositide metabolism due to acute ethanol administration. *J. Pharmacol. Exp. Ther.*, 245:120-123.

Deitrich R.A., Dunwiddie T.V., Harris R.A. and Erwin V.G., 1989, Mechanism of action of ethanol: Initial central nervous system actions. *Pharm. Rev.*, 41:489-537.

Erickson C.K. and Graham D.T., 1973, Alterations of cortical and reticular acetylcholine release by ethanol in vivo. *J. Pharmacol. Exp. Ther.*, 185:583-593.

Fowler C.J. and Tiger G., 1991, Modulation of receptor-mediated inositol phospholipid breakdown in the brain. *Neurochem. Int.*, 19:171-206.

Gonzales R.A. and Crews F.T., 1988, Effects of ethanol in vivo and in vitro on stimulated phosphoinositide hydrolysis in rat cortex and cerebellum. *Alcoholism: Clin. Exp. Res.*, 12: 94-98.

Gonzales R.A., Theiss C. and Crews, F.T., 1986, Effects of ethanol on stimulated inositol phospholipid hydrolysis in rat brain. *J. Pharmacol. Exp. Ther.*, 237:92-98.

Hoek J.B. and Taraschi T.F., 1988, Cellular adaptation to ethanol. *TIBS*, 13: 269-274.

Hoek J.B., Thomas A.P., Rubin R. and Rubin E., 1987, Ethanol-induced mobilization of calcium by activation of phosphoinositide-specific phospholipase C in intact hepatocytes. *J. Biol. Chem.*, 262: 682-691.

Hoek J.B., Thomas A.P., Rooney T.A., Higashi K. and Rubin E., 1992, Ethanol and signal transduction in the liver. *FASEB J.*, 6, 2386-2396.

Hoffman P.L. and Tabakoff B., 1990, Ethanol and guanine nucleotide binding proteins: a selective interaction. *FESEB J.*, 4, 2612-2622.

Hoffman P.L., Moses F., Luthin G.R. and Tabakoff B., 1986, Acute and chronic effects of ethanol on receptor-mediated phosphoinositol 4,5-bisphosphate breakdown in mouse brain. *Mol. Pharmacol.*, 30: 13-18.

Lee S.Y., Sim S.S., Kim J.W., Moon K.H., Kim J.H., Rhee S.G., 1990, Purification and properties of D-myo-inositol 1,4,5-trisphosphate 3-kinase from rat brain. *J. Biol. Chem.*, 265:9434-9440.

Lin T.A., Lin T-N., He Y.Y., Hsu C.Y., and Sun G.Y., 1992, Effects of focal cerebral ischemia on inositol 1,4,5-trisphosphate 3-kinase and 5-phosphatase activities in rat cortex. *Biochem. Biophys. Res. Commun.*, 184: 871-877.

Lin T.A., Navidi M., James W., Lin T.N. and Sun G.Y., 1993, Effects of acute ethanol administration on poly-phosphoinositide turnover and levels of inositol 1,4,5-trisphosphate in the mouse brain. *Alcoholism: Clin. Exp. Res.*, in press.

Rabin R.A., Wolfe B.B., Dibner M.D., Zahniser N.R., Melchior C.L. and Molinoff P.B., 1980, Effects of ethanol administration and withdrawal on neurotransmitter receptor systems in C57BL mice. *J. Pharmacol. Exp. Ther.*, 213:491-496.

Rubin T. and Hoek J.B., 1988, Ethanol-induced stimulation of phosphoinositide turnover and calcium influx in isolated hepatocytes. *Biochem. Pharmacol.*, 37:2461-2466.

Ritchie T., Kim H-S., Cole R., deVellis J., and Noble E.P., 1988, Alchol-induced alterations in phosphoinositide hydrolysis in astrocytes. *Alcohol*, 5: 183-187.

Sherman W.R., Munsell L.Y., Gish B.G. and Honchan M.P., 1985, The effect of systemic administration of lithium on phosphoinositide metabolism in rat brain kidney and testis. *J. Neurochem.* 44:798-807.

Simonsson P., Hansson, E., Alling C., 1989a, Ethanol potentiates serotonin stimulated inositol lipid metabolism in primary astroglial cell cultures. *Biochem. Pharmacol.*, 38:2801-2805.

Simonsson P., Sun G.Y., Vecsei L. and Alling C., 1989b, Ethanol effects on bradykinin-stimulated phosphoinositide hydrolysis in NG 108-15 neuroblastoma-glioma cells. *Cell Signalling*, 6:475-479.

Smith T.L., 1983, Influence of chronic ethanol consumption on muscarinic cholinergic receptors and their linkage to phospholipid metabolism in mouse synaptosomes. *Life Sci.*, 22:661-663.

Smith T.L., 1990, The effects of acute exposure to ethanol on neurotensin and guanine nucleotide-stimulation of phospholipase C activity in intact NIE-115 neuroblastoma cells. *Life Sci.*, 47:115-119.

Smith T.L., Yamamura H.I. and Lee L., 1986, Effect of ethanol on receptor-stimulated phosphatidic acid and polyphosphoinositide metabolism in mouse brain. *Life Sci.*, 39:1675-1684.

Sun, G.Y. and Sun, A.Y., 1983, Chronic ethanol administration induced an increase in phosphatidylserine in guinea pig synaptic plasma membranes. *Biochem. Biophys. Res. Commun.*, 113:262-268.

Sun, G.Y. and Sun, A.Y., 1985, Ethanol and membrane lipids. *Alcoholism: Clin. Exp. Res.*, 9:164-180.

Sun, G.Y. and Lin, T-N., 1989, Time course for labeling of brain membrane phosphoinositides and other phospholipids after intracerebral injection of [^{32}P]-ATP. Evaluation by an improved HPTLC method. *Life Sci.*, 44:689-696.

Sun, G.Y. and Lin T.N., 1990, Dynamic turnover of mouse brain phospholipids during normal aging and response to ischemia. *Upsala J. Med. Sci. (Suppl)*, 48:209-218.

Sun, G.Y., Huang, H-M., Lee, D-Z. and Sun, A.Y., 1984, Increased acidic phospholipids in rat brain membranes after chronic ethanol administration. *Life Sci.*, 35:2127-2133.

Sun, G.Y., Huang, H-M., Chandrasekhar, R., Lee, D.Z. and Sun, A.Y., 1987a, Effects of chronic ethanol administration on rat brain phospholipid metabolism. *J. Neurochem.*, 48:974-980.

Sun, G.Y., Huang, H-M., Lee, D-Z., Chung-Wang, Y-J., Wood, W.G., Strong, R. and Sun, A.Y., 1987b, Chronic ethanol effect on the acidic phospholipids of synaptosomes isolated from cerebral cortex of C57BL/6NNIA mice--a comparison with age. *Alcohol and Alcoholism*, 22:367-373.

Sun, G.Y., Chandrasekhar, R. and Huang, H-M., 1989a, Effects of acute and chronic ethanol administration on metabolism of brain acidic phospholipids. In "Molecular Mechanisms of Alcohol", Sun, G.Y., Rudeen, P.K., Wood, W.G., Wei, Y.H. and Sun, A.Y., eds., Humana Press, Clifton, New Jersey, pp 15-38.

Sun G.Y., Yoa F-G. and Lin T-N., 1990, Degradation of poly-phosphoinositides in brain subcellular membranes in response to decapitation insult. *Neurochem. Int.*, 17:529-535.

Sun G.Y., Navidi M., Yoa F.G. and Lin T.N., 1991, Effects of acute and chronic ethanol administration on phosphoinositide metabolism in mouse brain. *Alcohol and Alcoholism (Suppl 1)*, p215-219.

Sun, G.Y., Navidi M., Yoa F.G., Lin T.N., Orth O.E., Stubbs E.B. and MacQuarrie R.A., 1992, Lithium effects on inositol phosphate and inositol phospholipids in rat brain: studies with radiotracer technique and ion chromatography. *J. Neurochem.*, 58:290-297.

Sun, G.Y., Navidi, M., Yoa, F-G., Wood, W.G., and Sun, A.Y., 1993, Effects of chronic ethanol administration on poly-phosphoinositide metabolism in the mouse brain: Variance with age. *Neurochem. Int.*, 22:11-17.

Tabakoff B., Munoz-Marcus M. and Fields J.A., 1979, Chronic ethanol feeding produces an increase in muscarinic cholinergic receptors in mouse brain. *Life Sci.*, 25:2173-2180.

Takazawa K., Vandekerckhove J., Dumont J. and Erneux C., 1990, Cloning and expression in Escherichia coli of rat brain cDNA encoding a Ca^{2+}/calmodulin-sensitive inositol 1,4,5-trisphosphate 3-kinase. *Biochem J.*, 272:107-112.

Wood W.G., Armbrecht H.J. and Wise R.W., 1982, Ethanol intoxication and withdrawal among three age groups of C57BL/6NNIA mice. *Pharmacol. Biochem. Behav.*, 17:1037-1041.

Wood W.G. and Schroeder F., 1988, MiniReview: Membrane effects of ethanol: Bulk lipid versus lipid domains. *Life Sci.*, 43:467-475.

ETHANOL AND PHOSPHOLIPID DEPENDENT SIGNAL TRANSDUCTION:

THE VIEW FROM THE LIVER

Jan B. Hoek, Tomoyuki Nomura and Katsuyoshi Higashi

Department of Pathology and Cell Biology
Thomas Jefferson University
Philadelphia, Pa. 19107, USA

INTRODUCTION

Receptor-mediated signal transduction processes are common targets of acute or chronic ethanol treatment, both in brain and in peripheral tissues (reviewed in Hoffman and Tabakoff, 1990; Hoek et al, 1992). G protein-linked receptors coupled to adenylate cyclase were studied by several groups in different brain membrane preparations (Hoffman and Tabakoff, 1990). These studies identified the receptor-activated G_s protein α subunit as a likely target of acute ethanol actions in the central nervous system. The sensitivity to ethanol varied both with the receptor and the membrane preparation. Chronic ethanol treatment in vivo decreased the sensitivity to ethanol of adenylate cyclase activation (Hoffman and Tabakoff, 1990). Adenylate cyclase has also been recognized as an indirect target for ethanol in the studies of Diamond and coworkers (1991). Ethanol-induced changes in cellular adenosine uptake affected the extracellular level of this agonist; intracellular responses were determined by the type of adenosine receptors available on specific cells. Long-term ethanol treatment in vitro was associated with a decrease in the expression of the α_s subunit for activation of adenylate cyclase (Mochly-Rosen et al, 1988).

More recently, effects of ethanol on G protein-mediated activation of polyphosphoinositide-specific phospholipase C (PLC) were identified. Early studies on various brain preparations (Hoffmann et al, 1986; Smith et al, 1986; Gonzales et al, 1986) had indicated that ethanol did not significantly affect the basal PLC activity, but that acute or chronic ethanol treatment diminished the inositol phosphate accumulation in response to selected agonists. Hoffman et al (1986) reported that carbachol-induced inositol phosphate formation was sensitive to both acute and chronic ethanol treatment, whereas norepinephrine-induced inositol phosphate formation was not affected. An inhibition of agonist-induced inositol phosphate formation was also reported by Smith et al (1986) in liver cell preparations. The reason for the differential sensitivity to different agonists was not identified in these studies and the mechanism underlying these actions of ethanol remained uncharacterized.

Initial studies by our group (Hoek et al, 1987, 1988) on isolated liver cells detected a process that differed markedly from the findings reported earlier in brain preparations and other cells. Acute ethanol treatment induced a transient increase in cytosolic free Ca^{2+} concentration ($[Ca^{2+}]_{cyt}$), due to the rapid mobilization of intracellular Ca^{2+} stores; this was associated with an accumulation of inositol-1,4,5-trisphosphate ($InsP_3$) and a decrease in its precursor phosphatidylinositol 4,5-bisphosphate ($PtdInsP_2$). Kinetically, the rate of $InsP_3$ formation matched the increase in $[Ca^{2+}]_{cyt}$. Moreover, a vasopressin concentration that caused a similar rate of Ca^{2+} release also induced a comparable $InsP_3$ formation. Pretreatment of the cells with vasopressin or phenylephrine, which depleted the $InsP_3$-sensitive Ca^{2+} stores, prevented the ethanol-induced Ca^{2+} mobilization. Hence, the data indicated that ethanol activated PLC, even in the absence of other agonists. Similar results were identified in a number of other non-excitable cells, including human platelets, turkey erythrocytes and pancreatic acinar preparations (see Hoek et al (1992) for a recent review). Further studies on a membrane preparation from turkey erythrocytes that maintains adequate receptor-phospholipase C coupling, identified a direct action of ethanol on the G protein-activated PLC, where it probably affected the kinetics of GTP binding or GDP release from the G protein (Rooney et al, 1989). Chronic treatment of experimental animals with ethanol in vivo did not affect the activation of PLC by acute ethanol in isolated hepatocytes (Hoek et al, 1990). More recently, a rapid desensitization by ethanol of the responsiveness to hormonal stimuli was also detected in isolated hepatocytes and evidence was provided that this action of ethanol involved an activation of protein kinase C (Higashi and Hoek, 1991). Fig. 1 gives a schematic outline of the possible sites of action of ethanol on PLC in these cells.

The finding that ethanol activates PLC in hepatocytes and certain other peripheral cell types has prompted some recent studies to reevaluate the responses to ethanol in different resting and stimulated neuronal cell lines (Simonsson et al, 1989a,1989b,

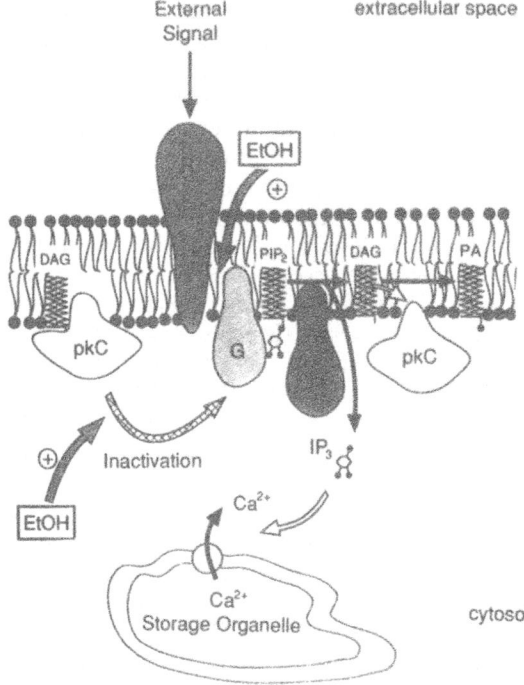

Figure 1. Schematic representation of the sites of action of ethanol in phospholipase C-linked signal transduction in the liver.

1991; Smith, 1990). Although these studies in general failed to identify a response to ethanol in resting cells, a considerable variety of responses was reported in agonist-stimulated cells. For instance, Simonsson et al (1989, 1991) reported that bradykinin-stimulated PtdInsP$_2$ hydrolysis was not affected by acute ethanol in NG108-15 neuroblastoma-glioma cells, but was inhibited transiently after 2-3 days of ethanol exposure. The same group observed an enhanced response to serotonin (but not to other agonists) in primary astroglial cell cultures (Simonsson et al (1989).

Is there a mechanistic basis for the different responses to ethanol of the polyphosphoinositide signalling system in peripheral cells and in different neuronal preparations? Does the apparent lack of an inositol phosphate response to acute ethanol exposure in several resting neuronal cells reflect a different make-up of the control of the PLC-linked signalling processes in brain? Is the reported inhibition by ethanol observed in some of the brain preparations stimulated with carbachol or other agonists related to the desensitization to hormonal responses in liver cells?

An answer to these questions will require a detailed comparative analysis, based on an understanding of the molecular components involved in the control of PLC activity in different cells; this kind of information is not yet available, even in a well-studied system such as the liver. Our studies have focussed on gaining further insights into the parameters that modulate the sensitivity of the cellular signalling systems to ethanol. During the course of our studies on liver cells, some of which will be detailed below, we have observed that the magnitude of the Ca^{2+} response to ethanol in vitro may vary substantially, depending on the incubation conditions or the pretreatment of the cells. These studies point to the feedback control of signalling processes by protein kinases and phosphatases as a crucial element that can alter the phenomenological response of a particular cell to a hormonal stimulus or to ethanol. Our studies suggest that the inhibition by ethanol of hormone-mediated responses can be altered independently from the conditions that allow PLC activation by ethanol. This inference is confirmed in studies on single cells, using microscopic fluorescence imaging techniques, where cell-to-cell variations in the response to ethanol also appeared to be common. We think these mechanisms reflect aspects of the control of PLC-mediated signal transduction that are of more general significance and may also be relevant for understanding the nature of the actions of ethanol on receptor-coupled processes in other cells systems, including neuronal cells.

CONTROL OF PHOSPHOINOSITIDE-LINKED SIGNAL TRANSDUCTION BY PROTEIN KINASES AND PHOSPHATASES

Most of our studies were carried out with isolated hepatocytes, stimulated with common G-protein-coupled agonists; vasopressin was routinely used. Vasopressin couples effectively to the activation of a phospholipase Cβ_1 in hepatocytes, mediated by a G protein of the G$_q$ subfamily. Similar to most G protein-coupled receptors that activate PLC in liver cells, the stimulation by subsaturating concentrations of vasopressin is inhibited by phorbol esters that activate protein kinase C, such 12-O-tetradecanoyl phorbol 13-acetate (TPA). Ethanol-induced PLC activation also shares this sensitivity to TPA: both the Ca^{2+} elevation and the InsP$_3$ formation by ethanol can be completely suppressed by TPA (Fig. 2). Since the activation of PLC itself causes formation of diacylglycerol (DAG) and activation of protein kinase C, this system could account for feedback inhibition of PLC that makes the stimulation transient, particularly with weak agonists, such as ethanol. Protein kinase C inhibitors, such as H7 or staurosporin, enhance the InsP$_3$ formation and the Ca^{2+} mobilization in response to both ethanol and low concentrations of vasopressin (Hoek et al, 1988 and unpublished data).

The site of action of protein kinase C has not been adequately established. In some instances (e.g., vasopressin) a high concentration of the agonists overcomes the inhibition, but with ethanol, or other weak agonists, there is almost complete suppression of both the $InsP_3$ formation and the Ca^{2+} response when cells are treated with maximally effective concentrations of TPA. The effects of TPA require phosphorylation, as indicated by the fact that the protein phosphatase inhibitor okadaic acid markedly potentiates the actions of TPA. A likely model is that one or more of the molecular components of the PLC-coupled signal transduction system is a target for protein kinase C, leading to uncoupling of the receptor - G protein - PLC complex.

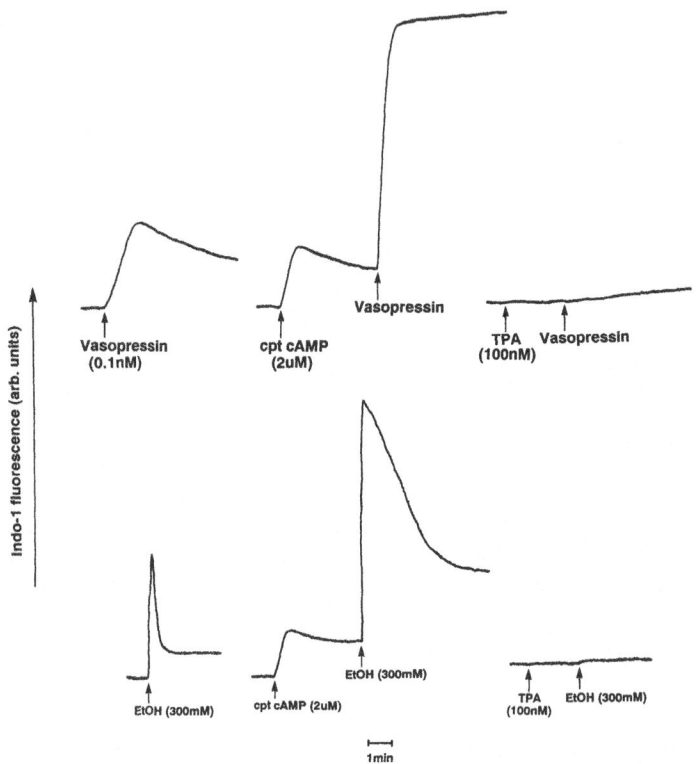

Figure 2. Effects of TPA (100 nM) and CPT-cAMP (2 μM) on the the intracellular Ca^{2+} change induced by ethanol (300 mM) and vasopressin (0.1 nM) in intact indo-1 loaded hepatocytes.

A similar feedback inhibition by protein kinase C of receptor-mediated PLC activation occurs in many other cells, including neuronal cell lines. In addition, there is evidence that not only PLC-coupled receptors are subject to this mode of control, but also voltage-gated or neurotransmitter-gated ion channels (Deitrich et al, 1989; Shearman et al, 1989). This control of ion channel activity by phosphorylation probably plays a role in some adaptive changes in channel activity. There is some evidence that adaptive responses to long-term ethanol treatment may involve protein kinase C (e.g., Messing et al, 1990).

A contrasting effect to that of protein kinase C is exerted in liver cells when protein kinase A is turned on, either by the activation of adenylate cyclase with glucagon or forskolin, or by the addition of a permeant cAMP analog, such as CPT-cAMP. Fig. 2 illustrates the effect of protein kinase A activation on the vasopressin-induced and ethanol-induced Ca^{2+} mobilization. The enhancement by CPT-cAMP results in a significant left-shift in the ethanol dose-response curve (Fig. 3).

222

At least two mechanisms may be involved. Firstly, protein kinase A potentiates the Ca^{2+} release from $InsP_3$-sensitive Ca^{2+} stores in the cell by making the receptor responsive to lower levels of $InsP_3$ (Burgess et al, 1991). This probably involves the phosphorylation of the $InsP_3$ receptor itself. Thus, a larger Ca^{2+} release occurs at low $InsP_3$ concentrations. Secondly, under certain conditions protein kinase A also potentiates the agonist-induced formation of $InsP_3$, i.e., it enhances the activation of phospholipase C by other agonists, although, by themselves, agents that increase cAMP levels do not cause much $InsP_3$ formation. Our recent data indicate that the latter effect of protein kinase A is most evident when phospholipase C is partly inhibited by phorbol esters, i.e., it may operate by decreasing the sensitivity of the signalling complex to protein kinase C (unpublished observations). Again, the response of the liver cells to ethanol is affected by protein kinase A in the same way as that to other agonists that activate PLC, i.e., there is a potentiation of $InsP_3$ formation and Ca^{2+} mobilization.

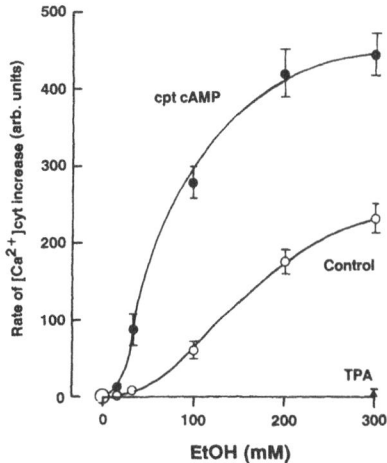

Figure 3. Effects of TPA (100 nM) and CPT-cAMP (2 μM) on the dose-response curve for ethanol-induced changes in intracellular Ca^{2+} concentration.

Although the actions of protein kinase C on the PLC systems are widely observed in many different cell types, including neuronal cells, the effects of protein kinase A are not generally shared by all these cells. Often, there is even an antagonistic effect of protein kinase A on the PLC system (e.g., Rubin and Hoek, 1990). The reason for these differences has not been clarified. One potential difference is at the level of the $InsP_3$ receptor, which comes in distinct neuronal and non-neuronal variants (Danoff et al, 1991). Whether this leads to a different sensitivity to protein kinase A has not yet been reported unequivocally. Nevertheless, the principle illustrated by the liver system is certainly more generally applicable, namely that there exists a set of intricate controls on the ability of the PLC system to respond to specific signals; protein kinases and phosphatases are important tools to exert these controls.

INHIBITION BY ETHANOL OF RECEPTOR-MEDIATED PLC ACTIVATION

In addition to the transient activation of PLC by ethanol in liver cells, this signal transduction system is also sensitive to inhibition by ethanol. This inhibition expresses itself in at least two ways in hepatocytes. One is in the very transient nature of the $InsP_3$ response to ethanol: our data indicate that ethanol itself helps suppress the activation of PLC. The other way is evident in a decreased response of ethanol-treated cells to vasopressin or other agonists that activate PLC. The extent of inhibition depends not only on the concentration of ethanol, but also on the agonist level, i.e., at saturating levels of vasopressin, the effect of ethanol is negligible, whereas at low hormone levels the sensitivity to ethanol is markedly enhanced. As indicated above, an inhibitory effect of ethanol on agonist-induced PLC activation has been reported in several other systems, including brain tissue and neuronal cell lines. By contrast, when the cells are treated with CPT-cAMP prior to the addition of ethanol, the inhibitory effect is diminished, even though the Ca^{2+} response to ethanol itself is enhanced. These actions of ethanol are illustrated in Fig. 4A, which shows the effects of CPT-cAMP treatment on the inhibition by ethanol and TPA of the initial rate of vasopressin-induced changes in cytosolic Ca^{2+} level.

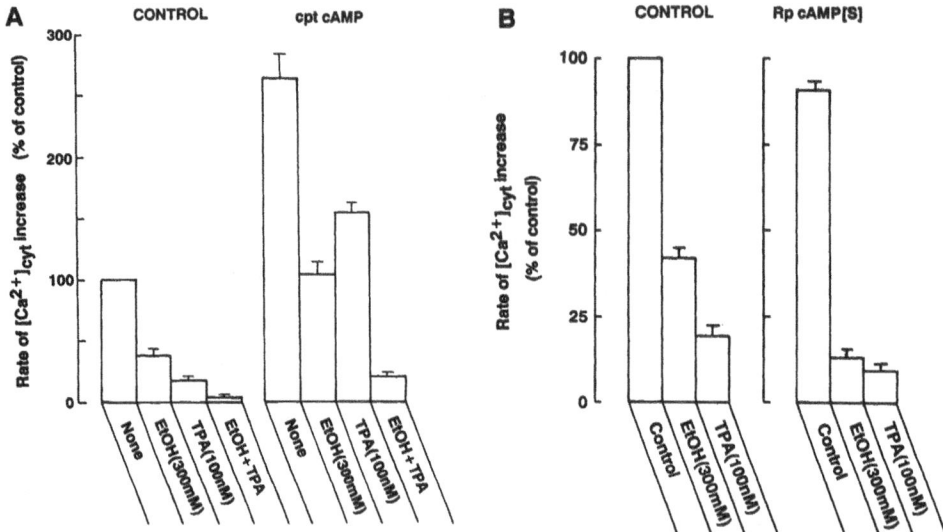

Figure 4. Effects of CPT-cAMP (A) and R_pcAMP[S] (B) on the inhibition by ethanol (300 mM) and TPA (100 nM) of the initial rate of vasopressin-induced changes in intracellular Ca^{2+} concentration. Cells were pretreated for 5 min with CPT-cAMP (2 μM) or R_pcAMP[S] (200 μM). Where indicated, ethanol (300 mM) or TPA (100 nM) was added 2 minutes prior to vasopressin.

We have characterized the inhibitory action of ethanol on liver cells as a desensitization, i.e., an inhibition of the signal transduction system that is secondary to the ethanol-induced activation of PLC (Higashi and Hoek, 1991). This designation implies that only those tissues that exhibit an ethanol-induced activation of PLC are subject to this kind of inhibition. This interpretation was based on our model that the activation of PLC would generate DAG and activate protein kinase C; consequently, processes in the cell that are sensitive to protein kinase C would be targets for an action of ethanol along this route. In support of this interpretation, inhibitors of protein kinase C partly prevented this action of ethanol. However, a number of recent observations relating to the involvement of protein kinase C and protein kinase A indicate that the inhibitory effect of ethanol is more complex than is suggested by this simple model.

As mentioned above, submaximal concentrations of agonists that activate PLC are almost invariably sensitive to TPA, and most of those (but not all) were also sensitive to pretreatment with ethanol. Protein kinase C inhibitors, such as H7 and staurosporin, counteracted the effects of ethanol, providing direct evidence for the involvement of this class of protein kinases in the inhibitory actions of ethanol (Higashi and Hoek, 1991). Contrary to expectation, however, when cells were treated with ethanol and a maximally effective concentration of TPA (100-400 nM), ethanol still markedly potentiated the actions of TPA. In Fig. 4A these combined effects of TPA and ethanol are shown for an intermediate concentration of vasopressin (0.4 nM). These experiments provide strong evidence that the action of ethanol is at least partly elsewhere, i.e., that ethanol enhanced the actions of protein kinase C itself, either directly, or indirectly. The synergistic effects of ethanol and TPA were evident even more when cells were treated with CPT-cAMP.

Two other classes of inhibitors could potentiate the effects of both TPA and ethanol in this system. In our earlier studies, we found that some agonists that activate PLC, but at the same time cause an increase in cAMP in the cells (e.g., the α_1-adrenergic agonist phenylephrine), were insensitive to ethanol (Higashi and Hoek, 1991). When protein kinase A was inhibited (by preincubating the cell with an inhibitory analog of cAMP, R_pcAMP[S]) the phenylephrine-induced phospholipase C activation became sensitive to ethanol. Thus, protein kinase A diminished the inhibitory action of ethanol, similar to the way it counteracted the inhibitory effect of TPA (Fig. 4A). Interestingly, as shown in Fig. 4B, inhibition of protein kinase A by R_pcAMP[S] also potentiated the effects of both TPA and ethanol, even with an agonist like vasopressin that, by itself, did not affect cAMP levels in the cell. Thus, even the baseline activity of protein kinase A in the cell can be sufficient to suppress some of the sensitivity of the system to ethanol and TPA.

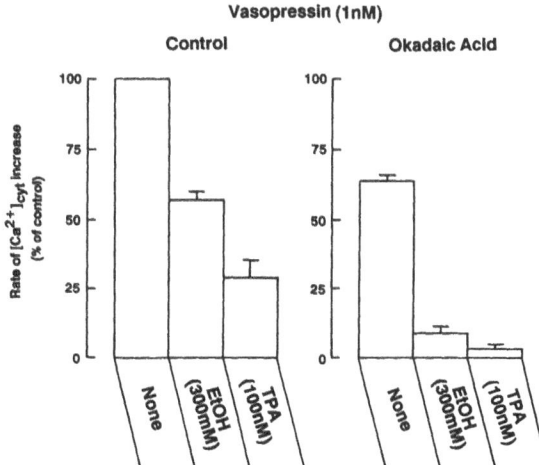

Figure 5. Synergistic effect of okadaic acid (1 μM) on the inhibition by ethanol and phorbol esters of vasopressin-induced increase in cytosolic Ca^{2+} level.

The actions of both protein kinase A and protein kinase C in intact cells are counteracted by protein phosphatases. Although the specific actions of these enzymes are still poorly understood, a series of specific inhibitors have recently become available, one of the most potent of which is okadaic acid. In view of the opposite actions of protein kinases A and C on the polyphosphoinositide-linked signalling system in liver, it was of interest to assess the effects of this compound, which inhibits the two most prominent classes of protein phosphatases, PP-1 and PP-2A. The data of Fig. 5 indicate that okadaic acid markedly enhanced the inhibitory effect of both TPA and ethanol. Thus, the effects of protein kinase C are in part counteracted by an ongoing dephosphorylation in the intact cell. By contrast, okadaic acid appeared to suppress the enhancement of PLC activation by protein kinase A (data not shown).

A summary of these opposing effects of protein kinase A and protein kinase C on the PLC activation by vasopressin is given in Table 1. Two overall messages emerge from this comparison: Firstly, ethanol always potentiates the effect of even saturating concentrations of phorbol esters (and vice versa). Secondly, ethanol decreases the effects of protein kinase A activation and conversely, its effects are enhanced when protein kinase A is inhibited. These data illustrate a counter-regulation of the PLC system by protein kinase A and protein kinase C. Ethanol interferes with these controls by shifting the balance in favor of the actions of protein kinase C. The mechanistic implications of these interactions of ethanol with the control of the PLC signal transduction system by protein kinase A and protein kinase C are not yet clear. However, recent experiments suggest that an important element in these controls are the protein phosphatases that counteract the actions of protein kinase C. Our data would be compatible with a model in which at least some of the effects of ethanol are mediated through a change in protein phosphatase activities. Further studies to analyze this system are currently in progress.

TABLE 1. Summary of effects of activators and inhibitors of protein kinase A, protein kinase C and protein phosphatase on the interactions of ethanol with PLC.

Agent	Action	Activation of Basal PLC	Inhibition of Hormonally Stimulated PLC
CPT-cAMP	PKA Activation	Potentiate	Diminish
R_pcAMP[S]	PKA Inhibition	No Effect	Potentiate
TPA	PKC Activation	Diminish	Potentiate
H7	PKC Inhibition	Potentiate	Diminish
Okadaic acid	PPase Inhibition	Diminish	Potentiate

EFFECTS OF ETHANOL ON THE PLC SIGNAL TRANSDUCTION SYSTEM IN INDIVIDUAL HEPATOCYTES

The previous set of experiments indicates that ethanol does more than just activating PLC and, thereby, turning on protein kinase C. In general, it appears that the susceptibility of the cellular signalling system to ethanol can be altered, irrespective of the strength of the ethanol-induced PLC activation. Conditions that enhance the PLC activation by ethanol may even suppress its ability to inhibit the response to vasopressin, and vice versa. These observations raise the question whether there is an obligatory connection between the ability of ethanol to activate PLC and to inhibit receptor-mediated control of the system, as we suggested previously (Higashi and Hoek, 1991).

In order to gain further insight into this relationship, we studied the effects of ethanol on the hormonal stimulation of individual hepatocytes by fluorescence microscopic imaging. During the past few years, the application of these techniques to the analysis of hormonally induced changes in cytosolic Ca^{2+} levels has opened up new insights into the control of intracellular signalling mechanisms. Woods et al (1986) were the first to demonstrate that hepatocytes stimulated with vasopressin or α_1-adrenergic agonists exhibit an oscillatory elevation of cytosolic Ca^{2+} levels. These oscillations do not show up in an analysis of cell suspensions, because the additive effects of a large number of cells that oscillate out of phase with each other give the impression of a smooth increase to a steady state elevation of $[Ca^{2+}]_{cyt}$.

These hormonally induced Ca^{2+} oscillations in single hepatocytes had several remarkable features. Firstly, cells responded to increasing doses of hormone with an increased frequency of oscillation of cytosolic Ca^{2+} levels, but did not increase the peak Ca^{2+} concentration of individual Ca^{2+} transients (i.e., there is frequency modulation rather than amplitude modulation (Thomas et al, 1991). More recent studies have demonstrated that each Ca^{2+} transient moves across the cell as a wave of Ca^{2+} elevation, originating in a particular area and spreading to other parts of the cell (Thomas et al, 1991). The hormone concentration has no effect on the rate of the Ca^{2+} wave spreading across the cell, but primarily affects the resting period between individual transients. Secondly, the response pattern of individual cells, even in the same microscope field, was often markedly different in the characteristics of oscillation frequency, peak Ca^{2+} concentration, and the length of the latency period before the onset of oscillations. Most strikingly, individual cells showed large differences in hormone sensitivity, i.e., low concentrations of hormone would stimulate only some of the cells; higher concentrations would trigger a higher frequency of oscillation, or even a sustained Ca^{2+} elevation in those cells, with other cells in the same field just starting to respond with a low frequency oscillation. Moreover, when the same cells, after a suitable recovery period, were stimulated again with the same hormone concentration, the characteristic features of that cell's Ca^{2+} oscillation could usually be repeated almost identically (Thomas et al, 1991). Figs. 6 A and B illustrate this for liver cells stimulated with vasopressin or phenylephrine. These features indicated that the factors that control the response to individual cells to a particular hormone concentration may vary widely, even under otherwise identical conditions of stimulation. This heterogeneity probably reflects individual differences between cells in the density, distribution and activity state of receptors, G proteins, PLC complexes, $InsP_3$ receptors, Ca^{2+} stores and protein kinases and phosphatases.

The response of individual cells to ethanol was notably different from those exhibited by hormonal stimuli (Fig. 6C). Instead of showing a repeated oscillation, ethanol usually induced only a single Ca^{2+} transient; the effect of ethanol concentration is reflected mostly in the number of cells responding. This is not entirely surprising, since the ethanol-induced increase in the $InsP_3$ level returns to basal levels within a few minutes, in contrast to hormonal stimuli, which cause a sustained steady state increase in $InsP_3$ concentration (Hoek et al, 1987). Moreover, after an initial stimulation by ethanol, a repeated stimulation of the same cells with ethanol did not generate a renewed Ca^{2+} response, even after a recovery period that is normally sufficient to regain hormonal responsiveness. The response to ethanol was also suppressed by pretreating the cells with other stimuli, including growth factors. Hence, cells that were cultured in a medium containing fetal calf serum often failed to respond to ethanol (unpublished observations). The reasons underlying this marked sensitivity of the ethanol-induced PLC activation are currently under investigation.

We have used the single cell Ca^{2+} analysis to gain further insight into the relationship between the ethanol-induced PLC activation and the ability to suppress hormonal responses in hepatocytes. The concept behind these experiments is as follows: if the ethanol-induced activation of PLC is a prerequisite for the desensitization, then the relationship between these phenomena should be evident on a cell-by-cell basis, i.e., a cell that exhibits an ethanol-induced inhibition of the response to vasopressin would be a cell that is capable of activating PLC (as indicated by the ability to cause an increase in $[Ca^{2+}]_{cyt}$). On the other hand, a cell that does not respond to ethanol by itself would not be expected to generate an inhibition of the response to vasopressin.

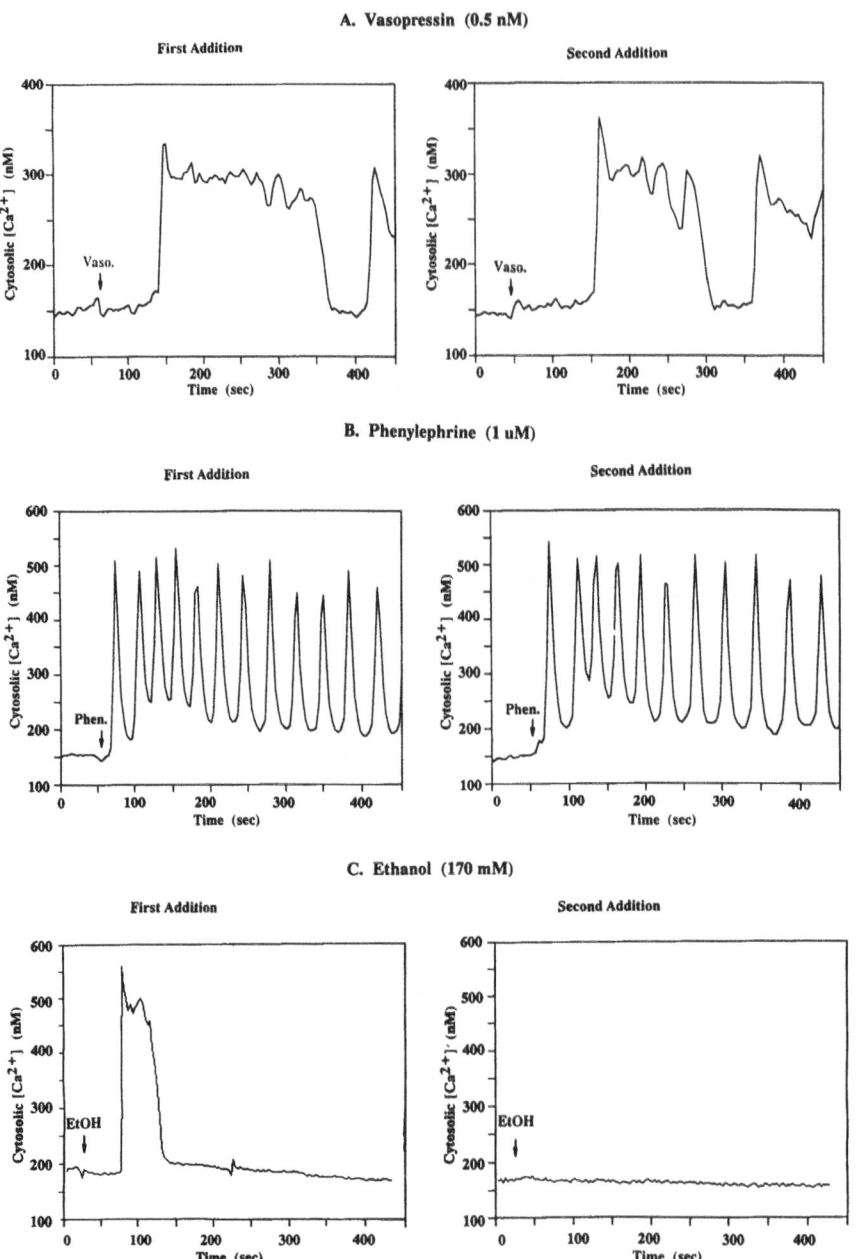

Figure 6. Oscillations in cytosolic Ca^{2+} level in single hepatocytes in response to 0.5 nM vasopressin (A), 1 μM phenylephrine (B) or 170 mM ethanol (C). Freshly isolated cells were plated for 2 h on polylysine-coated glass coverslips and loaded with fura 2. Changes in $[Ca^{2+}]_{cyt}$ were analyzed by digital fluorescence microscopy as described in detail elsewhere (see Thomas et al, 1991). After an initial stimulation, the agonist was removed and the cells were allowed to recover for 40 minutes prior to a second stimulation of the same field with the same agonist concentration.

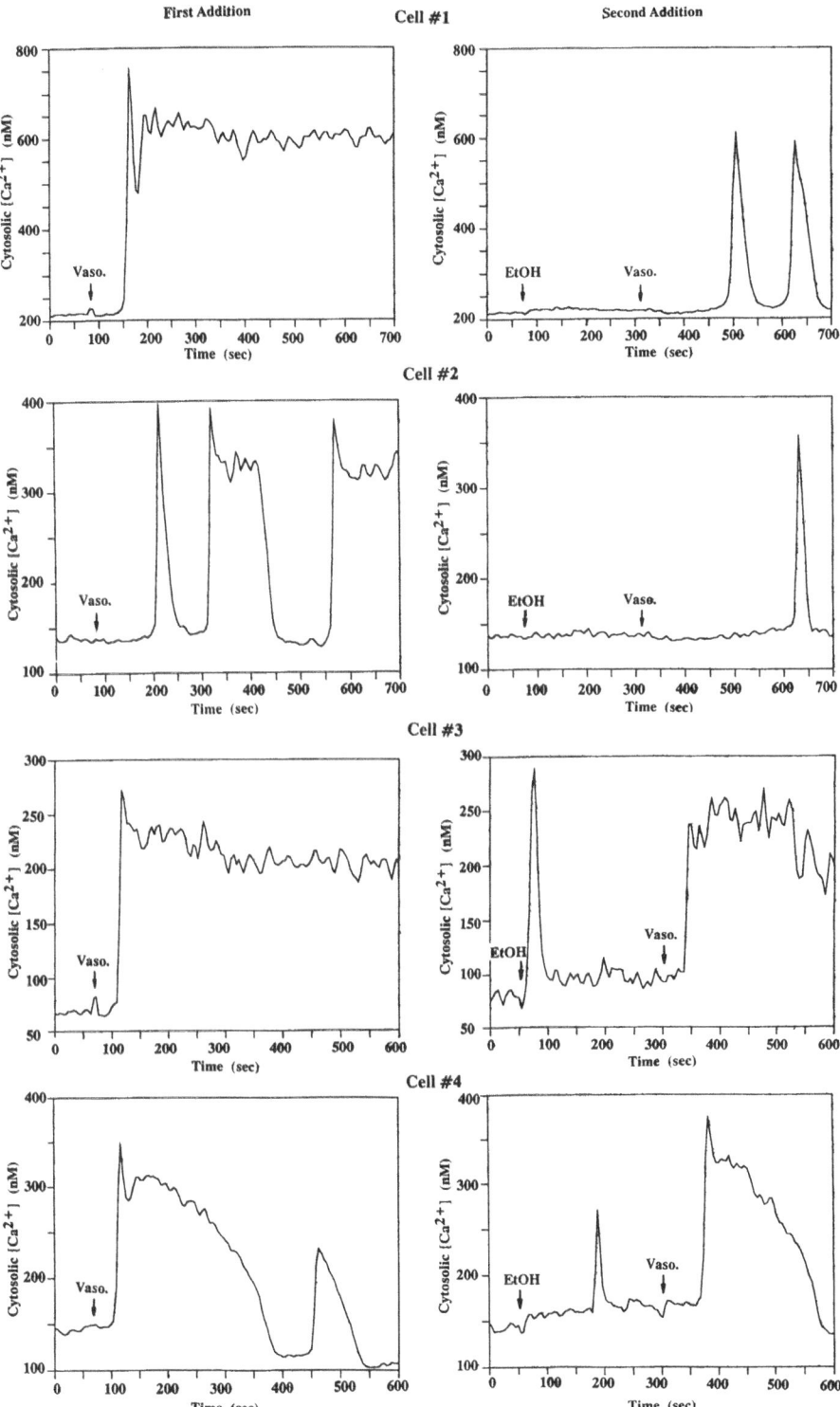

Figure 7. Effect of ethanol on the vasopressin-induced oscillations in cytosolic Ca^{2+} level in single hepatocytes. Cells were first stimulated with 0.5 nM vasopressin, allowed to recover for 40 minutes, and then treated with ethanol (170 mM), prior to a restimulation with vasopressin. Traces are representative examples of four individual cells from a single microscope field.

Results of a set of individual cells from a characteristic cell preparation are illustrated in Fig. 7.

In all these experiments, the cells were first stimulated by vasopressin (0.5 nM), to identify the cell-specific response to this agonist. After a 40 minute recovery period, cells were treated with ethanol (170 mM), followed 5 minutes later by vasopressin (0.5 nM). Control experiments were done to ensure that each cell responded the same way to a second stimulation with vasopressin, if ethanol was omitted (see Fig. 6). The data shown in Fig. 7 suggest clearly that there is no obligatory connection between PLC activation by ethanol and the inhibition of the vasopressin response. The figure shows two examples of cells that failed to show a Ca^{2+} elevation in response to ethanol (cells #1 and 2); yet, the subsequent response to vasopressin was severely suppressed, compared to the response in the same cell prior to ethanol treatment. By contrast, in two others (cells # 3 and 4), where ethanol generated a single Ca^{2+} peak, the response to vasopressin was almost unaffected.

A summary of the analysis of all the cells in a microscope field in these experiments is shown in Fig. 8. In this figure, the responses of individual cells were classified as negative (no response during the test period), oscillatory (i.e., a response that returned to baseline at least once during the test period) or saturated (i.e., a sustained elevation of cytosolic Ca^{2+} levels). The left two bars (A1 and A2) compare the response to two sequential additions of 0.5 nM vasopressin. The majority of cells in this field showed a sustained Ca^{2+} response and almost all cells maintained the same response pattern after a second challenge with vasopressin. The two bars on the right (B1 and B2) compare the response (in a different microscope field) to an initial challenge with vasopressin with that to a subsequent challenge by vasopressin, preceded by 170 mM ethanol. In this experiment, most of the cells showed a sustained Ca^{2+} elevation in response to vasopressin, but less than 20 % of the cells showed a detectable Ca^{2+} response to ethanol itself (not shown). There was no apparent correlation in individual cells between the ability to respond to ethanol and the suppression of the subsequent response to vasopressin. The data indicate that there is no obligatory connection between the capacity of ethanol to activate PLC and its ability to suppress the response to other agonists, i.e., the activation of PLC by ethanol and the enhancement of the feedback inhibition by protein kinase C may be mechanistically unrelated actions of ethanol.

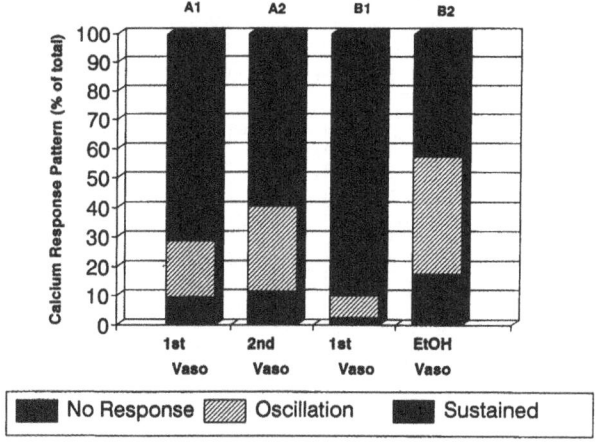

Figure 8. Effect of ethanol on Ca^{2+} response patterns to vasopressin stimulation in single hepatocytes. Conditions as described for Figs. 6 and 7. Results from 56 (A1 and A2) or 48 (B1 and B2) individual cells in the microscope field were grouped as either not responding, showing an oscillatory response, or showing a sustained Ca^{2+} elevation.

DISCUSSION

Ethanol and the control of PLC in the liver

The data discussed here and in our previous studies illustrate both similarities and significant differences in the mechanisms of activation of PLC by ethanol and by receptor-mediated agonists. The regulation of PLC activation by protein kinases A and C is essentially identical, whether the enzyme is activated by ethanol, or whether stimulation is by a hormone through a G protein-coupled receptor. The major difference between the PLC activation by ethanol and vasopressin is in switching off the enzyme, as reflected in the period during which an elevated InsP$_3$ level and an increased $[Ca^{2+}]_{cyt}$ is maintained. At the single cell level, this translates into a limited period in which an ethanol-induced Ca^{2+} oscillation is sustained. In most of our experiments, the Ca^{2+} response is limited to a single transient, rather than the repeated oscillations observed with hormonal stimulation, which continue for at least 15 minutes.

Our earlier studies indicated that PLC activation by ethanol was due to its action at the membrane level, possibly by interfering with protein-protein interactions involved in G protein activation (Hoek et al, 1992). However, the reason why the response to ethanol is much more transient, compared to hormonal stimuli, has not been adequately clarified. The data shown here and in earlier studies support the notion that the origin may lie in the synergistic effects of ethanol and protein kinase C activators, i.e., protein kinase C is much more potent in inhibiting PLC in the presence of ethanol than in its absence. Thus, the feedback inhibition of ethanol-induced PLC activity reflects the same actions of ethanol as its inhbition of hormone-induced PLC stimulation. This inhibitory action of ethanol does not have an obligatory link to the ethanol-induced activation of PLC. Neither of these actions of ethanol appears to depend on its oxidation, as most of these data were obtained in cells treated with 4-methylpyrazole, a potent inhibitor of alcohol dehydrogenase in liver.

A synergistic action of enhancing the effects of protein kinase C in this system is also exhibited by two other classes of agents, namely the protein phosphatase inhibitor okadaic acid and R$_p$cAMP[S], an inhibitor of protein kinase A. Okadaic acid enhances the action of protein kinase C, presumably by preventing the reversal of the phosphorylation of its target protein by an endogenous protein phosphatase. What this target protein is has not yet been defined. Protein kinase A counteracts the inhibitory effects of protein kinase C on PLC. The synergism between R$_p$cAMP[S] and TPA in the absence of protein kinase A activation indicates that even the endogenous activity is sufficient to contribute to this effect. Recent evidence in our laboratory suggests that the action of protein kinase A may also be in part mediated through a modulation of protein phosphatase activity in the cell (unpublished observations). Whether the synergistic effects of ethanol and protein kinase C activators are similarly related is currently under investigation. A recent study on cultured hepatocytes indicates that ethanol causes inhibition of adenylate cyclase, mediated through adenosine A$_1$ receptors (Nagy, 1992). However, in freshly isolated hepatocytes, we have been unable to detect a decrease in cellular cAMP levels in response to ethanol, nor have we found significant effects of adenosine deaminase or adenosine receptor antagonists on the effects of ethanol described here.

An important aspect that has emerged from studies of Ca^{2+} transients in single hepatocytes is the heterogeneity of the cellular responsiveness to various receptor-mediated agonists. However, a repeated hormonal stimulation of the same cells results in essentially reproducible Ca^{2+} oscillation patterns. Thus, the heterogeneity in response patterns in different cells probably reflects intrinsic differences in the control features expressed in individual cells, even when they are isolated and incubated under identical conditions and exposed to an identical stimulus.

The cells were also heterogeneous in their response to ethanol. This heterogeneity was expressed both in a variable capacity to exhibit an ethanol-induced Ca^{2+} transient and in a variable degree of inhibition by ethanol of the response to vasopressin. As with the heterogeneous hormonal responsiveness, the basis for the variability in the response to ethanol is not clear. Preliminary observations indicate that specific culture conditions, e.g., unidentified components in the fetal calf serum, suppress PLC activation by ethanol, without affecting the cell's ability to respond to vasopressin or phenylephrine. However, it is not clear why the response of the cells to the treatment with these factors is not uniform, or what the reason is for the slow and variable recovery from these (or other) suppressive factors. Currently, studies are in progress to identify these factors in more detail. Additionally, our experiments indicate that there is no cell-to-cell correlation between the heterogeneity in the ethanol-induced activation of PLC and the inhibition by ethanol of the hormone-induced Ca^{2+} oscillation. Thus, the factors responsible for either of these actions of ethanol independently exhibit cell-to-cell variability.

Are the effects of ethanol on the control of PLC in hepatocytes a model for other cell types?

Although there is so far no evidence of an ethanol-induced activation of PLC in brain tissue or in neuronal cell lines, there is considerable evidence that ethanol affects the response of these cells to other stimuli that activate PLC. Both inhibitory and stimulatory effects have been described, depending on the cell type, the agonist used and the time of exposure to ethanol (Hoffmann et al, 1986; Smith et al, 1986; Simonsson et al, 1989a; Simonsson et al, 1989b; Smith, 1990). Although there is evidence that G proteins may mediate some of these effects of ethanol (Simonsson, 1991), the mechanism of action of ethanol has not been characterized in these studies and there is little information to assess whether protein kinase C or protein phosphatases could play a role in mediating the effects of ethanol, either directly or indirectly. Several factors could be expected to modulate the consequences of ethanol exposure, due to the complexity of multiple simultaneous actions of this compound. Studies by Diamond and coworkers (1991) have identified ethanol-mediated actions on adenylate cyclase, caused by the accumulation of adenosine in the medium. It is not inconceivable that these effects could markedly modify the control of the PLC coupling to specific receptors. A careful analysis of the conditions under which ethanol acts should take into account the complex cross-talk mechanisms between different signal transduction systems operating in these cells. Since the details of cross-talk interactions are likely to be cell-specific and agonist-specific phenomena, the net outcome of ethanol exposure may well vary from one cell-type to another and may depend on the nature of the stimulus. The potential for cell-to-cell heterogeneity, as detected in liver cells, further adds to the complexity. Neverthelss, even though the phenomenology may vary from one experimental system to another, the underlying mechanistic basis for the actions of ethanol may well be closely related.

In conclusion, we feel that our studies on the liver system may point to actions of ethanol that have much wider consequences than presently realized. Actions of ethanol that interfere with the fundamental elements of the protein kinase/phosphatase-dependent regulatory system in the cell are bound to have a wide range of implications for metabolic control systems. The specific features of how these actions are expressed in an individual cell is likely to vary from one system to another.

Acknowledgements

This work was supported by US Public Health Service grants AA07186, AA07215 and AA08714.

REFERENCES

Burgess, G.M., Bird, G.St.J., Obie, J.F., and Putney, J.W. (1991) The mechanism for synergism between phospholipase C- and adenylate cyclase-linked hormones in liver. J.Biol.Chem. 266:4772-4781.

Danoff, S.K., Ferris, C.D., Donath, C., Fischer, G.A., Munemitsu, S., Ullrich, A., Snyder, S.H., and Ross, C.A. (1991) Inositol 1,4,5-trisphosphate receptors: Distinct neuronal and nonneuronal forms derived by alternative splicing differ in phosphorylation. Proc. Natl. Acad. Sci., U.S.A. 88:2951-2955.

Diamond, I., Nagy, L., Mochly-Rosen, D. and Gordon, A. (1991) The role of adenosine and adenosine transport in ethanol-induced cellular tolerance and dependence. Possible biologic and genetic markers of alcoholism. Ann. N.Y. Acad. Sci., 625:473-487.

Deitrich, R.A., Dunwiddie, T.V., Harris, R.A. and Erwin, V.G. (1989) Mechanism of action of ethanol: Initial central nervous system actions. Pharmacol. Rev. 41:489-537.

Gonzales, R.A., Thiess, C. and Crews, F.T. (1986) Effectys of ethanol on stimulated inositol phosphate hydrolysis in rat brain. J. Pharmacol. Exp. Therap. 237:92-98.

Higashi, K. and Hoek, J.B. (1991) Ethanol causes desensitization of receptor-mediated phospholipase C activation in isolated hepatocytes. J. Biol. Chem. 266:2178-2190.

Hoek, J.B., Thomas, A.P., Rubin, R. and Rubin, E. (1987) Ethanol-induced mobilization of calcium by activation of phosphoinositide-specific phospholipase C in intact hepatocytes. J. Biol. Chem. 262:682-691.

Hoek, J.B., Thomas, A.P. and Rubin, R. (1988) Ethanol-induced activation of phospholipase C activation is inhibited by phorbol esters in intact hepatocytes. Biochem. J. 251:865-871.

Hoek, J.B., Taraschi, T.F., Higashi, K., Rubin, E. and Thomas, A.P. (1990) Phospholipase C activation by ethanol in rat hepatocytes is unaffected by chronic ethanol feeding. Biochem. J. 272:59-64.

Hoek, J.B., Thomas, A.P., Rooney, T.A., Higashi, K. and Rubin, E. (1992) Ethanol and signal transduction in the liver. FASEB J. 6:2386-2396.

Hoffman, P.L. and Tabakov, B. (1990) Ethanol and guanine nucleotide binding proteins: a selective interaction. FASEB J. 4, 2612-2622.

Hoffman, P.L., Moses, F., Luthin, G.R. and Tabakoff, B. (1986) Acute and chronic effects of ethanol on receptor-mediated phosphatidylinositol 4,5-bisphosphate breakdown in mouse brain. Mol. Pharmacol. 30:13-18.

Messing, R.O., Sneade, A.B. and Savidge, B. (1990) Protein kinase C participates in up-regulation of dihydropyridine-sensitive calcium channels by ethanol. J.Neurochem. 55:1383-1389.

Mochly-Rosen, D., Chang, F.-U., Cheever, L., Kim, M., Diamond, I. and Gordon, A.S. (1988) Chronic ethanol causes heterologous desensitization by reducing α_s mRNA. Nature 333:848-850.

Nagy, L.E. (1992) Ethanol metabolism and inhibition of nucleoside uptake lead to increased extracellular adenosine in hepatocytes. Am. J. Physiol. 262:C1175-C1180.

Rooney, T.A., Hager, R., Rubin, E. and Thomas, A.P. (1989) Short-chain alcohols activate guanine nucleotide-dependent phosphoinositidase C in turkey erythrocyte membranes. J. Biol. Chem. 264:6817-6822.

Rubin, R. and Hoek, J.B. (1990), Inhibition of ethanol-induced platelet activation by agents that elevate cAMP. Thrombosis Res. 58:625-632.

Shearman, M.S., Sekiguchi, K. and Nishizuka, Y. (1989) Modulation of ion channel activity: A key function of the protein kinase C enzyme family. Pharmacol. Rev. 41, 211-237.

Simonsson, P., Sun, G.Y., Vecsei, L. and Alling, C. (1989a) Ethanol effects on bradykinin-stimulated phosphoinositide hydrolysis in NG108-15 neuroblastoma-glioma cells. Alcohol 6:475-479.

Simonsson, P., Hansson, E. and Alling, C. (1989b) Ethanol potentiates serotonin-stimulated inositol lipid metabolism in primary astroglial cell cultures. Biochem. Pharmacol. 38, 2801-2805.

Simonsson, P., Rodriguez, F.D., Loman, N. and Alling, C. (1991) G-proteins coupled to phospholipase C: molecular targets of long-term ethanol exposure.

Smith, T.L., Yamamura, H.I. and Lee, L. (1986) Effect of ethanol on receptor-stimulated phosphatidic acid and polyphosphoinositide metabolism in mouse brain. Life Sci. 39:1675-1684.

Smith, T.L. (1990) The effects of acute exposure to ethanol on neurotensin and guanine nucleotide stimulation of phospholipase C activity in intact NIE-115 neuroblastoma cells. Life Sci. 47:PL115-PL119.

Thomas, A.P., Renard, D.C. and Rooney, T.A. (1991) Spatial and temporal organization of calcium signalling in hepatocytes. Cell Calcium, 12:111-126.

Woods, N.M., Cuthbertson, K.S.R. and Cobbold, P.H. (1986) Repetitive transient rises in cytoplasmic free calcium in hormone-stimulated hepatocytes. Nature 319:600-602.

PHOSPHOLIPASE C COUPLED G-PROTEINS:
MOLECULAR TARGETS OF ETHANOL

Per Simonsson, F. David Rodríguez, Christer Larsson,
Niklas Loman and Christer Alling

Department of Psychiatry and Neurochemistry
Lund University
Box 638
S-220 09 Lund, Sweden

INTRODUCTION

The transfer of information into the intracellular compartment is an important step in neurotranmission. GTP-binding proteins (G-proteins) mediate this signalling and it is by now well established that ethanol differentially interacts with this set of proteins (reviewed by Hoffman and Tabakoff, 1990, Hoek et al., 1992). Most of the knowledge generated is connected to signalling through the adenylate cyclase system. Less is known about the role of phospholipase C coupled G-proteins. In analogy with the effect of ethanol on the cAMP system, it can be postulated that long-term ethanol may affect receptor-stimulated phosphoinositide hydrolysis by acting on the G-proteins coupled to this system. It is known that this is indeed the case in systems that are sensitive to acute exposure to alcohols (Rubin and Hoek, 1988, Rooney et al., 1989).

We have previously reported that ethanol inhibits bradykinin-stimulated phosphoinositide metabolism in NG 108-15 neuroblastoma x glioma hybrid cells (Simonsson et al., 1989b) and that this effect appears to be related to an effect on the G-proteins coupled to phospholipase C (Simonsson et al., 1991).

This report will focus on the role of G-proteins in ethanol-exposed cellular systems and their possible role in the maladaptation of this system to long-term exposure. Possible molecular mechanisms underlying these phenomenon will be discussed.

RECEPTOR-STIMULATED PHOSPHOINOSITIDE HYDROLYSIS IN NG 108-15 CELLS

NG108-15 cells were selected as experimental model as they are known to express bradykinin receptors coupled to phosphatidylinositol 4,5bisphosphate-specific phospholipase C (Yano et al., 1984). Using this approach we were able to reproduce the rapid hydrolysis of polyphosphoinositides and the subsequent formation of inositol 1,4,5)trisphosphate [$I(1,4,5)P_3$] (Fig 1). The formation of $I(1,4,5)P_3$ reached a maximum 10 sec after the addition of 10^{-6} M bradykinin, thus paralleling the hydrolysis

Alcohol, Cell Membranes, and Signal Transduction in Brain
Edited by C. Alling *et al.*, Plenum Press, New York, 1993

of phosphatidylinositol 4,5bisphosphate (PIP_2). Previous studies have revealed not only an increase in IP_3 but also in IP_2 and IP_1, although at a slower rate with a maximum after 30 sec (Simonsson et al., 1989a). The phospholipase C activation was not accompanied by an increase in free fatty acids, indicating that phospholipase A_2 does not interact in the signalling of the bradykinin receptor in this particular cell system (Simonsson et al., 1989a).

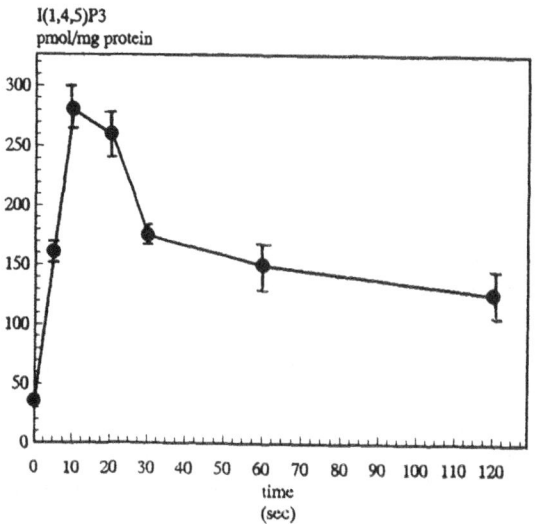

Fig 1. Bradykinin (10^{-6} M) stimulated the formation of I(1,4,5)P_3 in NG 108-15 cells.

ETHANOL AND PHOSPHOINOSITIDE HYDROLYSIS
Acute effects of ethanol

In contrast to the effects found on hepatocytes and on adenylate cyclase systems in brain cells, acute ethanol exposure did not significantly affect basal PIP_2 hydrolysis in NG 108 cells (Simonsson et al., 1989b). Although ethanol at very high concentrations (500 mM) reduced inositol phosphate formation 30 sec after the addition of bradykinin, no effect was seen with ethanol doses < 250 mM.

Long-term effects of ethanol

A significant inhibitory effect of long-term ethanol exposure was found on receptor-stimulated phosphoinositide hydrolysis (Simonsson et al., 1989b). The effect was detectable during the first 1-4 days of exposure to 100 mM ethanol. During this time period, ethanol inhibited the bradykinin-induced phosphoinositide hydrolysis and the formation of radiolabelled inositol polyphosphates. To obtain stable ethanol concentrations that were unaffected by evaporation, a protocol has been established that makes it possible to culture cells at constant levels of ethanol (Rodriguez et al., 1992a).

The previously reported findings have been confirmed in subsequent experiments where the receptor-mediated transduction was measured as the mass determination of I(1,4,5)P_3 formed within the first two minutes of bradykinin stimulation (Fig 2).

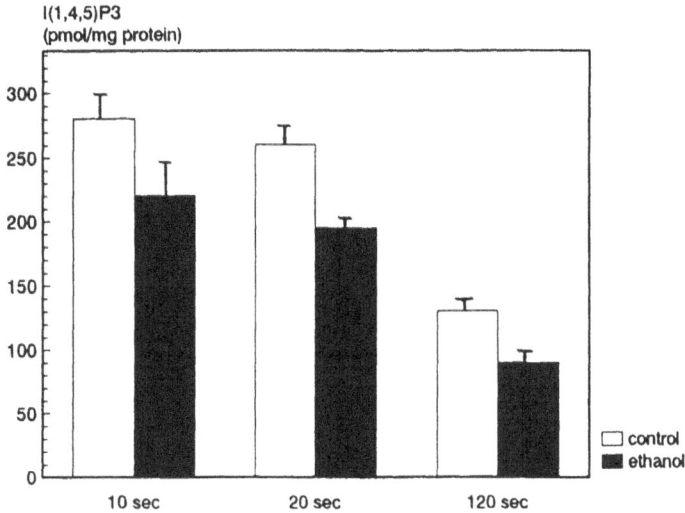

Fig 2. Exposure for 2 days to 100 mM ethanol inhibits the bradykinin-stimulated formation of I(1,4,5)P₃ in NG 108-15 cells.

MECHANISMS OF ACTION

Theoretically, ethanol could exhert its effects directly on any of the three major constituents of the signalling cascade, i.e. receptor protein, G-protein or phospholipase C. It is also possible that the effect was mediated via an alteration of the molecular communication between these interacting proteins in the plasma membrane or via an effect on the lipid-protein interaction. Experimental methods were therefore set up to approach the question on the molecular mechanism(s) underlying ethanol´s long-term inhibitory effects.

Receptor-binding

Basic pharmacological data revealed the presence of a high-affinity bradykinin receptor population with a K_d of approximately 1 nM (Simonsson et al., 1991). The profile of the binding studies are thus in line with the results published elsewhere (Imaizuma et al., 1989). Exposure for 4 days to 100 mM ethanol did not alter the specific binding of [³H]-bradykinin. However, which is of interest in the light of the GTP(S)-stimulation data discussed below, the expected reduction in ligand binding after the addition of GTP analogues was not found in ethanol-exposed cells. In conclusion, these data do not indicate that the binding protein of the transduction system in question was primarily influenced by ethanol.

Phosphatidylinositol 4,5bisphosphate-specific phospholipase C

An alteration in the expression or the function of the intracellular effector enzyme would yield an inhibition similar to the one observed after bradykinin-stimulation. An assay for the determination on phosphatidylinositol 4,5bisphosphate-specific phospholipase C was therefore established. By incubation of a membrane-rich preparation with [³H]-PIP₂ it was possible to follow the formation of water-soluble inositol phosphate metabolites formed after enzymatic cleavage of the radiolabelled phosphoinositide (Suzuki et al., 1989). However, no difference in the enzymatic activity of phosphatidylinositol 4,5bisphosphate-specific phospholipase C was found when comparing results obtained from control cells and ethanol-exposed cells (Simonsson et al., 1991). This makes it unlikely that the ethanol-inhibited signalling found in NG 108-15 cells was related to the intracellular effector enzyme.

G-proteins

G-proteins play a key role in the regulation of transmembrane communication. From a theoretical point of view it is therefore suggestive that ethanol´s effect on signal tranduction may indeed be mediated via an effect on these protein trimers (reviewed by Hoffman and Tabakoff, 1990). To date, this hypothesis has been tested extensively, particularily with regard to the adenylate cyclase system (Luthin and Tabakoff, 1984, Richelson et al., 1986, Diamond et al 1987, Valverius et al., 1987, 1989, Mochly-Rosen et al., 1988, Nagy et al., 1988). It thus appears that the adenylate cyclase-coupled G-proteins are mediating the effect of ethanol on this system, either by influencing the function or the turnover of this family of membrane-associated proteins. The negative findings with regard to the bradykinin binding protein and phosphatidylinositol 4,5bisphosphate-specific phospholipase C also suggest the role of the intermediary step in the mechanisms underlying ethanol´s effect on the bradykinin receptor complex.

The functional status of the G-proteins coupled to phospholipase C was evaluated by stimulating the G-proteins with the non-hydrolyzable GTPanalogue GTP(S) or, in some experiments, with NaF/AlCl$_3$. Originally, GTP(S) stimulations of NG 108-15 cells were performed after the addition of low doses of saponin (Rubin and Hoek, 1988). Following a lag period, phosphoinositide hydrolysis and inositol phosphate formation increased, reaching a maximum after 10 min. Using this method, it was shown that GTP(S)-stimulated phosphoinositide metabolism was significantly inhibited in cells exposed for 4 days to 100 mM ethanol (Simonsson et al., 1991).

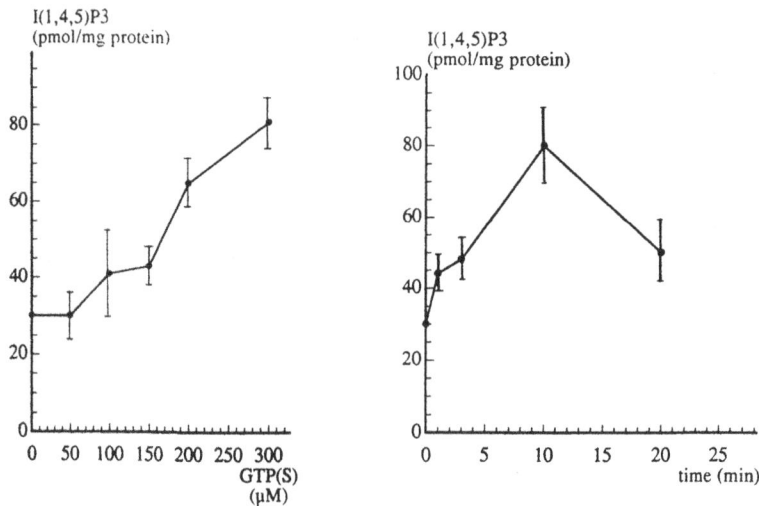

Fig 3. The time- and dose-dependent formation of I(1,4,5)P$_3$ induced by GTP(S) in NG 108-15 cells.

It is therefore suggestive to postulate that it is indeed the G-proteins that are primarily affected by long-term ethanol exposure. However, the data does not answer the question if the effect is mediated by a reduced expression or a inhibited function of the phospholipase C-coupled G-proteins. A definite answer can only be obtained following the characterization of the G-protein species associated to the bradykinin-receptor complex and the subsequent determination of this G-protein and the appropriate mRNA. Awaiting these results, this question can only be addressed using circumstantial evidence.

The time-course of the ethanol induced inhibition, with an effect only after 24 - 48 h of exposure, could suggest a role of altered expression and synthesis. However, the inhibition induced by ethanol disappeared rapidly after cessation of ethanol exposure. Normal signal transduction was found 1-2 hours after the end of ethanol exposure. It is unlikely that expression of the proteins in question would be normalized in this short period of time. It is therefore possible that the effect is rather on the function of the G-proteins. This hypothesis also gets some support from the data obtained from NaF/AlCl$_3$-stimulated cells. This approach is known to activate several forms of G-proteins, presumably by forming an AlF$_4^-$ complex that directly interacts with the G-protein. Although the conclusions from these experiments must be interpreted with caution due to the multitude of intracellular effects of these ions, they suggest that ethanol's principle mode of action is on the function of G-proteins. Using the NaF/AlCl$_3$ protocol for G-protein stimulations, no major effect of ethanol on phosphoinositide metabolism could be found (Simonsson et al., 1991). If GTP(S) and NaF/AlCl$_3$ indeed stimulate the same G-proteins, although at different sites and via different mechanisms, the inconsistency can be explained more easily by an effect on function rather than on expression.

The original method for GTP(S) stimulations suffers several drawbacks. These problems are mainly related to the permeabilization of the cells prior to stimulation. Although this is an effective method for introducing agents that do not readily penetrate the membrane, results may be biased by the experimental protocol. As the processes studied are related to the interaction of proteins in the lipid environment of the plasma membrane of the cell, it is possible that the permeabilization may in itself influence the proteins that are to be studied. In addition, permebilization drastically alters the intracellular environment and leads to altered basal levels of phospholipid and inositol phosphate turn-over. Although there were no differences between control cells and ethanol exposed cells, saponin-treatment more than doubled basal levels of IP$_1$ and IP$_2$ (Simonsson et al., 1991).

A modified method of GTP(S)-stimulation has therefore been developed and subsequently used. The protocol is based on the finding that high concentrations of

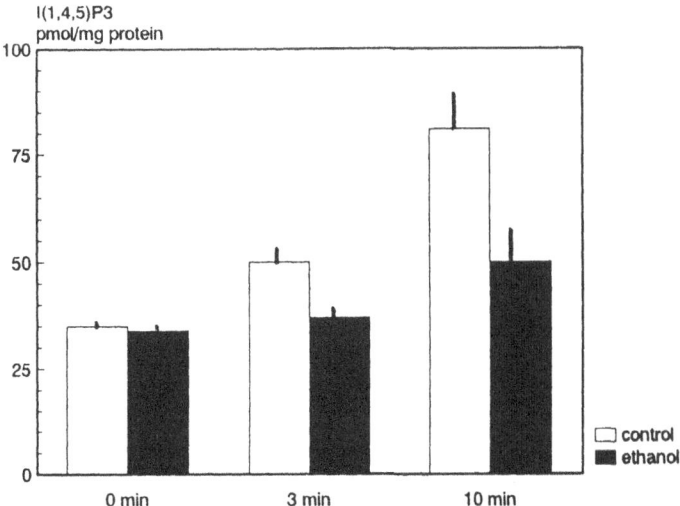

Fig 4. Exposure for 2 days to 100 mM ethanol reduced GTP(S)-stimulated I(1,4,5)P$_3$ in NG 108-15 cells.

GTP(S) (> 10^{-4}M) can activate phospholipase C, presumably by penetration or active uptake, in brain slices (Li et al., 1989) and in NG 108-15 cells (Chiang and Hauser, 1989). Although GTP-analogues have a low membrane permeability, it has been shown that the intracellular amount is sufficient for a significant activation of a N-ethylmaleimide-sensitive GTP-binding protein coupled to phospholipase C. The addition of GTP(S) to NG108-15 resulted in a time- and dose-dependant formation of endogenous $I(1,4,5)P_3$ (Fig 3). Maximal effect was observed at 3×10^{-4} M GTP(S) after 10 min of stimulation. Using this method, we could confirm our previous results that the GTP(S)-stimulated $I(1,4,5)P_3$ formation is indeed reduced after long-term ethanol exposure (Fig 4).

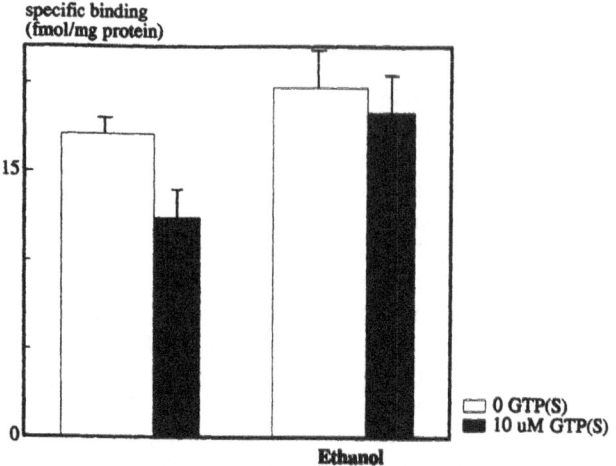

Fig 5. In NG 108-15 cells exposed for 4 days to 100 mM ethanol, the addition of GTP(S) did not reduce the specific binding of the bradykinin receptor.

In addition to these results, observations made during the study of the binding of [³H]-bradykinin indicate an alteration at the level of the G-protein. In ethanol-naive cells, the addition of GTP(S) reduced the specific binding of this receptor. However, in ethanol exposed cells, this phenomenon was not observed (Fig 5).

Taking these different observation into consideration, it appears that 2-4 days of ethanol exposure results in an inhibition of the phospholipase C signalling system by interfering with the G-proteins coupled to this system.

G-PROTEINS AND THE MALADAPTATION TO ETHANOL

Originally, the effect of ethanol was studied only during the first 4 days ef ethanol exposure. In recent studies, the time-span of exposure was increased, yeilding intriguing new data (Rodriguez et al., 1992b). In these experiments, the finding that 2-4 days of ethanol exposure inhibited signal transduction was reproduced. However, following 7 days of exposure to 100 mM ethanol, it was no longer possible to obtain an inhibition. Based on the knowledge of the role of G-proteins in the inhibitory effect of ethanol, the hypothesis that G-proteins participate in the adaptation, or rather maladaptation, was addressed.

GTP(S) stimulations once again resulted in reduced response after 2 or 4 days of ethanol treatment. However, after 7 days of exposure, no inhibition could be observed. In contrast, there was a clear trend towards increased inositol phosphate production in these cells at day 7. Binding studies were performed and it was concluded that after one week of ethanol exposure, the specific binding of [³H]-bradykinin to membranes of NG 108-15 cells was increased. Scatchard analysis of binding data revealed a shift to the left of the displacement curve, indicating a reduction in the low-affinity binding sites and an increase in the high-affinity population. This again can be interpreted in light of what is known concerning G-protein-receptor interaction. It thus appears that the ethanol exposure increased the functional capacity of the appropriate G-proteins, which, secondarily, lead to an adaptation of the signalling through this receptor complex. This observation is not new as parallel conclusions previously have been drawn from studies on the long-term effects of ethanol on cAMP-producing transduction systems (Valverius et al., 1987, 1989).

In conclusion, the recently published observations further stress the importance of the G-proteins in mediating the long-term effects of ethanol on cellular communication.

MOLECULAR BASIS OF ETHANOL-SENSITIVE G-PROTEINS

It is by now clear that ethanol exerts differential effects depending not only on receptor type but also on the cellular localization of the signalling system studied. This is not surprising considering the heterogeneity of proteins constituting receptor complexes. To date at least 20 different sub-types of Gα subunits have beeen isolated and characterized. It is also known that ethanol differentially affects the function of different Gα subtypes. For example, in studies on N1E-115 neuroblastoma cells, the inhibitory effect of ethanol was mediated via an increase in the inhibitory Gαi subunit while the same cell biological response was mediated via an inhibited expression of the stimulatory Gαs in NG 108-15 cells (Charness, Querimit and Henteleff, 1988).

Are ethanol sensitive G-proteins myristoylated?

The biochemistry underlying these differences in sensitivity to ethanol is unknown. However, the differences offer an approach for studying the more fundamental question regarding the existence of specific ethanol-sensitive phenomenon that could underly the ethanol sensitivity of different cellular processes. The well-established, but less thoroughly proven, hypothesis that ethanol exherts its effect via an interaction with membrane lipid-bound proteins offers one such possibility. Although all G-proteins must be associated to the membrane, it is possible that different G-proteins are differently regulated by the lipid environment. The observation that some membrane-associated proteins are covalently bound to fatty acids has recently been shown to regulate the interaction between the proteins and the membranes (Towler et al., 1988). Myristic acid is one of the fatty acids that most frequently has been shown to bind to these proteins. The myristoylation of specific Gα subunits is now established as a mechanism by which the stimulatory form of this G-protein is located to the membrane and thus activated (Jones et al., 1990, Mumby et al., 1990).

The fact that there is some correlation between ethanol sensitivity and the myristoylation among G-proteins makes it possible to hypothesize that myristoylated proteins are influenced by ethanol and that myristoylation is a mechanism related to ethanol´s effect on cellular signalling.

To test this hypothesis, NG 108-15 cells were cultured over night in the presence of hydroxy-myristate and N-myristoyl-glycinal diethylacetal, two known inhibitors of myristoylation. This regime reduced the incorporation of [¹⁴C]-myristate to cellular proteins as determined by fluorography of hydroxylamine-treated SDS-polyacrylamide gels. The most marked reductions were found with protein with approximate molecular weights of 43 and 80 kDa. Hydroxymyristate treatment reduced I(1,4,5)P₃ formation secondary to bradykinin stimulation but did not affects GTP(S) stimulations. In contrast,

N-myristoyl-glycinal-diethylacetal had no effect on bradykinin-induced response but inhibited GTP(S)-stimulated I(1,4,5)P$_3$ generation. Based on these inconsistencies, it was concluded that the bradykinin/G-protein/phospholipase C system studied in NG 108-15 cells did not contain myristoylated proteins. It thus seems unlikely that the ethanol sensitivity of this system was related to the presence of myristoylated G-proteins.

The role of differentiation

Morphological differentiation of cultured cells influences the efficacy of transmembrane signalling. This effect appears to be related to an altered expression of receptor-coupled G-proteins (Mullney et al., 1988, Barnett et al., 1990a,b, Mullaney and Milligan, 1990). The NG 108-15 cells used for these experiments are rapidly growing neuroblastoma-glioma hybrid cells and are thus different from the mature and differentiated cells found in brain. It is therefore likely that the transduction system studied differ from those found in differentiated cells. It is also possible that this fact may be responsible for the inhibitory effects observed after chronic ethanol exposure and that a different pattern may appear if differentiated cells are subjected to long-term ethanol exposure.

The findings of Balduini et al (Balduini and Costa, 1989, 1990, Balduini et al., 1991) attracted our attention as they indicate that the ethanol sensitivity of charbacol-stimulated phosphoinositide hydrolysis is correlated to the growth spurt of rodent brain. A selective inhibition of the muscarinic transduction was found only during a short time-span in growing rats. This period also coincided with maximal signal transduction of the muscarinic receptor.

To test the role of differentiation for signal transduction and ethanol´s effect on this system, NG 108-15 cells were differentiated by cultivation in medium containing 1 mM dibutyrylcAMP(dbcAMP). Cell growth was retarded and the morphology of the cells drastically altered after 4-6 days. Cells formed thick neurites, forming connections with surrounding cells in a bouton-like fashion. dbcAMP reduced bradykinin-stimulated I(1,4,5)P$_3$ formation already after 1 h of exposure. The inhibition increased during the following days and reached a maximum after 6 days.

The differentiated cells were exposed to 100 mM ethanol during the last 48 h of differentiation. However, this ethanol treatment did not inhibit bradykinin-stimulated I(1,4,5)P$_3$ formation. It thus appears that differentiated cells are resistent to the inhibitory effect seen in proliferating NG 108-15 cells. If, indeed, the effect was mediated via an altered expression of the appropriate G-proteins remains to be determined.

CONCLUSION

Long-term ethanol exposure leads to a transient inhibition of receptor-stimulated phosphoinositide hydrolysis, mediated via an effect on G-proteins. Following this period of inhibited transduction, the system is maladapted, presumably by an alteration at the level of the G-protein. However, the signal transduction systems are complex and heterogeneous, making it likely that these effects are not universally applicable to all phospholipase C-coupled signalling systems.

ACKNOWLEDGEMENT

The expert technical assistence of Berit Färjh and Monica Mihailescu is gratefully acknowledged. Myrisoylation inhibitors were kindly provided by Dr T. Saermark,

Panum Institute, Copenhagen, Denmark. Financial support was obtained from Medical Research Council, A. Påhlson Foundation, T. Nilsson Foundation, Swedish Alcohol Research Fund and the Medical Faculty, Lund University.

REFERENCES

Balduini W and Costa LG (1989) Effects of ethanol on muscarinic receptor-stimulated phosphoinositide metabolism during brain development. J Pharmacol Exp Ther 250:541-547.

Balduini W and Costa LG (1990) Developmental neurotoxicity of ethanol: in vitro inhibition of muscarinic receptor-stimulated phosphoinositide metabolism in brain from neonatal but not adult rats. Brain Res 512: 248-252.

Balduini W, Canduna SM, Manzo L, Cattabeni F and Costa LG (1991) Time-, concentration-, and age-dependent inhibition of muscarinic receptor-stimulated phosphophoinositide metabolism by ethanol in the developing rat brain. Neurochem Res 16:1235-1240.

Barnett JV, Shamah SM, Lassegue B, Griendling KK and Galper JB (1990a) Muscarinic cholinergic stimulation of inositol phosphate production in cultured embryonic atrial cells. Biochem J 271:437-442.

Barnett JV, Shamah SM, Lassegue B, Griendling KK and Galper JB (1990b) Development of muscarinic cholinergic stimulation of inositol phosphate production in cultured embryonic atrial cells. Biochem J 271:443-448.

Charness ME, Querimit LA and Henteleff M (1988) Ethanol differentially regulates G proteins in neural cultures. Biochem Biophys Res Commun 155:138-141.

Chiang C and Hauser G (1989) Effects of bradykinin, GTPγS, R59022 and N-ethylmaleimide on inositol phosphate production in NG 108-15 cells. Biochem Biophys Res Commun 165: 175-181.

Diamond I, Wrubel B, Estrin W and Gordon A (1987) Basal and adenosine receptor-stimulated levels of cAMP are reduced in lymphocytes from alcoholic patients. Proc Natl Acad Sci USA.84:11413-1416.

Hoek JB, Thomas AP, Rooney TA, Higashi T, Rubin E (1992) Ethanol and signal transduction in the liver. FASEB J 6:2386-2396.

Hoffman PL and Tabakoff B (1990) Ethanol and guanine nucleotide binding proteins: a selective interaction. FASEB J 4:2612-2622

Imaizumi T, Osugi T, Misaki N, Ushida S, Yoshida H (1989) Heterologous desensitization of bradykinin-induced phosphatidylinositol response and Ca^{2+} mobilization by neurotensin in NG 108-15 cells. Eur J Pharmacol 161:203-208.

Jones TLZ, Simonds WF, Merendino Jr JJ, Brann MR and Spiegel M (1990) Myristoylation of inhibitory GTP-binding protein α subunit is essential for its membrane attachment. Proc Natl Acad Sci USA 87:568-572

Li PP, Chiu AS and Warsh JJ (1989) Activation of phosphoinositide hydrolysis in rat cortical slices by guanine nucleotides and sodium flouride. Neurochem Int 14:43-48.

Luthin GR and Tabakoff B (1984) Activation of adenylate cyclase by alcohols reguires the nucleotide-binding protein. J Pharmacol Exp Ther 228:579-587.

Mochly-Rosen D, Chang F-H, Cheever L, Kim M, Diamond I and Gordon AS (1988) Chronic ethanol causes heterologous desensitization of receptors by reducing alpha, messenger RNA. Nature 333:848-850.

Mumbry SM, Heukeroth RO and Gordon JI and Gilman AG (1990) G-protein α-subunit expression, myristoylation, and membrane association in COS cells. Proc Natl Acad Sci USA 87:728-732.

Nagy LE, Diamond I and Gordon A (1988) Cultured lymphocytes from alcoholic subjects have altered cAMP signal transduction. Proc Natl Acad Sci USA 85, 6973-6976.

Richelson E, Stenstrom S, Forray C, Enloe L and Pfenning M (1986) Effects of chronic exposure to ethanol on the prostaglandin E1 receptor-mediated response and binding in a murine neuroblastoma clone (N1E-115). J Pharmacol Exp Ther 239:687-692.

Rodriguez FD, Simonsson P and Alling C (1992a) A method for maintaining constant ethanol concentration in cell culture medium. Alcohol Alcohol 27:309-313.

Rodriguez FD, Simonsson P, Gustavsson L and Alling C (1992b) Mechanisms of adaptation to the effects of ethanol on phospholipase C activation in NG 108-15 cells. Neuropharmacol 31:1157-1164.

Rooney TA, Hager R, Rubin E, and Thomas AP (1989) Short-chain alcohols activate nucleotide-dependent phosphoinositidase C in turkey erythrocyte membranes. J Biol Chem 264: 6817-6822.

Rubin R and Hoek JB (1988) Agonist-induced stimulation of phospholipase C in human platelets requires G-protein activation. Biochem J 254: 147-153.

Saito T, Lee JM and Tabakoff B (1985). Ethanol´s effects on cortical adenylate cyclase activity. J Neurochem 44: 1037-1044.

Saito T,, Lee JM, Hoffman PL and Tabakoff B (1987) Effects of chronic ethanol treatment on the beta-adrenergic receptor-coupled adenylate cyclase system of mouse cerbral cortex. J Neurochem 48:1817-1822.

Simonsson P, Sun GY, Aradottir S and Alling C (1989a) Bradykinin effects on phospholipid metabolism and its relation to arachidonic acid turnover in neuroblastoma x glioma hybrid cells (NG 108-15). Cell Signalling 1:587-598.

Simonsson P, Sun GY, Vecsei L and Alling C (1989b) Ethanol effects on bradykinin-stimulated phosphoinositide hydrolysis in NG 108-15 neuroblastoma x glioma hybrid cells. Alcohol 6:475-479.

Simonsson P, Rodriguez FD, Loman N and Alling C (1991) G proteins coupled to phospholipase C: Molecular targets of long-term ethanol exposure. J Neurochem 56:2018-2026.

Suzuki Y, Hraska KA, ReidI, Alvares UM and Avioli LV (1989). Characterization of phospholipase C activity of the plasma membrane and cytosol of an osteoblast-like cell line. Am J Med Sci 297:135-144.

Towler DA, Gordon JI, Adams SP, Glaser L (1988) The biology and enzymology of eukaryotic acylation. Ann Rev Biochem 57:69-99.

Valverius P, Hoffman PL and Tabakoff B (1987) Effects of ethanol on mouse cerebral cortical beta-adrenergic receptors. Mol Pharmacol 32:217-222.

Valverius P, Hoffman PL and Tabakoff B (1989) Hippocampal and cerebellar beta-adrenergic receptors and adenylate cyclase are diffrentially altered by chronic ethanol ingestion. J Neurochem 52:492-497.

SELECTIVE EFFECTS OF ETHANOL ON NEUROPEPTIDE - MEDIATED POLYPHOSPHOINOSITIDE HYDROLYSIS AND CALCIUM MOBILIZATION

Thomas L. Smith

Research Service (151)
Department of Veterans Affairs Medical Center
Tucson, AZ 85723

INTRODUCTION

The cell-surface receptor activation of phospholipase C and the subsequent production of 1,4,5 trisphosphate and diacylglycerol and intracellular calcium mobilization is a key signal transduction system utilized by brain as well as many peripheral tissues (for review, see Fisher et al, 1992). The role of phospholipase C as a mediator of ethanol actions in brain has been the subject of several investigations (Gonzales et al, 1986; Hoffman et al, 1986; Smith et al, 1986). These studies focused on the activation of PLC by classical neurotransmitters such as norepinephrine histamine, and acetylcholine. However, very little information is available regarding the possible effects of chronic ethanol on neuropeptide - stimulated PLC and subsequent calcium mobilization in neural tissues. This laboratory has recently reported that chronic ethanol significantly reduced the stimulation of[^{3}H]inositol phosphates production by bradykinin (Smith, 1991). In the present report, possible mechanisms for this down-regulated response were examined in N1E-115 neuroblastoma cells. Although neuroblastoma are transformed cells, they retain many neural properties (Bozou et al, 1986; Monck et al, 1990; Twombly et al, 1990; Milligan et al, 1990) and, therefore, represent a suitable model system to study alterations in signal transduction, thus avoiding the inherent complexity and heterogeneity of cell populations within brain tissues.

METHODS

Tissue Culture

Mouse neuroblastoma (N1E-115) cells were cultivated essentially as described previously (Smith, 1991). Briefly, cell passages 14-24 were grown in 75-cm^2 tissue culture flasks in 20ml Dulbecco's modified Eagle's medium (DMEM) supplemented with 10% (v/v) fetal bovine serum (Hyclone, UT), 50μg/ml streptomycin, and 50 units/ml penicillin (Sigma Chemical Co., St. Louis, MO) at 37°C in a humidified atmosphere of 93% air -7.2% CO_2. In the chronic ethanol experiments, 20ml of fresh medium containing 100mM ethanol was added daily to each flask for 7 days. Confluent cells from control and ethanol treated flasks were detached with modified Puck's D_1 solution and harvested immediately before each experiment. Cell viability for both control and ethanol treated cells was assessed by trypan blue exclusion and averaged greater than 85%.

[^{3}H]inositol Assay

The hydrolysis of polyphosphoinositides by phospholipase C was assessed by determining the accumulation of total [^{3}H]InP in the presence of lithium as originally described (Berridge et al., 1982). Briefly, confluent cells harvested from one flask (ca 7-8 X 10^6 cells) were washed twice with a 10mM HEPES buffer containing: NaCl, 110 mM; KCl 5.3mM; MgCl$_2$ 1 mM; sucrose 80mM; and glucose, 25mM, pH 7.4 and adjusted to 340 mOsmol (buffer A). Cells were resuspended in 2ml of the above buffer and preincubated with 50 to 75μCi[^{3}H]inositol (20Ci/mmol, NEN, Boston MA for 60 min at 37°C with gentle shaking. After three washes with buffer A, the cells were suspended to a final concentration of 7-8 X 10^5 cells/ml. Aliquots (220μl) of cells were incubated in triplicate in the presence of 10mM LiCl with or without ethanol for 10 min at 37°C, after which the reactions were initiated by the addition of the indicated concentration of neurotensin, bradykinin, or the stable GTP analog, GTP(S), in the total volume of 250μl and further incubated for 5 to 30 min. Reactions were terminated with the addition of 1 ml of $CHCl_3$/CH_3OH (1:2). After 30 min 0.35ml of $CHCl_3$ and 0.35ml of H_2O were added and aliquots of the resulting upper phase were applied to columns containing Dowex-1 resin (formate). [^{3}H]InP were routinely eluted with 3 ml of 0.1M formic acid/1.0 M ammonium formate and counted by liquid scintillation spectroscopy as described previously (Smith and Yamamura, 1987).

Intracellular Free Calcium [Ca^{2+}]$_i$ Determination

[Ca^{2+}]$_i$ was determined essentially as described previously (Henderson, et al., 1991). Harvested cells from one flask (8 - 12 X 10^6 cells) were washed twice with 5ml of a 10mM HEPES buffer containing: NaCl, 110mM; KCl, 5.3mM; MgCl$_2$, 1mM; sucrose, 80mM; and glucose, 25mM, pH 7.4 and adjusted to 340 mOsM (Buffer A). Washed cells (4-6 X 10^6) were resuspended in 1.25ml DMEM and incubated for 30 mins at 35°C with 9μl of a 1.25mM stock solution of fluo-3/AM (Molecular Probes, Eugene, OR) dissolved in DMSO containing 3.6% (w/v) pluronic acid. All washings and resuspensions were performed at 35°C.

Fluorescence determinations were made using a T-format series 300 spectrofluorometer (H&L Instruments, Burlingame, CA) with excitation wavelength set at 506 nm. An emmission wavelength of 526 nm was achieved with a narrow band pass filter (Microcoatings, Westford, MA). Fluorescence intensity, F, was monitored continuously with an on line IBM compatible personal computer. Immediately prior to each determination, a 200-250μl aliquot of the cell suspension was washed twice with 5ml of Buffer A, resuspended in md of Buffer A containing 1.8mM $CaCl_2$ and transferred to a quartz cuvette. After the cell suspension reached 35°C, indicated concentrations of drugs were added and the subsequent change in F recorded.

Free intracellular calcium was calculated as described previously (Tsien, et al., 1982) using a K_D of 450 nM for fluo-3/AM (Meyer, et al., 1990).

$$[Ca2+]_i = K_D \cdot \frac{F - Fmin}{Fmax - F}$$

Separate calibrations were performed on each sample. Fmin was determined in the presence of 0.1% SDS and excess EGTA. Fmax was then determined after the addition of a saturating concentration of calcium. Total fluorescence was corrected for both autofluorescence and dye leakage prior to the calculation of $[Ca^{2+}]_i$.

[³H]inositol Trisphosphate Specific Binding

[³H]Inositol 1,4,5-trisphosphate, InP_3, binding to N1E-115 cell membrane fragments was determined essentially as described previously (Worley, et al., 1987) with minor modifications (Smith, 1989). Freshly harvested cells from three pooled flasks (25-35 X 10^6 cells) were homogenized (Polytron, setting 7,15 secs.) in 4 ml ice-cold 50mM tris buffer (pH 8.3) containing: 20mM NaCl; 100mM KCl; 1mM EDTA. (buffer B). The tissue suspension was centrifuged at 35,000 x g for 10 mins and the pellet resuspended in the same buffer to provide a tissue concentration of approximately 2.5mg protein/ml. Assay conditions for [³H]InP_3 binding were as follows: 135µl (300 - 400µg protein) of membrane suspension was added in triplicate to polypropylene microcap tubes containing 1.2nM [³H]In (1,4,5)P_3 (40,000 dpm; 17 Ci/mmol, NEN) and increasing concentrations (1-200nM) of unlabeled InP_3 in buffer B to give a total volume of 500µl. Nonspecific binding was determined in the presence of 10µM unlabeled InP_3. After incubation on ice for 10 mins, the samples were centrifuged at 11,000 xg for 10 mins at 4° and the pellets washed twice with 500µl ice-cold buffer B. Solubilization of the pellet was achieved by the addition of 100µl tissue solubilizer (NCS, Amersham) and further incubated with shaking for 60 mins at 50°. Samples were transferred to scintillation vials and counted after 24 hours by liquid scintillation spectroscopy with a counting efficiency of 40%. K_D and Bmax values were obtained from Scatchard analysis of the InP_3 displacement curves (Worley, et al., 1987). The slope and intercept of the straight line were generated by least squares analysis.

RESULTS

The production of [³H]InP by maximally effective concentrations of neurotensin, bradykinin and the stable GTP analog, guanine 5' (- γ - thiotriphosphate), was determined in control cells and cells chronically exposed to 100mM ethanol for seven days. As shown in Fig.1 chronic ethanol exposure significantly reduced the stimulation of [³H]InP by bradykinin, but had no effect on this response to either neurotensin or GTP[S]. These results suggest a selective effect of ethanol on the coupling of neuropeptides to phospholipase C. The results obtained with GTP[S] also suggest that chronic ethanol does not alter the activity of phospholipase C, per se, nor does it affect the activation of phospholipase C by guanine nucleotide binding proteins.

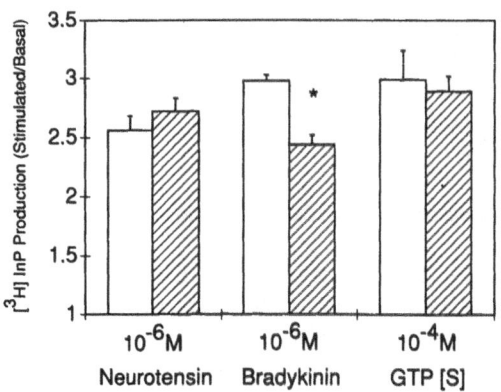

Figure 1. Effects of chronic ethanol exposure in the stimulation of total [³H] inositol phosphate formation by neurotensin, bradykinin and GTP[S]. Shaded bars indicate N1E-115 cells exposed to 100mM ethanol for seven days as described in Methods. Open bars indicate pair - cultured control cells. Results are the mean ±S.E.M. from 3-4 experiments performed in triplicate *p < 0.05. Student's t test (unpaired).

In order to establish whether or not the reduction of [³H]InP formation by bradykinin in cells chronically exposed to ethanol resulted in a parallel reduction in the ability of bradykinin to increase cytosolic free calcium, $[Ca^{2+}]_i$, the effects of ethanol on the resting and agonist stimulated $[Ca^{2+}]_i$ levels were determined. The results in Table 1 indicate that neither chronic nor acute ethanol exposures up to 400mM had any significant effect on resting $[Ca^{2+}]_i$. Acute ethanol up to 200 mM also had no effect on the rise in $[Ca^{2+}]_i$ stimulated by either bradykinin or neurotensin (results not shown).

In contrast to the acute studies, chronic ethanol exposure resulted in a highly significant reduction in the stimulation of $[Ca^{2+}]_i$ by bradykinin. However, the $[Ca^{2+}]_i$ response to neurotensin was unaltered in identically treated cells (Fig. 2).

Because the bradykinin - induced rise in $[Ca^{2+}]_i$ represents the release of intracellularly bound Ca^{2+} as well as Ca^{2+} influx across the plasma membrane, experiments were performed in the absence and presence of Ca^{2+} in the $[Ca^{2+}]_i$ assay buffer. The results are shown in Table 2. In the absence of external

Table 1. Effects of ethanol exposure on resting intracellular free calcium, $[Ca^{2+}]_i$.

Ethanol (mM)	0	100	200	400	chronic (7 days)
$[Ca^{2+}]_i$ (nM)	64 ± 2	55 ± 6	64 ± 2	68 ± 7	68 ± 3

Control N1E-115 cells were acutely exposed to the indicated concentrations of ethanol for 5 mins at 35°C prior to the determination of $[Ca^{2+}]_i$ as described in Methods. Cells chronically exposed to ethanol were assayed in the absence of ethanol in the assay buffer. Results represent the mean ± S.E.M. from 3-6 individual experiments.

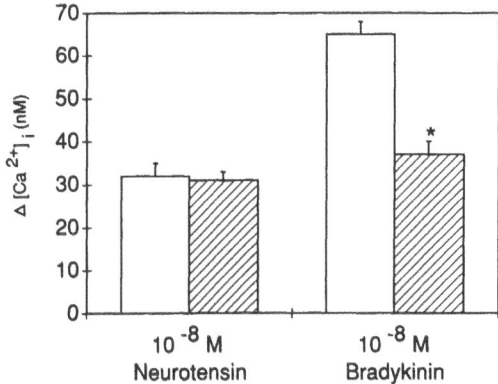

Figure 2. Agonist-stimulated rise in $[Ca^{2+}]_i$ in control (open bars) and cells chronically exposed to 100 mM ethanol for 7 days (shaded bars). $\Delta [Ca^{2+}]_i$ is defined as the difference between resting $[Ca^{2+}]_i$ and that calculated from the peak fluorescence intensity observed (10-15 sec) after the addition of agonists. Results represent the mean ± S.E.M. from 3-4 experiments. *p < 0.01 (Student's t test, unpaired).

Ca^{2+} the bradykinin stimulated rise in $[Ca^{2+}]_i$ was reduced by approximately 50% in control cells and 66% in cells chronically exposed to ethanol. However, even in the absence of external Ca^{2+}, the stimulation of $[Ca^{2+}]_i$ by bradykinin was still significantly less in chronically treated cells compared to that of control. These data indicate that chronic ethanol treatment inhibits the release of intracellularly bound Ca^{2+} by bradykinin, presumably due to the parallel reduction in bradykinin stimulated [³H]InP formation shown in Fig. 1. Another possible explanation for this phenomenon is that chronic ethanol exposure affects the binding characteristics of the inositol 1,4,5 triphosphate receptor in a manner similar to that reported previously in cerebella of mice made physically dependent to ethanol (Smith, 1987). Therefore, the specific binding of [³H]In (1,4,5)P₃ was determined in membrane fragments from control and ethanol exposed cells.

Table 2. Effects of extracellular Ca^{2+} on the magnitude of bradykinin stimulated rise in $[Ca^{2+}]_i$ in control and ethanol treated N1E-115 cells

buffer Ca^{2+}	$\Delta[Ca^{2+}]_i$	
	CONTROL	CHRONIC ETHANOL
+	65 ± 3	37 ± 3*
−	29 ± 3	12 ± 3*

Control and ethanol exposed (100mM, 7 days) cells were washed twice and resuspended in a HEPES-buffered salt solution either in the absence or presence of 1.8 mM Ca^{2+}. Stimulation of $[Ca^{2+}]_i$ was achieved with 10^{-8}M bradykinin as described in Methods. $\Delta[Ca^{2+}]$ is defined as in Fig. 2. Values represent the mean ± S.E.M. from four independent observations. *indicates p < 0.01 when compared to corresponding control values (Student's t test)

Figure 3 represents a typical [³H]In (1,4,5)P₃ binding displacement curve. Transformation of the binding data into Scatchard plots yielded corresponding K_D and Bmax values which are summarized in Table 3. It is clear from Table 3 that chronic ethanol treatment had no effect on the dissociation constant (K_D) or the binding capacity (Bmax) for the In (1,4,5)P₃ receptor.

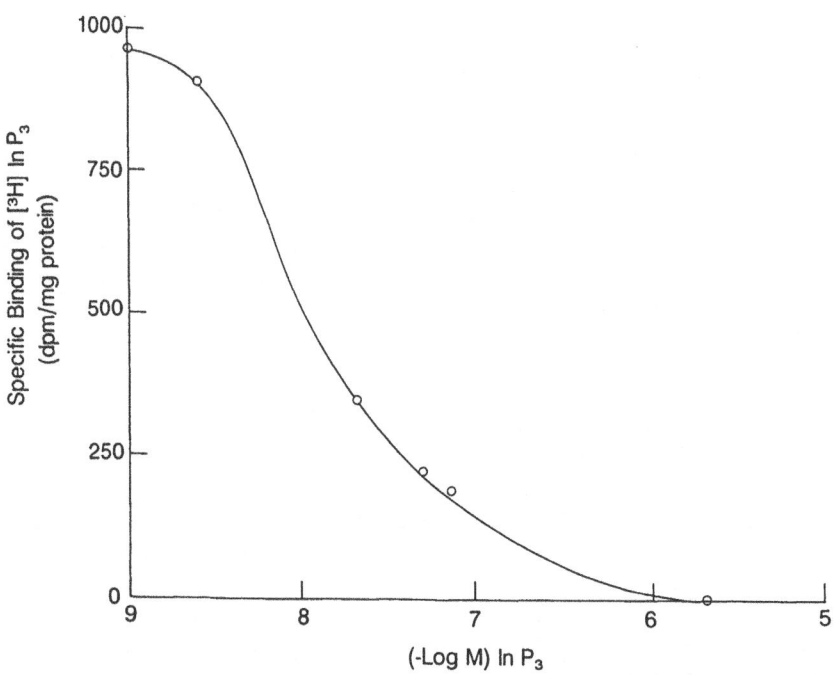

Figure 3. Representative displacement curve for [³H]In (1,4,5)P₃ specific binding to membrane fragments of control N1E-115 cells. Binding assays included 200-400 μg of washed membrane protein and 1.2nM [³H]In (1,4,5)P₃ as described in Methods. Nonspecific binding, determined in the presence of 2 μM unlabeled InP₃ was subtracted from the total binding to obtain specifically bound [³H]InP₃. Date points represent mean values from a typical experiment performed in triplicate.

Table 3. [³H]lnositol 1,4,5-trisphosphate specific binding in N1E-115 neuroblastoma membranes after chronic ethanol exposure.

Treatment	K_D (nM)	Bmax (fmoles/mg protein)
control (n=4)	31 ± 7	158 ± 22
Chronic ethanol (n=3) (100mM, 7 Days)	22 ± 6	174 ± 21

K_D and Bmax values were obtained by Scatchard analysis of the individual displacement curves as described in Methods. Value represents the mean ± S.E.M. from the indicated number of experiments performed in triplicate.

Because recent reports have demonstrated that protein kinase C (PKC) activity is increased in PC-12 cells after chronic ethanol treatment (Messing, et al., 1991) and that inhibition of PKC further enhanced the stimulatory effect of acute ethanol on serotonin stimulated [³H] inositol phosphate (Simonsson, et al., 1989), a series of experiments were performed in order to determine whether the coupling of both bradykinin and neurotensin to phospholipase C - stimulated $[Ca^{2+}]_i$ mobilization was regulated by PKC activation. The results are shown in Fig. 4. In the absence of extracellular Ca^{2+}, short term activation of PKC by phorbol 12-myristate 13 acetate (PMA) inhibited $[Ca^{2+}]$ mobilization by both bradykinin and neurotensin by approximately 50%. These results suggest that the activation of PKC is an unlikely mechanism for the selective effects of ethanol on bradykinin-mediated second messenger production.

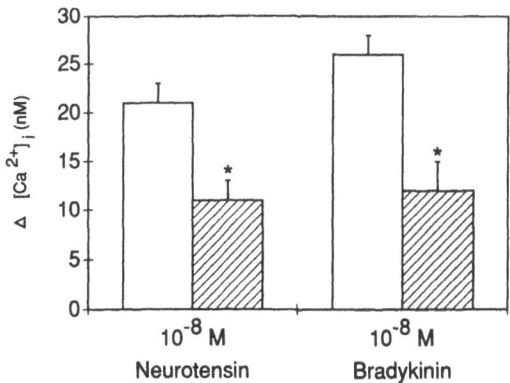

Figure 4. Effect of phorbol 12-myristate 13-acetate (PMA) on bradykinin and neurotensin stimulated $[Ca^{2+}]_i$ in the absence of extracellular Ca^{2+}. Control N1E-115 cells were exposed for 4-5 mins with 100nM PMA prior to stimulation with the indicated neuropeptides. $\Delta[Ca^{2+}]_i$ is defined as in Fig. 2. Values represent the mean ± S.E.M. from 3-4 experiments. *p < 0.01 when compared to control values. (Student's t test unpaired).

SUMMARY

The direct effects of chronic ethanol exposure on [^{3}H]InP formation and subsequent mobilization of [Ca^{2+}]$_i$ were determined in N1E-115 neuroblastoma, a cell line which provides a homogeneous population of neuronal-like cells. The results demonstrate that chronic ethanol has selective inhibitory effects on the generation of these second messengers by bradykinin. In contrast, chronic ethanol had no effect on the formation of [^{3}H]InP or [Ca^{2+}]$_i$ mobilization by neurotensin. The results further demonstrate that the ethanol induced inhibition of [Ca^{2+}]$_i$ mobilization by bradykinin was due, in part, to the reduced ability of bradykinin to release intracellularly bound Ca^{2+}, presumably a reflection of the parallel reduction in [^{3}H]In (1,4,5)P$_3$ formation, rather than on alteration in the specific binding parameters of the In (1,4,5)P$_3$ receptor. The results with GTP[S] also suggest that chronic ethanol exposure does not alter phospholipase C activity, per se, nor does it alter the normal activation of phospholipase C by guanine-nucleotide binding proteins. The possible activation of protein kinase C after chronic ethanol exposure does not represent a likely mechanism for the selective effects of ethanol since both bradykinin and neurotensin stimulated [Ca^{2+}]$_i$ mobilizations were similarly inhibited by protein kinase C activation. The results as a whole suggest that chronic ethanol exposure may partially disrupt the normal coupling efficiency between the bradykinin receptor and the G proteins associated with phospholipase C.

REFERENCES

Berridge, M.J., Downes, C.P. and Hanley, M.R., 1982, Lithium amplifies agonist-dependent phosphatidylinositol responses in brain and salivary glands. Biochem J. 206:587.

Bozou, J., Amar, S., Vincent J. and Kitabgi, P., 1986, Neurotensin-mediated inhibition of cyclic AMP formation in neuroblastoma N1E-115 cells: involvement of the inhibitory GTP-binding component of adenylate cyclase, Mol. Pharmacol. 29:489.

Fisher S.K., Heacock, A.M. and Agranoff, B.W., 1992, Inositol lipids and signal transduction in the nervous system: An update, J. Neurochem. 58:18.

Gonzales, R.A., Theiss, C. and Crews, F.T., 1986, Effects of ethanol on stimulated inositol phospholipid hydrolysis in rat brain, J. Pharmacol. Exp. Ther. 237:92.

Henderson, A.K., Yamamura, H.I., Maggi, C.A., Roeske, W.R., and Smith T.L., 1991, Neurokinin A receptors are coupled to calcium mobilization in transfected fibroblasts, Eur.J.Pharmacol. 206:251.

Hoffman, P.L., Moses F., Luthin, G.R. and Tabakoff, B., 1986, Acute and chronic effect of ethanol on receptor-mediated phosphatidyl 4,5-trisphosphate breakdown in mouse brain, Mol. Pharmacol. 30:13.

Messing, R.O., Peterson, P.J. and Henrich, C.J., 1991, Chronic ethanol exposure increases levels of protein kinase C δ and ϵ and protein kinase C -mediated phosphorylation in cultured neuronal cells, J. Biol. Chem. 266:23428.

Meyer, T., Wensel, T. and Stryer, L., 1990, Kinetics of calcium channel opening by 1, 4, 5-trisphosphate, Biochem. 29:32.

Milligan, D., McKenzie, F., McClue, S., Mitchell, F. and Mullaney, L., 1990, Guanine nucleotide binding proteins in neuroblastoma X glioma hybrid NG108-15 cells. Regulation of expression and function, Int. J. Biochem. 22:701.

Monck, J., Williamson, R., Rogulia, I., Fluharty, S. and Williamson, J., 1990, Angiotensin II effects on the cytosolic free Ca^{2+} concentration in N1E-115 neuroblastoma cells: Kinetic properties of the Ca^{2+} transient measured in single fura-2 loaded cells, J. Neurochem. 54:278.

Simonsson, P., Hansson, E. and Alling, C., 1989, Protein kinase C modulation of the ethanol effect on serotonin$_2$ receptor transduction in astrocytes, Alcohol & Alcoholism 24:401.

Smith, T.L., 1987, Chronic ethanol consumption reduces [^{3}H]inositol 1,4,5 trisphosphate specific binding in mouse cerebella membrane fragments, Life Sci. 41:2863.

Smith T.L., 1989, [^{3}H]inositol trisphosphate specific binding in two brain regions of long and short sleep mice, Pharmacol. Biochem. & Behav. 32:327.

Smith, T.L., 1991, Selective effects of acute and chronic ethanol exposure on neuropeptide and guanine nucleotide stimulated phospholipase C activity in intact N1E-115 neuroblastoma, J. Pharmacol. Exp. Ther. 258:410.

Smith, T.L., Yamamura, H.I. and Lee, L., 1986, Effects of ethanol on receptor-stimulated phosphatidic acid and polyphosphoinositide metabolism in mouse brain, Life Sci. 39:1675.

Smith, T.L., Yamamura, H.I., 1987, Competitive inhibition of phosphoinositide hydrolysis by the muscarinic receptor antagonist, AF-DX 116 is of low affinity in mouse cerebral cortex, Brain Res. 420:362.

Tsien, R.Y., Pozzan, T., Rink, T.J., 1982, Calcium homeostasis in intact lymphocytes: cytoplasmic free calcium monitored with a new, intracellularly trapped fluorescent indicator, J. Cell Biol. 94:325.

Twombly, P., Herman, M., Kye, C. and Narahashi, T., 1990, Ethanol effect on two types of voltage-activated calcium channels, J. Pharmacol. Exp. Ther. 254:1029.

Worley, P.F., Baraban, J.M., Supattapone, S., Wilson, V.S., and Snyder, S.H., 1987, Characterization of inositol trisphosphate receptor binding in brain, J. Biol. Chem. 262: 12132.

THE MUSCARINIC RECEPTOR-STIMULATED PHOSPHOINOSITIDE METABOLISM AS A POTENTIAL TARGET FOR THE NEUROTOXICITY OF ETHANOL DURING BRAIN DEVELOPMENT

Walter Balduini, Flaminio Cattabeni, Filippo Renò, and Lucio G. Costa[1]

Institute of Pharmacology and Pharmacognosy, University of Urbino, Urbino, Italy
[1]Department of Environmental Health, University of Washington, Seattle, WA, USA

The underlying molecular events by which EtOH induces its profound deleterious effects on the developing central nervous system (CNS) are still poorly understood. In the past, the accepted scientific view of the effect of EtOH was that it altered neuronal activity by virtue of a non specific interaction with neuronal membranes. However, recent developments suggest that its action may be directed toward pre- and/or post-synaptic events, including synthesis, turnover, release, binding of neurotransmitters to receptor sites, and second messenger systems. In the present chapter, we present data showing a selective interaction of EtOH with muscarinic receptor-stimulated phosphoinositide hydrolysis during postnatal development in the rat suggesting, therefore, that this system may represent a specific and relevant target for the developmental neurotoxicity of EtOH.

Ethanol and Brain Development

Excessive ethanol use during pregnancy results in a set of fetal malformations and behavioral deficits that have been collectively named Fetal Alcohol Syndrome (FAS) (Streissguth et al. 1980). Principal features of this syndrome include: central nervous system dysfunction (mild to moderate mental retardation, microencephaly and brain malformations, irritability in infancy, possible hyperactivity in childhood), growth deficiency (of prenatal onset), and particular facial features (short, upturned nose, small eyeballs, altered facial symmetry). Microencephaly represents one of the most indicative diagnostic symptoms of FAS and has been found to occur in over 80% of confirmed cases. However, cognitive and intellectual deficits in children born to alcoholic mothers have been described even in the absence of typical characteristics of FAS (Streissguth et al. 1980). EtOH exerts different effects depending upon the time of administration (West and Pierce 1986). One of the most vulnerable periods for the developing brain is that of the brain growth spurt (BGS) which in humans begins with the last trimester of pregnancy and continues throughout the first 18-24 months. Although the timing of the BGS with respect to birth varies across species (Dobbings and Sands, 1979; Gottlieb et al. 1977), it is always characterised by proliferation of glial cells and maturation of neurones that develop dendrites, axons and synaptic contacts (Dobbings and Sands, 1979). In the rat the BGS is completely postnatal (peaking around postnatal days 6 and 10; Dobbings and Sands, 1979) and the neurogenetic events occurring during this period are comparable to those occurring during the third trimester in humans (Samson and Diaz, 1982). The administration of EtOH to rats during the BGS (from postnatal day 4 to day 10) causes microencephaly, learning deficits and other behavioral abnormalities without alteration of body weight. Therefore, these animals may represent a useful experimental model to study the molecular events underlying the neurotoxic effects of EtOH during this stage of brain development.

Muscarinic Receptors, Phosphoinositide Turnover and Brain Development

Activation of muscarinic receptors (particularly the M_1 subtype) causes the hydrolysis of membrane phosphoinositides with the generation of two important second messengers: inositol 1,4,5-trisphosphate which releases calcium from intracellular stores, and diacylglycerol which stimulates protein kinase C. In the rat, the density of cholinergic muscarinic receptors, as well as other parameters of the cholinergic system (choline acetyltransferase, acetylcholinesterase), gradually increases during postnatal development and reaches adult values at about two months of age. The developmental profile of muscarinic receptor-stimulated phosphoinositide hydrolysis, however, differs from that of the other cholinergic parameters: it is relatively low at birth, peaks between 6 and 10 days, and gradually declines to adult values (Figure 1; Balduini et al. 1987; 1990; 1991a; Heacock et al. 1987). This developmental profile is peculiar for activation of muscarinic receptors in that it was not observed for three other neurotransmitters (norepinephrine, serotonin, histamine) which are coupled to phosphoinositide metabolism in rat brain (Balduini et al. 1987; 1990; 1991). The striking resemblance of the temporal pattern of muscarinic receptor-stimulated phosphoinositide hydrolysis with the curve of the BGS led us to hypothesize a role for this biochemical response in brain development. Recent studies have provided further support to this hypothesis.

For example, Hohman et al. (1988) have shown that neonatal lesions of the basal cholinergic neurones result in abnormal cortical development. Van Hoof et al. (1988) have shown that activation of muscarinic receptors in nerve growth cones isolated from neonatal rat brain can stimulate the phosphorilation of the nerve-specific protein B-50 (GAP43, a major substrate for PKC), which is presumed to have a relevant role in modulation of neurite outgrowth and maturation of the growth cone into a synaptic terminal. Furthermore, Ashkenazi et al. (1989) have shown that activation of muscarinic receptor subtypes which elicit phosphoinositide hydrolysis stimulates, in an age-dependent fashion, DNA synthesis in prenatal rat brain-derived primary astrocytes. The correlation between phosphoinositide response and DNA synthesis and its age-dependence led the authors to conclude that activation of phosphoinositide hydrolysis by acetylcholine is involved in its mitogenic action and is an important factor in astroglial cell growth during brain development. Although the number of muscarinic receptors is low in the neonatal brain (Figure 1; Balduini et al. 1987), acetylcholine is present in significant amounts ($\approx$ 30% of adult levels). It is conceivable, therefore, that acetylcholine could have profound influences on brain development through its "amplified" phosphoinositide response during the brain growth spurt. As a consequence, an interaction of EtOH with the phosphoinositide system coupled to muscarinic receptors during this critical period of brain development could lead to alteration in brain cytoarchitecture, reduced axons and neurites extension, reduced synaptogenesis and impaired astrocyte proliferation. These indeed are among the reported developmental neurotoxic effects of EtOH (West, and Pierce, 1986).

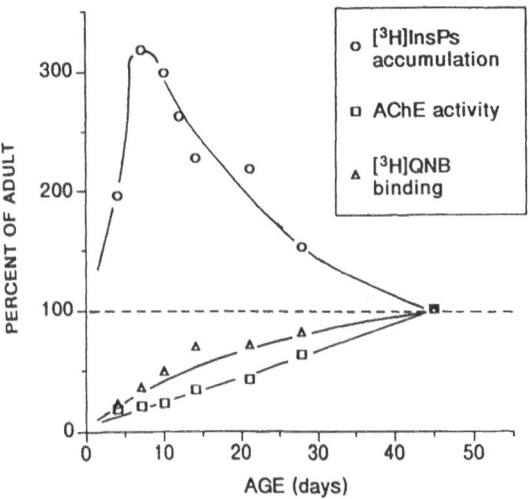

Figure 1. Carbachol-stimulated [^{3}H]inositol phosphates accumulation, [^{3}H]QNB binding and acetylcholinesterase activity in rat cerebral cortices as a funtion of age. Results are expressed as percentage of adult values.

The Interaction of Ethanol with Muscarinic Receptor-Stimulated Phosphoinositide Hydrolysis During Brain Development

The cholinergic system is particularly sensitive to the effects of EtOH. In adult rats the levels and the release of acetylcholine has been shown to be altered by EtOH exposure (Hunt and Dalton, 1976), as well as the density of muscarinic receptors in various brain areas (Tabakoff et al. 1979; Nordberg and Wahlstrom, 1982; Smith, 1983). EtOH has also been found to interact with phosphoinositide hydrolysis (for a review, see Hoek and Rubin, 1990). We have therefore investigated whether EtOH exposure during the BGS at a dose level known to induce long-lasting microencephaly would cause any alteration of muscarinic receptor-stimulated phosphoinositide hydrolysis. EtOH was diluted to 20% (w/v) in water and administered by gavage, according to the procedure of Sonderegger et al. (1982) and Bruns et al. (1984), at the dose of 4 g/kg of body weight in two 2 g/kg doses three hours apart. A control group received the caloric equivalent of a sucrose solution. In initial experiments, animals were treated with EtOH or sucrose from postnatal day 4 (PN 4) to 10 and sacrificed on day 7, 10, 12, 20, and 45 of age. Blood EtOH concentrations (BAC), measured on day 4, 7 and 10, 45 min after the second 2 g/kg dose, were 248, 237 and 283 mg/dl, respectively. EtOH administration had no effect on body weights, as compared to sucrose-fed pups. However, the brain weights of EtOH treated rats were significantly lower than those of controls, starting from day 12 (16% decrease) to day 45 (4% decrease). Muscarinic receptor-stimulated phosphoinositide hydrolysis, measured in cerebral cortex slices (Balduini et al., 1987), was significantly decreased in 7 and 10 day-old rats, but not at the other ages tested (Figure 1; Balduini and Costa, 1989). This effect was specific for the phosphoinositide system coupled to muscarinic receptors, since the stimulation elicited by norepinephrine, histamine and serotonin did not differ from controls. Similar effects were also found in the hippocampus. Binding experiments did not reveal any difference in muscarinic receptors number or affinity between controls and EtOH treated rats (Balduini and Costa, 1989).

The presence of an alteration of muscarinic receptor-stimulated phosphoinositide hydrolysis only on days 7 and 10 but not at the other ages, after termination of exposure, suggested that the effect was probably linked to the presence of high blood and brain concentrations of EtOH. Therefore, an identical EtOH exposure of adult rats would be expected to cause a similar effect. However, treatment of 75 day-old rats with the same protocol of EtOH administration, which resulted in similar BAC levels (283 mg/dl), had no effect on the same system. These results indicate that during the period of its maximal efficacy (i.e., during the BGS) the phosphoinositide system coupled to muscarinic receptors is more sensitive to the effect of EtOH. In vitro experiments, performed in cortical slices of neonatal and adult rats confirmed the results obtained with in vivo EtOH exposure (Balduini and Costa, 1990). These studies indicated that EtOH has a time-, concentration-, age-, brain region- and neurotransmitter-specific inhibitory effect on phosphoinositide hydrolysis stimulated by muscarinic agonists. In vitro, by increasing the time of incubation, the inhibitory effect of EtOH was significant at a concentration of 50 mM, which is similar to that reached in the in vivo experiments (Balduini and Costa, 1989; Balduini et al. 1991b). The effect of EtOH was more pronunced in the cerebral cortex, hippocampus and cerebellum (Balduini and Costa, 1990; Balduini et al. 1991b), three brain areas thought to be more sensitive to the developmental neurotoxicity of EtOH. Other aliphatic alcohols, which cause microencephaly when administered to rat during the BGS (Grant and Samson, 1982; 1984), also inhibited muscarinic receptor-stimulated phosphoinositide hydrolysis in neonatal rats (Candura et al. 1991).

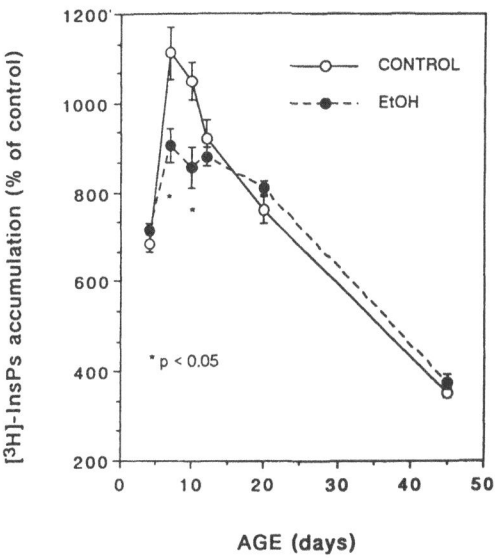

Figure 2. Carbachol-stimulated phosphoinositide hydrolysis in cerebral cortex slices from control and EtOH treated rats at different ages during postnatal development.

In another series of in vivo experiments we evaluated the sensitivity of the phosphoinositide system coupled to muscarinic receptors during different stages of the BGS. The protocol of EtOH exposure and the results obtained have been summarized in Table 1. Significant reductions of muscarinic receptor-stimulated phosphoinositide hydrolysis and brain weights have been observed after EtOH exposure on days 6-7-8 or 10-11-12 (i.e. during the peak of the BGS), but not after exposure on days 2-3-4 (i.e. during the initial stage of the BGS) or 13-14-15 (i.e. during the declining stage of the BGS). Also a single day of treatment during the peak of the BGS (day 7) was without effect. It should be noted that when muscarinic receptor-stimulated phosphoinositide metabolism was not inhibited, no differences in brain weights were observed (Table 1).

These findings confirm the results of in vitro experiments indicating that the period of the highest sensitivity to EtOH of the phosphoinositide system coupled to muscarinic receptors is at the peak of the BGS (Balduini et al. 1991b), and suggest a direct relationship between

Table 1. Effect of EtOH administration at different stages of the BGS on muscarinic receptor-stimulated phosphoinositide hydrolysis and brain weight.

Days of Treatment (age)	Maximal Stimulation (% of basal)[1]		Brain Weight[2]	BAC[3]
	Control	EtOH		
2 - 3 - 4	597 ± 11	626 ± 95	NS	303 ± 38
6 - 7 - 8	1000 ± 19	834 ± 24[*]	$p < 0.05$	261 ± 15
10 - 11 - 12	881 ± 20	769 ± 29[*]	$p < 0.05$	312 ± 46
13 - 14 - 15	784 ± 36	793 ± 37	NS	298 ± 11
7	1006 ± 22	1030 ± 32	NS	308 ± 17

[1]Different groups of neonatal rats were treated on selective days during the period of the BGS with 4 g/kg EtOH (in two 2 g/kg three hours apart) and sacrificed 45 min after the second dose on the last day of administration. Results represent the mean (± S.E.M.) of six to nine samples assayed in separate experiments. Maximal stimulation was that observed in the presence of 1 mM carbachol and is expressed as percentage of basal values. Basal accumulation (in the absence of carbachol) did not differ between control and EtOH treated rats. [*]$p < 0.05$

[2]Brain weights were measured on PN 12 for groups treated on days 2-3-4, 6-7-8 or 7, and on PN 21 for groups treated on days 10-11-12 and 13-14-15. NS, not statistically different from controls (sucrose treated animals).

[3]Blood EtOH concentrations (BAC) were determined 45 min after the second 2g/kg dose on the last day of administration.

developmental neurotoxicity of EtOH and inhibition of this second messenger system. Since the phosphoinositide-linked signal transduction system plays a critical role in cell activity, it is possible that its inhibition by EtOH during this critical stage of brain development may alter the interaction among neurones and/or neurones and glial cells which express muscarinic receptors and may cause delays in cell growth and synaptic contact formation.

The molecular mechanism(s) responsible for this effect of EtOH remains a matter of speculation. A direct interaction of EtOH with the muscarinic receptor recognition site seems unlikely since no effect on radioligand binding was observed in vitro or after in vivo administration. A plausible hypothesis is that EtOH might interfere with the mechanism(s) responsible for the age-dependent differential ability of muscarinic agonists to stimulate inositol metabolism. It has been suggested that the higher accumulation of inositol phosphates induced by activation of muscarinic receptors during brain development might be due to an enhanced coupling of the receptor with phospholipase C (Balduini et al. 1987), probably mediated by a novel GTP binding protein . This protein, therefore, might represent a potential molecular target of EtOH. In tune with this hypothesis, we have found that in a membrane preparation in which muscarinic agonists stimulate phosphoinositide hydrolysis only in the presence of GTP, the inhibitory effect of EtOH was more pronounced in neonatal (where it was significant at concentrations as low as 50mM) than in adult rats (Candura et al. 1992).

Other factors might also play a role in the developmental profile of muscarinic receptor-stimulated phosphoinositide metabolism and in the differential effects of EtOH, such as the developmental changes in the activities of phospholipase C, 3 kinase and 5-phosphatase (Moon et al. 1989), or protein kinase C (particularly the gamma subtype; Sposi et al. 1989) and the relative distribution and/or coupling of muscarinic receptor subtypes. Finally, an interesting hypothesis to be considered is the formation of phosphatidylethanol, an "abnormal" phospholipid found in large amounts in brains of acutely and chronically treated rats (Alling et al. 1984) which might activate protein kinase C (Asaoka et al. 1988) and exert a feedback inhibition on muscarinic receptor stimulated activation of phospholipase C.

ACKNOWLEDGMENTS

This study was supported in part by a grant from NIAAA (AA-08154).

REFERENCES

Alling, C, Gustavsson, L., Mansson, J.E., Benthin, G., and Auggard, E., 1984, Phosphatidyl-ethanol formation in rat organs after ethanol treatment, *Biochem Biophys Acta* 793: 119-122.

Asaoka, Y., Kikkawa, U., Sekiguchi, K, Shearman, M.S., Kosaka, Y, Nakano, Y., Satoh, T. and Nishizuka, Y., 1988, Activation of brain-specific protein kinase C subspecies in the presence of phosphatidylethanol, *FEBS Lett* 231: 221-224.

Ashkenazi, A., Dramachandran, J., and Capon, D.J., 1989, Acetylcholine analogue stimulates DNA synthesis in brain derived cells via specific muscarinic receptor subtypes, *Nature* 340: 146-150.

Balduini, W., Murphy, S.D., and Costa, L.G., 1987, Developmental changes in muscarinic receptor-stimulated phosphoinosiitde metabolism in rat brain, *J Pharmacol Exp Ther* 241: 421-427.

Balduini, W., and Costa, L.G., 1989, Effects of ethanol on muscarinic receptor-stimulated phosphoinositide metabolism during brain development, *J Pharmacol Exp Ther* 250: 541-547.

Balduini, W, and Costa, L.G., 1990, Developmental neurotoxicity of ethanol: in vitro inhibition of muscarinic receptor-stimulated phosphoinositide metabolism in brain from neonatal but not adult rats, *Brain Res* 512: 248-252.

Balduini, W., Murphy, S.D., and Costa, L.G., 1990, Characterization of cholinergic muscarinic receptor-stimulated phosphoinositide metabolism in brain from immature rats, *J Pharmacol Exp Ther* 253: 573-579.

Balduini, W., Candura, S.M., and Costa, L.G., 1991a, Regional development of carbachol-, glutamate-, norepinephrine-, and serotonin-stimulated phosphoinositide metabolism in rat brain, *Dev Brain Res* 62: 115-120.

Balduini, W., Candura, S.M., Manzo, L., Cattabeni, F., and Costa, L.G., 1991b, Time-, concentration-, and age- dependent inhibition of muscarinic receptor-stimulated phosphoinositide metabolism by ethanol in the developing rat brain, *Neurochem Res* 16:1235-1240.

Bruns, E.M., Kruckeberg, T.W., Stibler, H., Cerven, E., and Borg, S., 1984, The effects of ethanol exposure during the brain growth spurt in rats, *Teratology* 29: 251-258.

Candura, S.M., Balduini, W., and Costa, L.G., 1991, Interaction of short chain aliphatic alcohols with muscarinic receptor-stimulated phosphoinositide metabolism in cerebral cortex from neonatal and adult rats, *NeuroToxicology* 12: 23-32.

Candura, S.M., Manzo, L., and Costa, L.G., 1992, Inhibition of muscarinic receptor and G-protein-dependent phosphoinositide metabolism in cerebrocortical membranes from neonatal rats by ethanol, *Neurotoxicology* 13: 281-288.

Dobbing, J., and Sands, J., 1979, Comparative aspects of the brain growth spurt, *Early Human Develop* 3: 79-83.

Gottlieb, A., Keydar, I., and Epstein, H.T., 1977, Rodent brain growth stages: an analytical review, *Biol Neonate* 32: 166-176.

Grant, K.A., and Samson, H.H., 1982, Ethanol and tertiary butanol-induced microencephaly in the neonatal rat: comparison of brain growth parameters, *Neurobehav Toxicol Teratol* 4: 315-321

Grant, K.A., and Samson, H.H., 1984, *n*-Propanol-induced microencephaly in the neonatal rat, *Neurobehav Toxicol Teratol* 6: 165-169.

Heacock, A.M., Fisher, S.K., and Agranoff, B.W., 1987, Enhanced coupling of neonatal muscarinic receptors in rat brain to phosphoinositide turnover, *J Neurochem* 48: 1904-1911.

Hoek, J.B., and Rubin, E., 1990, Alcohol and membrane-associated signal tranduction, *Alcohol & Alcoholism* 25: 143-156.

Hohman, C.F., Brooks, A.R., and Coyle, J.T., 1988, Neonatal lesions of the basal forebrain cholinergic neurons results in abnormal cortical development, *Dev Brain Res* 42: 253-264.

Hunt, W.A., and Dalton, T.K., 1976, Regional brain acetylcholine levels in rats acutely treated with ethanol or rendered ethanol-dependent, *Brain Res* 109: 628-631.

Moon, K.H., Lee, S.Y., and Rhee, S.G., 1989, Developmental changes in the activities of phospholipase C, 3-kinase and 5-phosphatase in rat brain, *Biochem Biophys Res Comm* 164: 370-374.

Nordberg, A., and Wahlstrom, G., 1982, Tolerance, physical dependence and changes in muscarinic receptor binding sites after chronic ethanol treatment in the rat, *Life Sci* 31: 277-287.

Samson, H.H., and Diaz, J., 1982, Effects of neonatal ethanol exposure on brain development in rodents. *in*: "Fetal Alcohol Syndrome. Vol III. Animal Studies". Abel E.L. ed., Boca Raton, CRC Press, pp 131-150.

Smith, T.L., 1983, Influence of chronic ethanol consumption on muscarinic cholinergic receptors and their linkage to phospholipid metabolism in mouse synaptosomes *Neuropharmacology* 22: 661-663.

Sonderegger, T.L., Colbern, D., Calmes, H., Corbitt, S., and Zimmerman, E., 1982, Methodological note: intragastric intubation of ethanol to rat pups, *Neurobehav Toxicol Teratol* 4: 477-481.

Sposi, N.M., Bottero, L., Cossu, G., Russo, G., Testa, U., and Peschile, C., 1989, Expression of protein kinase C genes during ontogenic development of the central nervous system, *Mol Cell Biol* 9: 2284-2288.

Streissguth, A.P., Landesman-Dwyer, S., Martin, J.C., and Smith D.W., 1980, Teratogenic effects of alcohol in humans and laboratory animals, *Science* 209: 353-361.

Tabakoff, B., Munoz-Marcus, M., and Fields, J. Z., 1979, Chronic ethanol feeding produces an increase in muscarinic cholinergic receptors in mouse brain, *Life Sci* 25: 2173-2180.

Van Hoof, C.O.M., De Graan, P.N.E., Oestriecher, A.B., and Gispen, W.H., 1989, Muscarinic receptor activation stimulates B50/GAP43 phosphorylation in isolated nerve growth cones, *J Neurosci* 9: 3753-3759.

West, J.R., and Pierce, D.R., 1986, Perinatal alcohol exposure and neuronal damage, *in*: Alcohol and Brain Development, J.R. West, ed., Oxford University Press, New York, pp. 120-157.

EFFECTS OF ETHANOL ON RECEPTOR-MEDIATED PHOSPHOLIPASE D ACTIVITY IN HUMAN NEUROBLASTOMA CELLS

María del Carmen Boyano-Adánez and
Lena Gustavsson

Department of Psychiatry and Neurochemistry
University of Lund
PO Box 638
S-220 09 Lund
Sweden

INTRODUCTION

During recent years, activation of phospholipase D by several receptor agonists has been demonstrated in a variety of cells, indicating a role for this enzyme in signal transduction cascades. The mechanisms of activation involve guanine nucleotide binding proteins as well as protein kinase C (for review see Exton, 1990; Billah and Anthes, 1990; Shukla and Halenda, 1991). Phospholipase D catalyses phosphatidic acid (PA) formation by hydrolysis of the terminal phosphodiester bound of glycerophospholipids (Kanfer 1980). Furthermore, this enzyme catalyses a transphosphatidylation reaction by which the phosphatidyl moiety of the phospholipid substrate is transferred to primary, short-chain alcohols to produce the corresponding phosphatidylalcohols (Yang et al., 1967; Dawson, 1967). In the presence of ethanol, transphosphatidylation via phospholipase D consequently leads to phosphatidylethanol (PEth) production (Gustavsson and Alling, 1987; Kobayashi and Kanfer, 1987).

Because of its ability to use ethanol as a substrate, it is possible that phospholipase D is involved in some of the ethanol-induced effects on cell function. A significant amount of PEth is accumulated in cell membranes in the presence of ethanol (Alling et al., 1984; Benthin et al., 1985; Lundqvist et al., 1993) and this amount is further increased upon stimulation with receptor agonists or activators of protein kinase C (Exton, 1990; Billah and Anthes, 1990; Shukla and Halenda, 1991). Simultaneously with PEth formation in the presence of ethanol, the normal receptor-induced production of PA is inhibited. Thus, ethanol causes a net decrease in the initial rate of phospholipase D-mediated PA formation, an effect that is compensated by the accumulation of PEth. The degree of PEth formation and inhibition of PA are dependent on the ethanol concentration (Gustavsson and Hansson, 1990). Because of its unique ability to use short-chain alcohols as substrate, phospholipase D may be involved in the ethanol-induced changes in cell function.

Muscarinic receptors are coupled to guanine nucleotide binding proteins to modulate several different signal transduction pathways, including stimulation of phospholipases A_2 and C (Conklin et al., 1988; Richards, 1991). Among the different receptors reported to be coupled to phospholipase D, muscarinic receptors are the most well studied in the nervous system. It has been demonstrated that muscarinic receptors activate phospholipase D in synaptosomes (Qian and Drewes, 1989), astrocytoma 1321N1 cells (Martinson et al., 1989), LA-N-2 human neuroblastoma cells (Sandmann and Wurtman, 1991) and astroglial cells in primary culture (Gustavsson et al., 1993).

The neuroblastoma cell SH-SY5Y displays muscarinic receptors coupled to poly-phosphoinositide-specific phospholipase C (Serra et al., 1988; Mei et al., 1989; Lambert and Nahorski, 1990). In the present study, we demonstrate that stimulation of muscarinic receptors in this cell type also leads to activation of phospholipase D. Both PA production and PEth formation were increased after addition of muscarinic agonists. However, ethanol induced only minor changes in the PA formation, indicating that the major part of PA was formed by another enzyme, probably via a phospholipase C-catalysed pathway.

MATERIALS AND METHODS

Materials

SH-SY5Y cells were a gift from Dr. S. Påhlsson, department of Pathology, Uppsala University, Sweden. PA, carbachol, atropine and mecamylamine were obtained from Sigma chemical co. (U.S.A.). HPTLC-plates (silica gel 60) were obtained from Merck (Germany). Tissue culture dishes were from Costar (U.S.A). Cell culture medium (Eagle's minimal essential medium with L-glutamine and Earl's salt) with and without Hepes (4-(2-hydroxyethyl) 1-piperazineethanesulfonic acid) and fetal calf serum were from GIBCO (Scotland). Streptomycin and penicillin were from ASTRA Pharmaceuticals (Sweden). [^{3}H]-palmitic acid, specific radioactivity 60.0 Ci/mmol, was obtained from Dupont NEN Product (U.S.A.). PEth standard was synthesized essentially according to the method by Eibl and Kovatchev (1981) from egg yolk phosphatidylcholine by using peanut phospholipase D (Sigma chemical co., U.S.A). Ready-Safe scintillation liquid was from Beckman, Beckman Instruments (U.S.A.).

Cell Culture

SH-SY5Y cells were cultured in 35 mm diameter plastic dishes containing Eagle's minimal essential medium with L-glutamine and Earl's salts supplemented with 10% fetal calf serum, 10 IU/ml penicillin and 10 μg/ml streptomycin. Cells were allowed to grow for 1 week after passage in an incubator under humidified atmosphere containing 5% CO_2/95% air. The medium was changed once during this time. Cells used were from passages 45-60.

Cell Incubations

Cells were labelled with [^{3}H]palmitic acid (4 μCi/ml medium) for 6 hours. The labelling was performed without serum or antibiotics in the medium. 15 minutes before the treatment with ethanol, carbachol or antagonists, the medium was changed to cell culture medium buffered with Hepes, pH 7.4. Ethanol was added 5 minutes before the addition of carbachol. When the antagonists, atropine (100 nM) and mecamylamine (10 μM), were used, they were added 10 minutes before carbachol. The reaction was stopped by addition of 500 μl of ice-cold isopropanol, washing the dishes with water and scraping the cells into chloroform/methanol 1:2 (by volume).

Lipid Extraction and Separation

The lipids were extracted according to Bligh and Dyer (1959). The lipid containing chloroform-phase was evaporated under N_2 and the lipids were redissolved in chloroform:methanol 2:1 before application on HPTLC-plates. The solvent system used for PEth isolation was ethylacetate:isoctane:acetic acid 90:50:20 (by volume). The plates were stained with iodine. The spots corresponding to individual phospholipids were scraped into scintillation vials and the radioactivity analysed in a scintillation counter.

RESULTS

The formation of [³H]PA and [³H]PEth was measured in SH-SY5Y cells, prelabelled with [³H]palmitic acid, at different times after stimulation with the cholinergic receptor agonist carbachol (1mM). In the presence of ethanol (150 mM), carbachol induced a rapid increase in PEth formation reaching maximum after 1 minute of stimulation (Fig. 1a). Similarly, there was a rapid increase in PA formation after carbachol stimulation (Fig. 1b). When ethanol was present there was a slight inhibition of PA formation (Fig. 1). PEth formation was dependent on the concentration of carbachol, reaching a maximum at 1 mM with a half maximal effective concentration of approximately 3×10^{-5} M (Fig. 2).

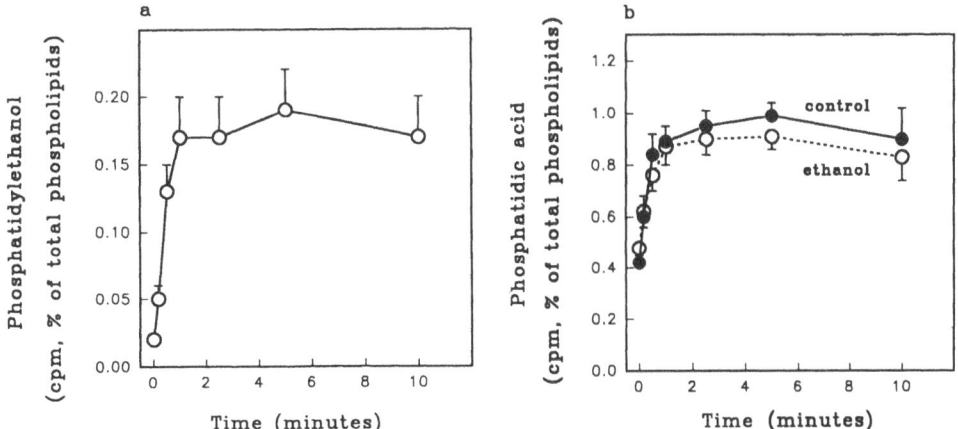

Figure 1. Time-course of carbachol-induced PEth (a) and PA (b) formation. SH-SY5Y cells prelabelled with [³H]-palmitic acid were stimulated with carbachol (1 mM) for different periods of time. The stimulations were performed both without ethanol (●) and in the presence of 150 mM ethanol (○). The data are presented as means ± S.E.M. (a: n=8, b: n=12).

To further characterize the carbachol-induced increase in phospholipase D activity the effects of the antagonists, atropine and mecamylamine were studied (Fig. 3). As described above carbachol induced PEth formation and this effect was blocked by atropine, a muscarinic receptor antagonist (Fig. 3). Atropine caused a reduction in PEth formation by 94%. On the other hand, mecamylamine, a nicotinic receptor antagonist, inhibited the carbachol-induced increase in the amount of PEth to a lesser extent (40%).

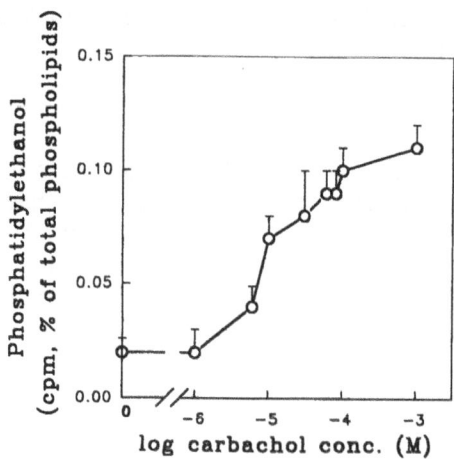

Figure 2. Concentration-response curve for carbachol-induced PEth formation. SH-SY5Y cells prelabelled with [³H]-palmitic acid were stimulated with carbachol at different concentrations for 30 seconds in the presence of ethanol (150 mM). The data are presented as means ±S.E.M. (n=20).

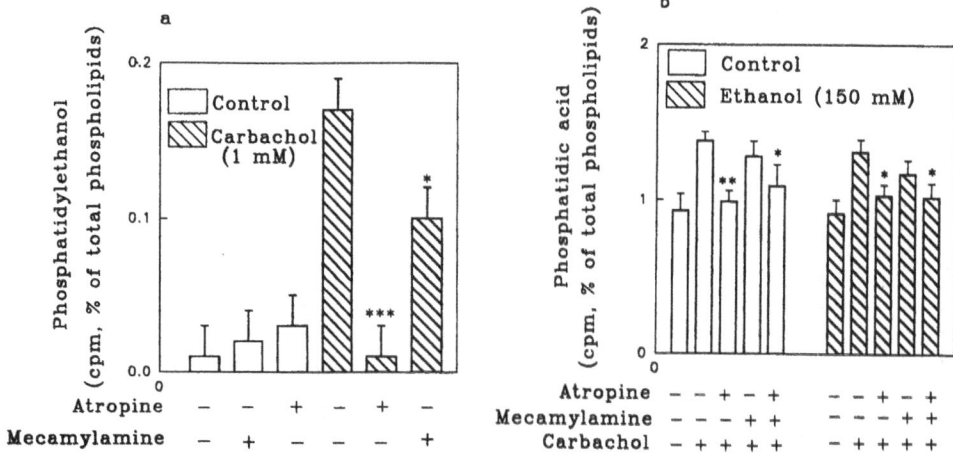

Figure 3. Effects of cholinergic receptor agonists and antagonists on PEth (a) and PA (b) formation. SH-SY5Y cells prelabelled with [³H]-palmitic acid were stimulated with carbachol (1 mM) for 30 seconds. PEth formation (a) was studied in the presence of 150 mM ethanol whereas PA formation (b) was studied both in the presence and absence of ethanol (150 mM) as indicated in the figure. The antagonists, atropine (100 nM) and mecamylamine (10 μM) were added 10 minutes before carbachol. The results are presented as means ± S.E.M. (a: n=15, b: n=12). *** $p < 0.001$, ** $p < 0.01$, * $p < 0.05$ compared to carbachol-induced product formation in the absence of antagonists.

In the same experiments PA formation was analysed. When atropine was added, the carbachol-induced PA formation was inhibited by 87%. Mecamylamine tended to decrease PA formation but this effect was not statistically significant. Preincubation with both atropine and mecamylamine resulted in the same degree of inhibition as did atropine alone.

The production of PEth and PA was dependent on the concentration of ethanol. The amount of PEth formed after carbachol stimulation was enhanced with increasing concentrations of ethanol (Fig. 4). PEth was measurable at ethanol concentrations as low as 10 mM. Ethanol caused a slight inhibition of the PA formed with a reduction of 16 % with 200 mM ethanol (Fig. 5). On the other hand, ethanol did not affect the carbachol-induced diacylglycerol formation (data not shown).

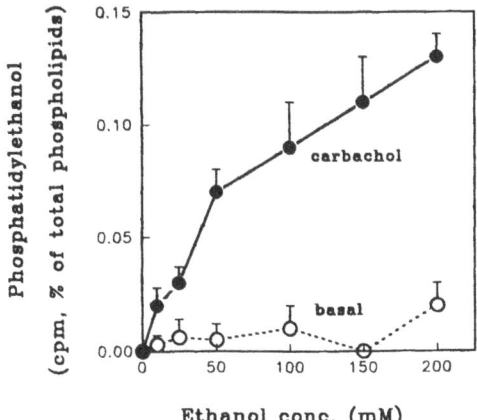

Figure 4. Effect of different ethanol concentrations on carbachol-induced PEth formation. SH-SY5Y cells prelabelled with [^{3}H]-palmitic acid were stimulated with carbachol (1 mM) for 30 seconds at different concentrations of ethanol (●). Basal levels correspond to ethanol alone (○). Ethanol was added 5 minutes before carbachol. The data are presented as means ± S.E.M. (n=6).

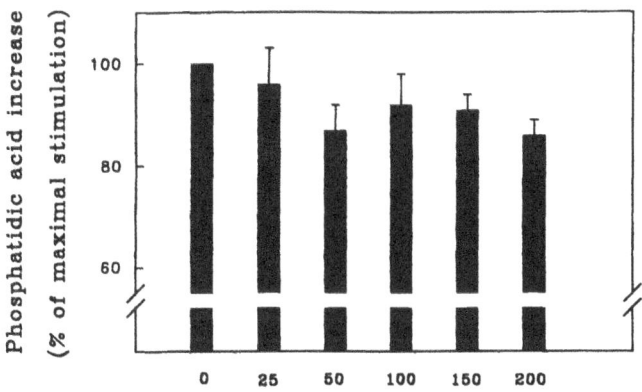

Figure 5. Effect of different ethanol concentrations on carbachol induced PA formation. SH-SY5Y cells prelabelled with [^{3}H]-palmitic acid were stimulated for 30 seconds with carbachol (1 mM) at different concentrations of ethanol. Ethanol was added 5 minutes before carbachol. The data are presented as means ± S.E.M. (n=6).

DISCUSSION

In the present study, we have demonstrated that muscarinic acetylcholine receptors are coupled to phospholipase D activity in the human neuroblastoma cell line SH-SY5Y. Carbachol induced an increase in PA formation, and when ethanol was present, production of PEth. The carbachol-induced responses were completely inhibited by a muscarinic receptor antagonist, atropine, but only slightly affected by a nicotinic receptor antagonist, mecamylamine. This indicates that the activation of phospholipase D by carbachol was mainly mediated via muscarinic receptors. These results are in agreement with studies from other cell types and tissues originating from brain which have pointed to a coupling between muscarinic acetylcholine receptors and phospholipase D (Martinsson et al., 1989; Qian and Drewes, 1989; Sandmann and Wurtman, 1991; Gustavsson et al., 1993).

Five types of cloned muscarinic acetylcholine receptors have been described: m_1, m_3 and m_5 coupled to phosphoinositide metabolism and calcium mobilization, whereas m_2 and m_4 are coupled (negatively) to adenylate cyclase (for review see Richards, 1991). Which of these subtypes that are coupled to phospholipase D activity in SH-SY5Y cells is still not known. Phospholipase D activation has been analysed in human embryonic kidney cells transfected with DNA coding for different muscarinic receptor subtypes. Although activation could be induced by all receptor subtypes, the highest efficiency was obtained with cells displaying m_1 and m_3 subtypes (Sandmann et al., 1991). Moreover, phospholipase D was demonstrated to be coupled to the m_1 subtype but not m_2 in HeLa cells transfected with different muscarinic acetylcholine receptor genes (Pepitoni et al., 1991). Based on pharmacological evidence it has been suggested that SH-SY5Y cells mainly express muscarinic receptors of the m_1 subtype and that these receptors are coupled to polyphosphoinositide breakdown (Serra et al., 1988; Mei et al., 1989). On the other hand, Lambert et al. (1989) demonstrated binding sites only for a muscarinic receptor subtype defined pharmacologically as m_3 on SH-SY5Y cells. Also the m_3 receptors found on SH-SY5Y cells were suggested to be linked to phosphoinositide metabolism and changes in intracellular calcium (Lambert and Nahorski, 1990). Therefore, it is probable that the carbachol-induced increase in phospholipase D activity in SH-SY5Y cells found in the present study is mediated either via m_1 and/or m_3 receptor subtypes, although this needs to be further investigated.

In the presence of ethanol, phospholipase D catalyses the formation of PEth at the expense of PA, leading to an almost complete inhibition of agonist-induced PA formation at optimal ethanol concentrations (Gustavsson et al., 1993). In the present study, ethanol only slightly inhibited the carbachol-induced PA formation. This indicates that PA is formed mainly by another mechanism than phospholipase D activation in SH-SY5Y cells. Furthermore, the initial rate of [^{3}H]PA formation was substantially higher (approximately 4-fold) than the initial increase in the amount of [^{3}H]PEth. This discrepancy further points to the involvement of another enzyme activity in the PA production following muscarinic stimulation of SH-SY5Y cells. In addition to the phospholipase D pathway, PA may be formed through the sequential action of phospholipase C and diacylglycerolkinase. Muscarinic activation of polyphosphoinositide-specific phospholipase C has been demonstrated in SH-SY5Y cells (Serra et al., 1988; Mei et al., 1989; Lambert and Nahorski, 1990). Moreover, receptor-mediated activation of a phospholipase C specific for phosphatidylcholine has been indicated during recent years (Exton, 1990). Involvement of such a phospholipase C in the muscarinic receptor agonist-induced formation of 1,2-diacylglycerol was demonstrated to occur in SK-N-SH cells, the parent human neuroblastoma of the clone SH-SY5Y (Lee et al., 1991). However, whether phosphatidylcholine breakdown via phospholipase C occurs also in SH-SY5Y cells and contributes to the carbachol-induced increase in PA is still not clear.

We conclude that stimulation of muscarinic receptors activates phospholipase D in SH-SY5Y cells, leading to PA formation, and when ethanol is present production of PEth. Ethanol only slightly inhibited the carbachol-induced PA formation, indicating that the major part of PA formed after muscarinic stimulation is formed by another pathway which probably involves activation of a phospholipase C.

ACKNOWLEDGEMENTS

This study was supported by grants from Wenner-Gren Center Foundation, the Albert Påhlsson Foundation, the Medical Faculty, University of Lund, the Swedish Alcohol Research Fund, the Swedish Medical Research Council (proj. No. 03P-08895 and 21x-05249) and Alcalá University, Spain.

REFERENCES

Alling, C., Gustavsson, L., Månsson, J-E, Benthin, G., and Änggård E., 1984, Phosphatidylethanol formation in rat organs after ethanol treatment. *Biochim. Biophys. Acta* 793:119.

Benthin, G., Änggård, E., Gustavsson, L., and Alling C., 1985, Formation of phosphatidylethanol in frozen kidneys from ethanol-treated rats. *Biochim.Biophys.Acta* 835:385.

Billah, M.M., and Anthes, J.C., 1990, The regulation of cellular functions of phosphatidylcholine hydrolysis. *Biochem. J.* 269:281.

Bligh, E.G., and Dyer, W.S., 1959, A rapid method of total lipid extraction and purification. *Can. J. Biochem. Physiol.* 37:911.

Conklin, B.R., Brann, M.R., Buckley, N.J., Ma, A.L., Bonner, T.I., and Axelrod, J., 1988, Stimulation of arachidonic acid release and inhibition of mitogenesis by cloned genes for muscarinic receptor subtypes stably expressed in A9 L cells. *Proc. Natl. Acad. Sci. U.S.A.* 85:8698.

Dawson, R.M., 1967, The formation of phosphatidylglycerol and other phospholipids by the transferase activity of phospholipase D. *Biochem.J.* 102:205.

Eibl, H., and Kovatchev, S., 1981, Preparation of phospholipids and their analogs by phospholipase D. *Methods Enzymol.* 72:632.

Exton, J.H., 1990, Signalling through phosphatidylcholine breakdown. *J. Biol. Chem.* 265:1.

Gustavsson, L., and Alling, C., 1987, Formation of phosphatidylethanol in rat brain by phospholipase D. *Biochem. Biophys. Res. Commun.* 142:958.

Gustavsson, L., and Hansson, E., 1990, Stimulation of phospholipase D activity by phorbol esters in cultured astrocytes. *J. Neurochem.* 54:737.

Gustavsson, L., Lundqvist, C., Hansson, E., Rodríguez, D., Simonsson, P., and Alling, C., 1993, Ethanol-induced changes in signal transduction via formation of phosphatidylethanol. This volume.

Kanfer, J.N., 1980, The base-exchange enzymes and phospholipase D of mammalian tissue. *Can.J.Biochem.* 58:1370.

Kobayashi, M., and Kanfer, J.N., 1987, Phosphatidylethanol formation via transphosphatidylation by rat brain synaptosomal phospholipase D. *J. Neurochem.* 48:1597.

Lambert, D.G., Ghataorre, A.S., and Nahorski, S.R., 1989, Muscarinic receptor binding characteristics of a human neuroblastoma SK-N-SH and its clones SH-SY5Y and SH-EP1. *Eur.J.Pharmacol.* 165:71.

Lambert, D.G., and Nahorski, S.R., 1990, Muscarinic-receptor-mediated changes in intracellular Ca^{2+} and inositol 1,4,5-trisphosphate mass in a human neuroblastoma cell line, SH-SY5Y. *Biochem.J.* 265:555.

Lee, C., Fisher, S.K., Agranoff, B.W., and Hajra, A.K., 1991, Quantitative analysis of molecular species diacylglycerol and phosphatidate formed upon muscarinic receptor activation of human SK-N-SH neuroblastoma cells. *J.Biol.Chem.* 266:22837.

Lundqvist, C., Rodríguez, F.D., Simonsson, P., Alling, C., and Gustavsson, L., 1993, Phosphatidylethanol affects inositol 1,4,5-triphosphate levels in NG108-15 neuroblastoma x glioma hybrid cells. *J. Neurochem.* 60:in press.

Martinson, E.A., Goldstein, D., and Brown, J.H., 1989, Muscarinic receptor activation of phosphatidylcholine hydrolysis. *J. Biol. Chem.* 264:14748.

Mei, L., Roeske, W.R., and Yamamura, H.I., 1989, The coupling of muscarinic receptors to hydrolysis of inositol lipids in human neuroblastoma SH-SY5Y cells. *Brain Res.* 504:7.

Pepitoni, S., Mallon, R.G., Pai, J.-K., Borlowski, J.A., Buck, M.A., and McQuade, R.D., 1991, Phospholipase D activity and phosphatidylethanol formation in stimulated HeLa cells expressing the human m1 muscarinic acetylcholine receptor gene. *Biochem. Biophys. Res. Commun.* 176:453.

Qian, Z., and Drewes, L.R., 1989, Muscarinic acetylcholine receptor regulates phosphatidylcholine phospholipase D in canine brain. *J. Biol. Chem.* 264:21720.

Richards, M.H., 1991, Pharmacology and second messengers interactions of cloned muscarinic receptors. *Biochem. Pharmacol.* 42:1645.

Sandmann, J., Peralta, E.G., and Wurtmann, R.J., 1991, Coupling of transfected muscarinic acetycholine receptors subtypes to phospholipase D. *J. Biol. Chem.* 266:6031.

Sandmann, J., and Wurtman, R.J., 1991, Stimulation of phospholipase-D activity in human neuroblastoma (LA-N-2) cells by activation of muscarinic acetylcholine receptors or by phorbol esters-relationship to phosphoinositide turnover. *J. Neurochem.* 56:1312.

Shukla, S.D., and Halenda, S.P., 1991, Phospholipase D in cell signalling and its relationship to phospholipase C. *Life Sci.* 48:851.

Serra, M., Mei, L., Roeske, R.R., Lui, G.K., Watson, M., and Yamamura, H.I., 1988, The intact human neuroblastoma cell (SH-SY5Y) exhibits high-affinity [³H]Pirenzepine binding associated with hydrolysis of phosphatidylinositols. *J. Neurochem.* 50:1513.

Yang, S.F., Freer, S., and Benson, A.A., 1967, Transphosphatidylation by phospholipase D. *J. Biol. Chem.* 242:477.

ETHANOL AND CELL TYROSINE KINASE

Shivendra D. Shukla, Cindy Y. Zhu, Ilsa I. Rovira and Archie W. Thurston, Jr.

Department of Pharmacology
University of Missouri-Columbia
School of Medicine
One Hospital Drive
Columbia, MO 65212

INTRODUCTION

Virtually no part of the body is spared by the effects of ethanol. Cells are the primary targets (Alcohol and Health, NIAAA Report, 1990). Ethanol has long been known to affect cell metabolism and enzyme activities (Hawkins and Kalant, 1972). Ethanol is metabolized in cells predominantly to acetaldehyde which can cause damaging effects. One of the first interactions of ethanol on a cell is with the cell membrane (Goldstein and Chin, 1981; Sun and Sun, 1985; Taraschi and Rubin, 1985). There is evidence that ethanol in the concentration range 50-500 mM has influences on membrane fluidity (Harris and Schroeder, 1981). Further, the effects of ethanol on membranes can be more discrete as it modifies lateral and vertical lipid domains (Wood et al., 1989).

Another important effect of ethanol is at the transmembrane signalling steps (Hoek et al., 1992). This aspect has drawn considerable attention since a variety of cell functions and responses are controlled by these signalling mechanisms. In the past few years, it has been demonstrated that multiple signalling pathways and receptor functions are modulated by ethanol. Among the signalling pathways, the phospholipase C mediated hydrolysis of phosphoinositides seems to be particularly sensitive to ethanol (Crews et al., 1989; Hoek et al., 1992). For example, ethanol attenuated receptor mediated phosphatidylinositol-4,5 bisphosphate breakdown in mouse brain (Hoffman et al., 1986). Ethanol decreases the bradykinin-stimulated IP_3 production in NG108 cells (Simonsson et al., 1989). On the other hand, ethanol itself causes activation of phospholipase C in platelets (Rubin et al., 1988) and in the liver (Hoek et al., 1992). Mobilization of Ca^{2+} also occurs in cells exposed to ethanol (Hoffman et al., 1989; Hoek et al., 1992). Most of these effects of ethanol are noted in the range of 50 to 500 mM concentrations. Ethanol has also been proposed to modulate the efficacy of the G-protein α_s subunit in stimulating adenylate cyclase (Nagy et al., 1990; Hoek et al., 1992), adenylate cyclase activity (Bode and Molinoff, 1988; DePetrillo and Swift, 1992) and cGMP levels (Hunt et al., 1977; Hoffman et al., 1989). Acidic phospholipids in the brain, i.e. polyphosphoinositides, are also influenced by ethanol, particularly through a phospholipase A_2 pathway (Sun et al., 1989, 1991). Phospholipase

Alcohol, Cell Membranes, and Signal Transduction in Brain
Edited by C. Alling et al., Plenum Press, New York, 1993

D (PLD) mediated pathways (Shukla and Halenda, 1991) are also suggested to be influenced by ethanol. PLD catalyzes the transphosphatidylation reaction and in the presence of ethanol a novel phospholipid, namely phosphatidylethanol (PEth) is formed. The level of PEth increases in ethanol treated cells, e.g. in NG108 cells (Lundquist *et al.*, 1992) or in the brain of alcoholics (Alling *et al.*, 1984).

Receptor-gated ion channel activities are also modulated by ethanol. Recent developments in this area are quite illustrative of the complexities of ethanol actions. GABA (γ-aminobutyric acid) is the major inhibitory neurotransmitter in mammalian brain. Acute *in vitro* exposure to ethanol activates GABA-stimulated chloride channels (Allan and Harris, 1987). On the other hand, voltage-dependent calcium and sodium channels in the brain are inhibited by acute ethanol exposure (Harris and Allan, 1989). Ethanol inhibits calcium flux and cyclic-GMP production stimulated by n-methyl-D-asparate (NMDA) (Hoffman *et al.*, 1989; Michaelis *et al.*, 1992). It is, therefore, obvious that ethanol actions on ion transport are both stimulatory and inhibitory depending on the type of receptor and cell system. At present, there is no unitary hypothesis to tie together these effects of ethanol on cell membrane and receptor signalling, and it remains an exciting area of research.

Tyrosine kinase has emerged as one of the components involved in cell signal transduction, cell growth, development and differentiation (Hunter and Cooper, 1985; Yarden and Ullrich, 1988; Aaronson, 1991; Cantley *et al.*, 1991; Dhar and Shukla, 1990, 1991). Ethanol is known to cause fetal alcohol syndrome (FAS) or fetal alcohol effects (FAE) which cause prenatal and postnatal growth retardation. As a consequence, CNS dysfunction and major organ system malfunction occur (Jones and Smith, 1973; Ouellette *et al.*, 1977; Rudeen *et al.*, 1989). Exposure of the fetus to ethanol results in the reduction of neuronal migration and synaptogenesis in specific brain regions. Differentiation of neurons is also effected by ethanol. Alcohol also alters the process of maturation and replication of lymphocytes in the spleen (Grossman *et al.*, 1988). These observations demonstrate the marked effect of ethanol on cell growth and development (see also Snyder *et al.*, 1992). The fact that tyrosine kinase plays a role in the aforementioned processes prompted the study of the effect of ethanol on tyrosine kinase. Several receptors (e.g. EGF, PDGF, FGF) have intrinsic tyrosine kinase activity. Additionally, there are soluble tyrosine kinases. Several oncogene products, e.g. $pp60^{c-src}$, fyn, lyn, lck, yes, etc., are also tyrosine kinases (Collet and Erikson, 1978; Anderson *et al.*, 1990). Furthermore, a link is emerging between tyrosine kinase and cell signalling components (transducer/effector relationship) (Ullrich and Schlessinger, 1990; Cantley *et al.*, 1991; Shukla, 1992). For example, EGFR tyrosine kinase phosphorylates phospholipase C-γ_1 involved in IP_3 production. PI-3-kinase and GAP (GTPase activating protein) are also substrates of tyrosine kinase (Anderson *et al.*, 1990; Cochet *et al.*, 1991; Walker *et al.*, 1988). The question whether ethanol affects tyrosine kinase is, therefore, quite relevant to explain several actions of ethanol at the cellular level and in alcoholism.

In order to ascertain the effect of ethanol on cell tyrosine kinase, several cell models were considered. One of the initial requirements to perform this study was to select a cell for which background literature information on receptor-coupled tyrosine kinase and signalling responses was available. In this respect, A431 cells, a human epidermoid carcinoma cell is one of the readily identifiable cells with such properties. These cells have an abundance of EGF receptor tyrosine kinase (Carpenter *et al.*, 1979; Kawamoto *et al.*, 1983; Buthrow *et al.*, 1983). EGF stimulation of tyrosine kinase has been shown to phosphorylate PLC-γ_1. This phosphorylation causes translocation of the PLC-γ_1 to the membrane and its activation (Nishibe *et al.*, 1990; Rhee, 1991).

Effect of Ethanol on Tyrosine Kinase Activity

In this study, A431 cell membranes were utilized to examine the influence of ethanol on the membrane associated tyrosine kinase. A431 cells were obtained from American Tissue Culture Collection (ATCC, Rockville, MD) and maintained in a humidified incubator with 5% CO_2/95% air at 37°C. Cells were grown in Dulbecco's Modified Eagles Medium (DMEM) with 4.5 g/L glucose supplemented with 10% heat inactivated fetal calf serum and 100 U/mL penicillin. Cells were subcultured weekly into 150 mm dishes at a density of 0.4-0.45×10^6 cells per mL. Media was replaced every 2-3 days. Cells were harvested for membrane preparations upon confluency. Membranes from A431 cells were isolated by a modification of published procedure (Carpenter et al., 1979). A431 cells were harvested and resuspended in a sodium free lysis buffer, buffer A, (5 mM $MgCl_2$, 2 mM EGTA and 10 mM Tris-HCl pH 7.0) at a concentration of 2.5×10^6 cells/mL. The cells were lysed by repeated freezing in liquid nitrogen and thawing. The thawed suspension was then centrifuged at 1000 × g for 15 minutes to remove unbroken cells and nuclei. The suspension was then centrifuged at 20,000 × g for 30 minutes at 4°C. The pellet was resuspended in one volume of buffer A and layered onto 9 volumes of 35% (w/w) sucrose dissolved in buffer A. The sucrose gradient was then centrifuged at 40,000 × g for 45 minutes at 4°C. The interface between the sucrose and buffer was collected and diluted with 9 volumes of buffer A. The resulting suspension was centrifuged at 75,000 × g for 30 minutes at 4°C. The resulting pellet was resuspended in PBS with 10 mM $MgCl_2$ at a protein concentration of 3-5 µg/µL.

In order to assay the tyrosine kinase, an *in vitro* assay system was employed in these studies. Briefly, 10 µg of the isolated membrane preparations plus the agonist or vehicle in the absence or presence of ethanol was diluted to 9 µL with water and the required agent was incubated for 30 minutes at 37°C with 10 µL of the synthetic substrate in 2X assay buffer (60 nM HEPES - pH 7.4, 20 mM $MgCl_2$, 0.2 mM DDT, 50 µg/mL BSA, 0.3% (v/v) Nonidet P-40, 140 nM sodium orthovanadate, 120 nM ATP and 1 mM peptide substrate) and 1 µL of [γ-^{32}P]-ATP at 1 µCi/µL. The synthetic substrate is a 13 amino acid peptide (Arg-Arg-Leu-Ile-Glu-Asp-Ala-Glu-Tyr-Ala-Ala-Arg-Gly) which contains the phosphorylation site sequences of the pp60[c-src] and contains only one tyrosine residue and no serine or threonine residues. The reaction was terminated by transfer of one volume of

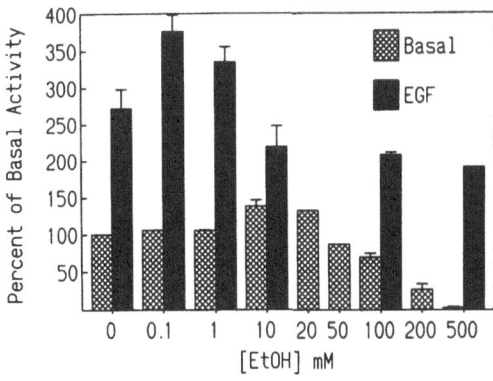

Fig. 1. Ethanol modulation of basal and EGF stimulated tyrosine kinase. Experimental details are described in the text. Tyrosine kinase activity is presented as percent increase over the basal activity.

sample into 6 volumes of 5% TCA at 4°C. After incubation on ice for 1 hour the sample was then centrifuged for 15 minutes in a tabletop microfuge. The supernatant is then spotted onto phosphocellulose discs and washed twice for 5 minutes with 250 mL 1% acetic acid (w/w) and twice with 250 mL of water. Radioactivity in phosphocellulose discs was measured using liquid scintillation spectrophotometry. Non-specific radioactivity bound to the filter was measured by using a control tube without substrate and was subtracted from all values to calculate the specific tyrosine kinase activity (Thurston and Shukla, 1992b).

A431 cell membranes have a basal tyrosine kinase activity that is stimulated 2-3 fold by 1.5 µM EGF. Concentrations of ethanol up to 1 mM had no effect on the basal tyrosine kinase activity in A431 cell membranes. When the concentration was increased further, a small but significant rise in the tyrosine kinase activity was observed which was maximum at 10 mM ethanol (Fig. 1). However, ethanol at higher concentrations showed an inhibition of the activity which declined to lower than basal levels at 200-500 mM ethanol. In the presence of EGF, the pattern was different. Ethanol at low concentrations potentiated the EGF-stimulated tyrosine kinase. At 0.1 mM ethanol, EGF-stimulated tyrosine kinase was 30 to 50% higher when compared to EGF alone. However, as the concentration of ethanol increased, a gradual decrease in EGF-activated tyrosine kinase was noted. It is clear from these results that the tyrosine kinase response to ethanol was bimodal. At ethanol concentrations above 100 mM, an inhibition in both basal and EGF-stimulated tyrosine kinase was apparent. Although low concentrations of ethanol (up to 1 mM) had no effect on basal tyrosine kinase, EGF-stimulated tyrosine kinase was quite sensitive to it (Thurston and Shukla, 1992 a&b). An important implication of these results is that low concentrations of ethanol can modulate agonist-coupled tyrosine kinase. At these low concentrations of ethanol, the basal tyrosine kinase is unaffected. Such modulations can thus influence agonist-induced responses. At higher ethanol concentrations, both the basal and EGF-stimulated tyrosine kinase are inhibited, and this can further affect the cell responses and physiological status of cells.

Effect of Different Chain Length Alcohols on Tyrosine Kinase

As discussed earlier, 10 mM ethanol increased the basal tyrosine kinase activity in A431 cell membranes. In order to determine the specificity of ethanol, the effect of increasing chain length alcohols on tyrosine kinase was also studied. Using the above concentration, it was observed that methanol (CH_4O) had little effect on the activity, while ethanol (C_2H_6O) stimulated the tyrosine kinase. 1-Propanol (C_3H_8O) and 1-butanol ($C_4H_{10}O$) had an inhibitory effect, the former causing more inhibition than the latter one. On the other hand, in the presence of different alcohols (10 mM) EGF-stimulated tyrosine kinase was seen highest in the presence of methanol. Increasing chain length decreased the stimulation; butanol showed the lowest effect (Thurston and Shukla, 1992b). These studies indicate that increasing the alcohol chain length decreases the effect on tyrosine kinase. Such a relationship has been reported for various other responses as well (Hawkins and Kalant, 1972; Goldstein and Chin, 1981).

Ethanol Modulation of Tyrosine Kinase Mediated Phosphorylation of Phospholipase Cγ_1

In A431 cells, several proteins are substrates for EGF receptor tyrosine kinase. Phospholipase Cγ_1 is one such protein. This issue has been examined in intact cells and by immunoprecipitation of PLC-γ_1 using published protocols (Kim et al., 1991). A431 cells grown in 35 mm dishes were placed in serum free DMEM for 18 hours and then changed to phosphate deficient DMEM for 1 hour. The media was aspirated and replaced with fresh phosphate free DMEM supplemented with 300 µCi/mL carrier-free ^{32}P for 1 hour. During the last 20 minutes of labelling, vanadate and molybdate (1 mM each) were added as well as the respective concentrations of ethanol. The media was aspirated and replaced with 1 mL of medium supplemented with vanadate and molybdate (1 mM each) containing EGF (200 ng/mL) or vehicle in the presence or absence of ethanol for 5 minutes. The assay was terminated by aspiration of media, addition of 1 ml lysis cocktail (20 mM HEPES pH 7.2; 1% Triton-X-100; 10% glycerol; 50 mM NaF; 1 mM PMSF; 1 mM vanadate and 10µg/mL leupeptin) and freezing in liquid nitrogen. The thawed lysate was transferred to microfuge tubes and precleared by incubation with normal mouse serum (30 minutes) and heat fixed killed *Staphylococcus aureus* precoated with Protein A and Protein A linked Agarose (45 minutes). The resulting supernatant was treated with serum containing monoclonal antibodies to PLC-γ_1 overnight (kindly supplied by Dr. S.G. Rhee) and precipitated with the *Staph A* protein A and the protein A linked agarose. The immunoprecipitate was then washed three times with buffer (1% Triton X-100; 1% deoxycholate; 1% SDS; 150 mM NaCl and 50 mM Tris; pH 8.5). The immunoprecipitate was solubilized with cocktail as described by Laemmli (1970) and electrophoresed on discontinuous SDS-PAGE and then autoradiographed.

Ethanol at 1 mM concentration stimulated the phosphorylation of PLC-γ_1 by about 30 to 40% over the basal level whereas concentrations above 100 mM decreased the basal ^{32}P-radioactivity of PLC-γ_1 by about 15 to 20%. EGF-stimulated phosphorylation of PLC-γ_1 by 2- to 3-fold. Ethanol at 1, 100 and 200 mM caused a graded inhibition of PLC-γ_1 phosphorylation by EGF. In the presence of 100 mM ETOH or above, EGF had no effect on PLC-γ_1 phosphorylation. These studies indicate that the ethanol has a profound effect on basal and EGF-stimulated phosphorylation of PLC-γ_1 (Thurston and Shukla, 1992b).

CONCLUSIONS

Both basal and EGF-stimulated tyrosine kinase in A431 cell membranes is modulated by ethanol in a biphasic manner. In contrast to basal tyrosine kinase the EGF-stimulated activity is very sensitive to ethanol. In membranes, the possibility of several tyrosine kinases exist. It is, therefore, possible that ethanol may have a varying degree of effects on different tyrosine kinases. In the case of EGF, it is most likely that EGF receptor tyrosine kinase activity is affected by ethanol. In alcoholism, the ethanol concentrations above 20 mM are considered intoxicating, and those in the patho-physiologically relevant range of 20 to 200 mM could have detrimental effects on cells. Despite this, the effect of ethanol at low concentrations can be of significance as well. The fact that agonist (EGF)-stimulated tyrosine kinase is highly susceptible to ethanol leads to the suggestion that even at low levels of ethanol, stimulus-coupled tyrosine kinase activity is modulated. This, in turn, would affect the cellular responses elicited by the stimulus. Further, higher concentrations of ethanol inhibit tyrosine kinase and this may have adverse consequences on cells or tissues.

The mechanism for the biphasic effects of ethanol on EGF receptor tyrosine kinase is not known. In a recent report it was shown that ethanol at 5% (about 855 mM) caused a 47% decrease in [^{125}I]EGF binding to membrane preparations of buccal mucosa (Wang *et al.*, 1992) resulting from the decrease of binding sites rather than the affinity of high affinity receptors. These investigators suggested that ethanol decreased EGF receptor binding through modifications of the receptor molecules thereby impairing receptor kinase activity. Results using lower ethanol concentrations were not presented. Protein kinase assays using [^{32}P]ATP showed an EGF-stimulated phosphorylation of 170 KDa protein in the control but not in the ethanol-treated sample. Although in the studies by Wang *et al.* (1992) very high concentrations of ethanol were used, nonetheless, the inhibition of tyrosine kinase is similar to the observations with A431 cell membranes discussed above. However, it must be pointed out here that under basal conditions too (i.e., in the absence of any EGF), the tyrosine kinase activity is affected by ethanol in a biphasic manner. Thus, the ethanol's effect on tyrosine kinase must also involve other mechanisms than those proposed by Wang *et al.* (1992). For example, ethanol may affect tyrosine kinase by interfering with the ATP binding site, or the catalytic (active) site or the substrate binding site. In this context, it will also be relevant to determine the influence of ethanol on protein phosphotyrosine phosphatases (Tonks and Charbonneau, 1989). These issues warrant further investigation.

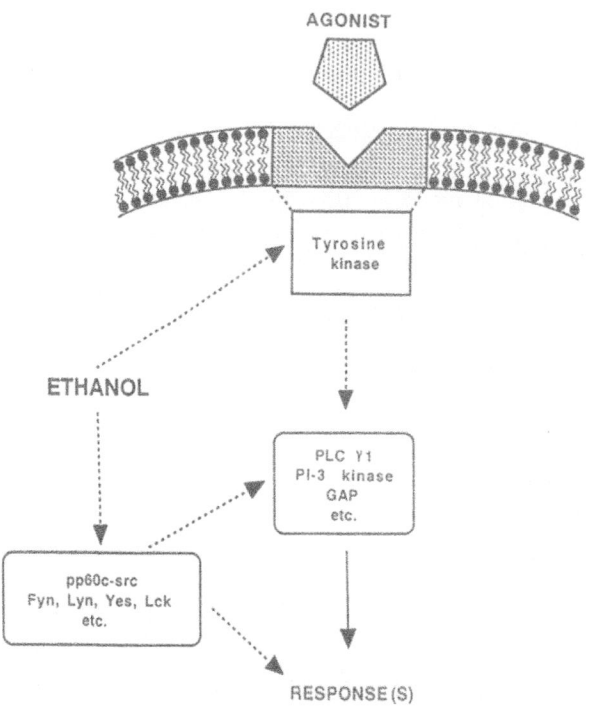

Fig. 2. A hypothetical model depicting possible relationships between ethanol and tyrosine kinase.

Tyrosine kinase can phosphorylate several membrane-bound and endogenous substrates. It has been observed that the phosphorylation of PLC-γ_1, a substrate for EGFR tyrosine kinase, was inhibited by ethanol. It is known that EGF causes phosphoinositide turnover in A431 cells via PLC-γ_1. An expected net result of ethanol effect, therefore, would be inhibition of the EGF-stimulated phosphoinositide turnover. It is relevant to note that 5-hydroxy-tryptamine stimulated inositol phosphate formation is inhibited in platelets from alcoholics (Simonsson and Alling, 1988). Chronic ethanol treatment was also demonstrated to inhibit receptor stimulated phosphoinositide hydrolysis in rat liver slices (Gonzales and Crews, 1991).

Ethanol modulation of tyrosine kinase would have multiple consequences including those on cell signalling, growth and differentiation. As depicted schematically in Fig. 2, ethanol can affect membrane tyrosine kinase as well as soluble tyrosine kinases of various types. Several tyrosine kinases and their target substrates are implicated in cellular mechanisms leading to a response. Low concentrations of ethanol have significant effects on agonist-sensitive activity of tyrosine kinase. These concentrations of ethanol (0.1 mM-10 mM), which are below the commonly accepted intoxicating (or damaging) concentrations, are, therefore, worthy of attention. Higher concentrations of ethanol exert a substantial inhibition of both basal and EGF-stimulated tyrosine kinase. Thus, as far as the sensitivity of tyrosine kinase is concerned, the effects of acute exposure to both low and high concentrations of ethanol are relevant in assessing the patho-physiology of ethanol. Future developments in this field will have a significant impact towards understanding the mechanisms involved in short-term and long-term effects of ethanol and on cell functions mediated by tyrosine kinase.

Acknowledgements

We are grateful to Professor Grace Sun for her expert advice. We also appreciate the skillful assistance of Judy Richey in the typing of the manuscript.

REFERENCES

Aaronson, S.A., 1991, Growth factors and cancer, Science 254:1146.

Alcohol and Health: Seventh Special Report to the U.S. Congress on Alcohol and Health, 1990, U. S. Department of Health and Human Services, PHS-NIAAA.

Allan, A.M. and Harris, R.A., 1987, Acute and chronic ethanol treatments alter GABA receptor operated chloride channels, Pharmacol. Biochem. Behav. 27:665.

Alling, C., Gustavsson, L., Momsson, J. E., Benthin, G. and Anggard, E., 1984, Phosphatidylethanol formation in rat organs after ethanol treatment, Biochim. Biophys. Acta 793:119.

Anderson, D., Koch, C.A., Grey, L., Ellis, C., Moran, M.F. and Pawson, T., 1990, Binding of SH2 domains of phospholipase C-γ1, GAP, and src to activated growth factor receptors, Science 250:979.

Bode, D.C. and Molinoff, P.B., 1988, Effects of ethanol *in vitro* on the β-adrenergic receptor-coupled adenylate cyclase system, J. Pharm. Exp. Therap. 246:1040.

Buhrow, S. W., Cohen, S., Garbers, D. L. and Staros, J.V., 1983, Characterization of the interaction of 5'-p-fluorosulfonylbenzoyl adenosine with the epidermal growth factor receptor/protein kinase in A431 cell membranes, J. Biol. Chem. 258:7824.

Cantley, L.C., Auger, K.R., Carpenter, C., Duckworth, B., Graziani, A., Kapeller, R. and Saltoff, S., 1991, Oncogenes and signal transduction, Cell 64:281.

Carpenter, G., King, Jr., L. and Cohen, S., 1979, Rapid enhancement of protein phosphorylation in A-431 cell membrane preparations by epidermal growth factor, J. Biol. Chem. 254:4884.

Cochet, C., Filhol, O., Payrastre, B., Hunter, T. and Gill, G.N., 1991, Interactions between the EGF receptor and phosphoinositide kinase, J. Biol. Chem. 266:637.

Cohen, S., Ushiro, H., Stosheck, C. and Chinkers, M., 1982, A native 170,000 epidermal growth factor receptor-kinase complex from shed plasma membrane vesicles, J. Biol. Chem. 257:1523.

Collet, M.S. and Erikson, R.L., 1978, Protein kinase activity associated with the avian sarcoma virus src gene product, Proc. Natl. Acad. Sci. USA 75:2021.

Crews, F.T., Pontzer, NJ. and Chandler, L. J., 1989, Effect of ethanol on receptors coupled to phosphoinositide hydrolysis in brain. In: *Molecular Mechanisms of Alcohol: Neurobiology and Metabolism*, G. Y. Sun *et al*, eds., Humana Press, p. 39.

DePetrillo, P. B. and Swift, R. M., 1992, Ethanol exposure results in a transient decrease in human platelet cAMP levels: evidence for a protein kinase C mediated process. Alcoholism: Clinical and Experimental Research 16:290.

Dhar, A., Paul, A. and Shukla, S.D., 1990, Involvement of tyrosine kinase in PAF stimulation of phsopholipase C in rabbit platelets: Studies with Genistein and monoclonal antibody of phosphotyrosine, Mol. Pharmacol. 37:519.

Dhar, A. and Shukla, S.D., 1991, Involvement of pp60^{c-src} in PAF stimulated platelets: Evidence for translocation from cytosol to membranes, J. Biol. Chem. 266:18797.

Goldstein, D.B. and Chin, J.H., 1981, Interaction of ethanol with biological membranes, Fed. Proc. 40:2073.

Gonzales, R. A. and Crews, F. T., 1991, Chronic ethanol inhibits receptor-stimulated phosphoinositide hydrolysis in rat liver slices, Alcohol 8:131.

Grossman, C.J., Mendenhall, C.L. and Roselle, G.A., 1988, Alcohol and immune regulation: *In vivo* effects of ethanol on con-A sensitive thymic lymphocyte function, Int. J. Immunopharm. 10:187.

Harris, R.A. and Schroeder, F., 1981, Ethanol and the physical properties of brain membranes: Fluorescence studies, Mol. Pharmacol. 20:128.

Harris, R.A. and Allan, A.M., 1989, Alcohol intoxication: Ion channels and genetics. FASEB J. 3:1689.

Hawkins, R. D. and Kalant, H., 1972, The metabolism of ethanol and its metabolic effects, Pharmacol. Rev. 24:67.

Hoek, J.B., Thomas, A. P., Rooney, T. A., Higashi, K. and Rubin, E., 1992, Ethanol and signal transduction in the liver, FASEB J. 6:2386.

Hoffman, P.L., Moses, F., Luthin, G. R. and Tabakoff, B., 1986, Acute and chronic effects of ethanol on receptor-mediated phosphatidylinositol 4,5-bisphosphate breakdown in mouse brain, Mol. Pharmacol. 30:13.

Hoffman, P.L., Rabe, C.S., Moses, F. and Tabakoff, B., 1989, N-methyl-D-aspartate receptors and ethanol: Inhibition of calcium flux and cyclic GMP production. J. Neurochem. 52:1937.

Hunt, W.A., Redos, J.D., Dalton, T.K. and Catravas, G.N., 1977, Alterations in brain c-GMP levels after acute and chronic treatment with ethanol, J. Pharm. Exp. Therap. 201:103.

Hunter, T. and Cooper, J.A., 1985, Protein tyrosine kinases, Ann. Rev. Biochem. 54:897.

Jones, K. L. and Smith, D. W., 1973, Pattern of malformation in offspring in chronic alcoholic mothers, Lancet 1:1267.

Kawamoto, T., Sato, J. D., Le, A., Polikoff, J., Sato, G. H. and Mendelsohn, J., 1983, Growth stimulation of A431 cells by epidermal growth factor: identification of high affinity receptors for epidermal growth factor by an anti-receptor monoclonal antibody, Proc. Natl. Acad. Sci. 80:1337.

Kim, U.H., Fink, Jr., D., Kim, H.S., Park, D.J., Contreras, M.L., Guroff, G. and Rhee, S.G., 1991, Nerve growth factor stimulates phosphorylation of phospholipase C-γ in PC-12 cells, J. Biol. Chem. 266:1359.

Laemmli, U. K., 1970, Cleavage of structural proteins during the assembly of the head of bacteriophage T$_4$, Nature (Lond.) 227:680.

Lundquist, C., Rodriguez, F. D., Simonsson, P., Alling, C. and Gustavsson, L., 1992, Phosphatidylethanol and ethanol increase inositol 1,4,5-trisphosphate levels in NG108-15 neuroblastoma x glioma hybrid cells. Alcohol and Alcoholism 27:61.

Michaelis, E.K., Michaelis, M.L., Kumar, K.N., Tilakaratne, N., Joseph, D.B., Johnson, P.S., Babcock, K.K., Aistrup, G.L., Schowen, R.L., Minami, H., Sugawara, M., Odashima, K. and Umezawa, Y., 1992, Purification, reconstitution and cloning of an NMDA receptor-ion channel complex from rat brain synaptic membranes: implication for neurobiological changes in alcoholism, Ann. N.Y. Acad. Sci. 654:7.

Nagy, L. E., Diamond, I., Casso, D. J., Franklin, C. and Gordon, A., 1990, Ethanol increases extracellular adenosine by inhibiting adenosine uptake via the nucleoside transporter, J. Biol. Chem. 265:1946.

Nishibe, S., Wahl, M. I., Hernandez.Sotomayor, S.M.T., Tonks, N. K., Rhee, S. G. and Carpenter, G., 1990, Increase of the catalytic activity of phospholipase C-gamma 1 by tyrosine phosphorylation, Science 250:1253.

Ouellette, E. M., Henry, R. L., Rosman, N. P. and Weiner, L., 1977, Adverse effects on offspring of maternal alcohol abuse during pregnancy, N. Engl. J. Med. 297:528.

Rhee, S. G., 1991, Inositol phopsholipid-specific phospholipase C: interaction of the γ1 isoform with tyrosine kinase, Trends in Biochem. Sci. 16:297.

Rubin, R., Ponnappa, B. C., Thomas, A. P. and Hoek, J. B., 1988, Ethanol stimulates shape change in human platelets by activation of phosphoinositide-specific phospholipase C, Arch. Biochem. Biophys. 260:480.

Rudeen, P. K. and Creighton, J. A., 1989, Mechanisms of central nervous system alcohol-related birth defects. In *Molecular Mechanisms of Alcohol, Neurobiology, and Metabolism*, Ed., G. Y. Sun *et al.*, p. 147.

Shukla, S.D., 1992, Platelet activating factor receptor and signal transduction mechanisms, FASEB J. 6:2296.

Shukla, S.D. and Halenda, S.P., 1991, Phospholipase C and D in receptor signalling. Life Sci. 48:851.

Simonsson, P. and Alling, C., 1988, The 5-hydroxytryptamine stimulated formation of inositol phosphate is inhibited in platelets from alcoholics, Life Sci., 42:385.

Simonsson, P., Sun, G. Y., Vecsei, L. and Alling, C., 1989, Ethanol effects on bradykinin-stimulated phosphoinositide hydrolysis in NG108-15 neuroblastoma-glioma cells. Alcohol 6:475.

Snyder, A. K., Singh, S. P. and Ehmann, S., 1992, Effects of ethanol on DNA, RNA and protein synthesis in rat astrocyte cultures, Alcoholism: Clinical and Experimental Research 16:295.

Sun, G. Y., Chandrashekhar, R. and Huang, H., 1989, Effects of acute and chronic ethanol administration on metabolism of brain acidic phospholipids, in *Molecular Mechanisms of Alcohol*, Ed., G. Y. Sun, *et al.*, Humana Press, p. 15.

Sun, G. Y., Navidi, M., Yoa, F.-G. and Lin, T. N., 1991, Effects of acute and chronic ethanol administration on phosphoinositide metabolism in mouse brain, Alcohol and Alcoholism, Suppl. 1:215.

Sun, G.Y. and Sun, A.Y., 1985, State of the art review: Ethanol and membrane lipids, Alcoholism: Clin. Exp. Res. 9:164.

Taraschi, T. F. and Rubin, E., 1985, Effects of ethanol on the chemical and structural properties of biologic membranes, Lab. Invest. 52:120.

Thurston, Jr., A.W. and Shukla, S.D., 1992a, Ethanol modulates basal and epidermal growth factor (EGF)-stimulated tyrosine kinase in A431 cell membranes, Alcohol and Alcoholism 27:68.

Thurston, Jr., A.W. and Shukla, S.D., 1992b, Ethanol modulates epidermal growth factor-stimulated tyrosine kinase and phosphorylation of PLC-γ_1. Biochem. Biophys. Res. Commun. 185:1062.

Tonks, N.K. and Charbonneau, H., 1989, Protein tyrosine dephosphorylation and signal transduction, Trends in Biochem. Sci. 14:497.

Ullrich, A. and Schlessinger, J., 1990, Signal transduction by receptors with tyrosine kinase activity, Cell 61:203.

Walker, D.H., Dougherty, N. and Pike, L.J., 1988, Purification and characterization of a PI-kinase from A431 cells, Biochemistry 27:6504.

Wang, S., Jacober, L., Wang, C., Slomiany, A. and Slomiany, B.L., 1992, Ethanol-induced structural and functional alterations of epidermal growth factor receptor in buccal mucosa. Int. J. Biochem. 24:85.

Wood, W.G., Gorka, C. and Schroeder, F., 1989, Acute and chronic effects of ethanol on transbilayer membrane domains, J. Neurochem. 52:1925.

Yarden, Y. and Ulrich, A., 1988, Growth factor receptor tyrosine kinase, Ann. Rev. Biochem. 57:443.

ETHANOL

Michael F. Miles, Gregory Gayer and Michael Sganga

Ernest Gallo Clinic and Research Center and
Department of Neurology
University of California at San Francisco
San Francisco General Hospital
San Francisco, CA 94110

INTRODUCTION

Much of the pathology caused by alcoholism can be attributed to the continued use of large quantities of ethanol. Alcoholics can expose themselves to extraordinarily large concentrations of ethanol due to an acquired resistance to the drug, termed tolerance. Resistance to ethanol can occur as dispositional (pharmacokinetic) or functional (pharmacodynamic) tolerance (Jaffe, 1985). Dispositional tolerance refers to an alteration in the rate of metabolism, absorption, distribution, or secretion. Functional tolerance refers to adaptive changes that occur within the central nervous system upon chronic ethanol exposure.

It is clear that changes in the disposition rate of ethanol in the body, although they occur, are not sufficient to account for the tolerance observed in humans. For example, alcoholics have been reported to survive or even remain sober with blood ethanol levels above 100 mM (Lindblad and Olsson, 1976; Charness et al., 1989). In contrast, blood ethanol levels above 65 mM are usually associated with death or severe central nervous system depression in naive individuals. This suggests that alterations in brain function provides a large role in the adaptation to chronic ethanol use. The occurrence of withdrawal symptoms in alcoholics upon cessation of drinking further supports the idea that significant functional changes in the CNS accompany chronic exposure to ethanol.

It is our working hypothesis that ethanol-induced alterations in CNS gene expression could underlie the development of tolerance and dependence. We and several other laboratories have recently documented that ethanol can cause increases (Parent et al., 1987; Charness et al., 1988; Kolber et al., 1988; Miles et al., 1991) or decreases (Dave et al., 1986; Mochly-Rosen et al., 1988) in the abundance of specific mRNA's in cultured neural cells or in the CNS of intact animals. Interestingly, some of our studies have shown that ethanol can induce highly coordinate changes in the expression of a number of genes in neural cells (Miles et al., 1992). This suggests that similar mechanisms may underlie the regulation of ethanol-responsive genes (EtRGs). Our studies have directly shown, that at least with the Hsc70 chaperonin gene, ethanol-induced increases in mRNA abundance are due to increases in Hsc70 gene transcription (Miles, Diaz et al., 1991).

This chapter will summarize some of our recent findings regarding the regulation of neural gene expression by ethanol. In addition to describing the scope and specific examples of ethanol-responsive genes (EtRGs), we will discuss our studies on the mechanism(s) underlying regulation of gene expression by ethanol. Specifically, we have attempted to identify ethanol-responsive *cis*-acting elements in the promoter regions of EtRGs. It is our long-term goal that through the identification of such ethanol-responsive elements and their cognate DNA-binding proteins, we can obtain a more thorough understanding of the mechanisms underlying CNS adaptation to ethanol.

IDENTIFICATION OF EtRGs

Two-Dimensional Gel Studies

Our initial studies used high resolution 2-D gel electrophoresis of in vitro translation products to provide a comprehensive analysis of changes in mRNA abundance induced by ethanol exposure (Miles, Diaz et al., 1992). These and all the other studies described in this chapter utilized the NG108-15 neuroblastoma x glioma cell line as a cell culture model for chronic ethanol exposure. These cells have been extensively used to study neuronal function and the biochemistry of cellular adaptation to ethanol (Charness et al., 1983; Charness et al., 1986; Gordon et al., 1986; Miles, Diaz et al., 1991; Miles, Diaz et al., 1992).

Chronic ethanol treatment (1-2 days) did not cause significant changes in total RNA abundance per cell or in the ability of RNA to program in vitro protein synthesis in a reticulocyte lysate (data not shown). However, detailed quantitative analysis of 2-D gels identified ethanol-induced increases and decreases in mRNA abundance. In these studies we chose 100 well-resolved, random proteins on each autoradiogram to obtain a representative picture of

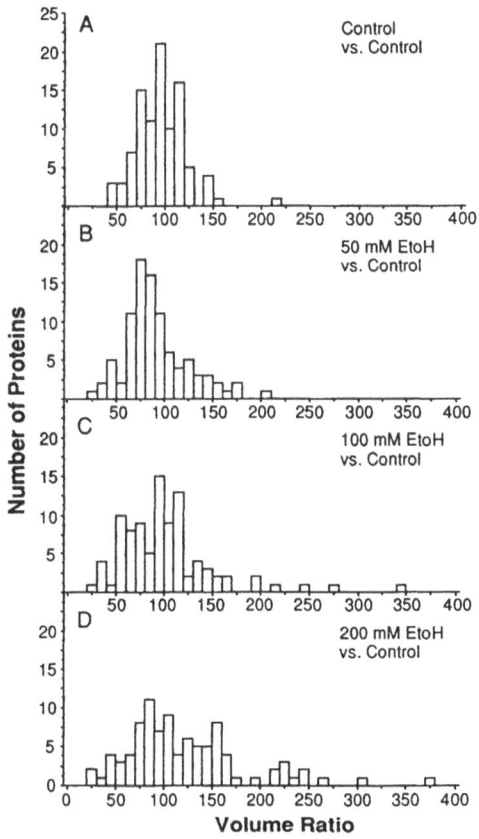

Figure 1. Quantitative 2-D Gel Analysis of Ethanol-induced Changes in Gene Expression in NG108-15 Cells. Total RNA was isolated from control or ethanol-treated (50-200 mM, 48 hours) NG108-15 cells followed by in vitro translation. Proteins were analyzed by 2-D gel electrophoresis and computerized densitometry. Histograms reflect the ratio of homologous protein volumes from pairs of control gels (panel A) or ethanol-treated samples (panels B-D) compared to control. Panel A represents the comparison of separate flasks of control cells. The y-axis in each case shows the number of proteins in each volume ratio grouping. Panel A represents the comparison of 98 matched proteins while panels B-D contain data from 94 proteins each. Results are representative of experiments repeated three times. Data is reproduced with permission from (Miles, Diaz et al., 1992).

ethanol-responsive gene expression. Analysis of autoradiograms from duplicate or triplicate control cultures confirmed that the quantitative procedures were highly reproducible. There was a very tight clustering of points when the ratio of peak volumes from homologous peaks of two control gels were plotted as a histogram (Fig. 1A).

If ethanol treatment produces a significant number of increases or decreases in mRNA abundance, this should increase the spread of points about the mean in the histograms shown in Fig. 1. When control gels were compared to ethanol-treated samples there was indeed a significant increase in the spread of points around the mean (Figs. 1B-D). The distribution of data points suggests that both decreases and increases in mRNA abundance are caused by ethanol treatment, although the results with 200 mM ethanol suggest that there may be a prevalence of gene inductions (Fig. 1, panel D). Dose response curves for individual proteins (mRNA) suggest that there may be families of EtRGs that respond in a coordinate fashion with either increases or decreases in mRNA abundance (see Fig. 3 in Miles, Diaz et al., 1992). This raises the possibility that similar mechanisms might underlie the regulation of these EtRGs.

Subtractive Hybridization Cloning

Recently we have directly isolated EtRGs using a modified subtractive hybridization protocol (Miles et al., 1992). This has confirmed our 2-D gel electrophoresis studies and provided further support for our hypothesis that changes in gene expression underlie CNS adaptation to ethanol.

Subtractive hybridization was performed using a modified version of the phenol-enhanced reassociation technique (PERT) described by Travis and Sutcliffe (Travis and Sutcliffe, 1988). [^{32}P]-labeled cDNA (cDNA$_{etoh}$) was prepared using poly(A)$^+$ RNA from ethanol-treated cells. Driver DNA was generated from the control λ-Zap cDNA library by "mass excision" of the phage library. Hybridization (PERT) of the cDNA$_{etoh}$ with a large excess of driver DNA was followed by isolation of unhybridized ssDNA$_{etoh}$ by hydroxylapatite chromatography. This material was then used as a hybridization probe for screening a cDNA$_{etoh}$ library or for construction of a subtracted cDNA library.

Our subtractive hybridization cloning identified 7 cDNA clones as possible EtRGs as confirmed by Northern blot hybridization (data not shown). The ethanol dose response for representative EtRGs (Fig. 2) shows that several of these genes reach their maximum induction levels (2-4 fold x control) at ethanol concentrations as low as 25 mM. This strongly suggests that these inductions would have physiological relevance since the average blood ethanol concentration in a large study of "sober" ethanol users was above 50 mM (Urso et al., 1981).

STUDIES ON IDENTIFIED EtRGs

In addition to the 2-dimensional gel and subtractive cloning studies described above, we have also identified EtRGs through an analysis of various "candidate genes". We directed these studies toward two specific groups of genes; those known to be responsive to cyclic AMP and members of the so-called "stress protein response".

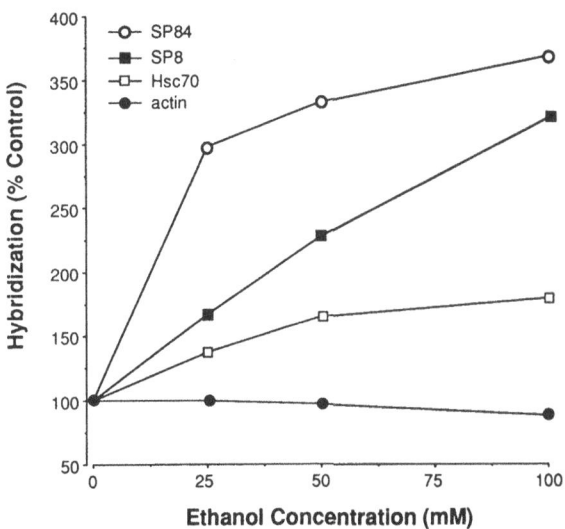

Figure 2. Ethanol Dose Response Curves for Representative EtRGs Isolated by Subtractive Hybridization. NG108-15 cells were grown for 24 hours with the indicated concentrations of ethanol. Mock-treated control cultures were treated identically. Total RNA was then analyzed by Northern blot using ^{32}P-labeled probes prepared from EtRGs (SP84 and SP8) isolated by subtractive hybridization cloning (Miles, Elliot et al., 1992). Control hybridizations were done with a known ethanol-responsive (Hsc70) (Miles, Diaz et al., 1991) and ethanol-unresponsive gene (β-actin). Autoradiograms were quantitated by densitometry.

The analysis of cAMP-responsive genes was initiated due the large amount of experimental data documenting that acute and chronic ethanol treatment causes significant increases and decreases, respectively, in the levels of receptor-stimulated cAMP (Charness, Simon et al., 1989). Furthermore, it is well established that changes in cyclic AMP levels will modulate the expression of a wide variety of genes (Maurer, 1981; Miles et al., 1981; Jungmann et al., 1983; Waterman et al., 1985; Montminy et al., 1986).

We chose tyrosine hydroxylase (TH) as a candidate cAMP-responsive gene since this gene encodes the rate limiting enzyme in catecholamine biosynthesis and a number of studies have documented changes in catecholamines with acute or chronic ethanol exposure. For example, following acute ethanol exposure, transcerebral dialysis in freely moving rats shows increased dopamine release in the nucleus accumbens and dorsal caudate nucleus (Di-Chiara and Imperato, 1985). Interestingly, agents that interrupt adrenergic neurotransmission by destroying adrenergic neurons, blocking catecholamine synthesis, or preventing receptor activity prevent the development of tolerance (Ritzmann and Tabakoff, 1976) and aggravate ethanol withdrawal-induced seizures (Goldstein, 1973).

Stress protein gene expression was also chosen as a candidate target for ethanol regulation. The stress protein or so-called heat shock response encompasses a number of genes, many of which are structurally related, which are induced following a variety of different metabolic or toxic stressors (Welch, 1987; Schlesinger, 1990; Hightower, 1991). Furthermore, some authors have previously reported the induction of at least some stress proteins following exposure of eucaryotic cells to very high ethanol concentrations (>0.8 M) (Li and Hahn, 1978; Li and Werb, 1982; Li, 1983). We reasoned that perhaps chronic exposure to much lower ethanol concentrations, as seen in alcoholics, might induce a specific subset of stress proteins.

Tyrosine Hydroxylase Gene Expression

Tyrosine hydroxylase (TH) catalyzes the initial, rate-limiting step in monoamine synthesis by hydroxylation of tyrosine to form dihydroxyphenylalanine (Nagatsu et al., 1964). Elevations in intracellular cAMP produce increases in TH gene transcription in PC12 cells through a cAMP regulatory element (CRE) located at position -44 to -37 (TGACGTCA) (Lewis et al., 1987) of the TH promoter.

We have recently shown that basal levels of TH mRNA and protein are increased in N1E-115 neuroblastoma cells following exposure to ethanol (50-100 mM) for 24-48 hours (Gayer et al., 1991). Figure 3 shows the results of slot blot hybridization analysis of TH mRNA levels compared with various control RNA species. It is clear that ethanol produces substantial increases in basal TH expression. TH mRNA levels were 240% of control in cells exposed to 100 mM ethanol. These inductions are accompanied by corresponding changes in TH protein (Gayer, Gordon et al., 1991).

Hsc70 Gene Expression

Hsc70 is an abundant cellular protein that is a constitutively expressed member of the 70 kDa stress protein gene family (Welch, 1987). This gene responds only weakly, if at all, to heat shock or other typical stress protein inducers. Hsc70 has been shown to function as a molecular chaperonin which binds to nascent polypeptide chains on polyribosomes and is required for their transport across the endoplasmic reticulum or mitochondrial membrane (Chirico et al., 1988; Deshaies et al., 1988; Beckmann et al., 1990). Hsc70 has

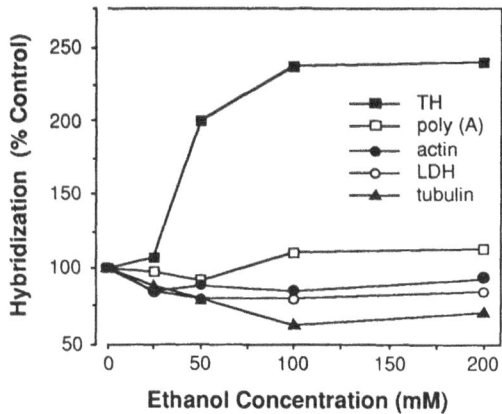

Figure 3. Slot blot analysis of ethanol-induced changes in TH mRNA levels in N1E-115 cells. Levels of TH, actin, LDH, and β tubulin mRNAs from control cells and cells exposed to ethanol at the indicated concentration for 3 days were quantitated by slot blot hybridization and computerized densitometry. Data are expressed as a percent of control (untreated) and are normalized to poly(A) RNA content. The values for poly(A) content are included and show no significant change with ethanol over this concentration range. Results are representative of experiments repeated at least three times. Data is reproduced with permission from (Gayer, Gordon et al., 1991).

also been shown to act as an uncoating enzyme for the removal of clathrin triskelia from clathrin-coated vesicles (Schmid and Rothman, 1985; Chappell et al., 1986). Thus, this protein plays a crucial role in protein and vesicular trafficking.

Figure 4 shows a dose response for induction of Hsc70 mRNA by ethanol in NG108-15 neuroblastoma x glioma cells. In contrast to TH (see Fig. 3), Hsc70 mRNA increased steadily with higher ethanol concentrations. Furthermore, we have observed that Hsc70 mRNA levels remain elevated as long as ethanol is present, for periods of up to 4 weeks (Miles, Diaz et al., 1991).

Other investigators have previously shown that very high concentrations of ethanol (>800 mM) could induce Hsp70, the highly inducible 70 kDa stress protein that is closely related to the constitutively expressed Hsc70. We have recently shown that concentrations of ethanol observed in alcoholics (50-100 mM) are capable of inducing Hsc70 gene expression without any alteration in Hsp70 expression (Miles, Diaz et al., 1991). Other known stress protein inducing agents cause an induction of Hsp70 with or without moderate increases in Hsc70 expression. This result, coupled with data from our 2-D gel studies (Miles, Diaz et al., 1992), suggests that ethanol causes a novel induction of a subset of stress proteins. We might thus expect that ethanol would regulate Hsc70 expression through a mechanism that does not involve the transcription factors involved in heat shock protein induction (HSF).

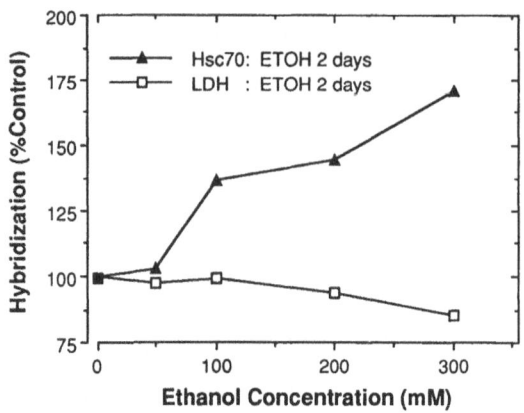

Figure 4. Northern Blot Analysis of Hsc70 Induction by Ethanol. NG108-15 cells were exposed to ethanol at the indicated concentrations for 2 days and then analyzed for LDH (□) or Hsc70 (▲) mRNA. Autoradiograms were quantitated by computerized densitometry. Data is expressed as a percent of control (untreated). Values are the mean of quadruplicate determinations and are representative of independent experiments repeated 3-6 times. Data is adopted with permission from (Miles, Diaz et al., 1991).

MECHANISMS OF EtRG INDUCTION BY ETHANOL

Our discussion above documents that coordinate changes in gene expression occur in neural cells exposed to ethanol for prolonged periods. The dose response curves of EtRGs in our 2-D gel studies and those seen with induction of TH and Hsc70 are remarkably similar. This suggests that similar mechanisms may underlie the induction of EtRGs. Our studies on Hsc70 had shown that ethanol-induced increases in Hsc70 mRNA are due to increased Hsc70 gene transcription, as determined by nuclear runoff analysis (Miles, Diaz et al., 1991). The identification of TH and Hsc70 as EtRGs enables us to perform detailed mechanistic studies on the induction of these genes by ethanol since the promoter regions of these genes have previously been cloned. We have therefore sought to determine the *cis*-acting elements in the promoter regions of the TH and Hsc70 genes which confer ethanol-responsiveness on these two EtRGs. Figure 5 confirms that such ethanol-responsive elements (EtREs) should exist since proximal portions of the TH and Hsc70 promoters conferred dose-dependent ethanol-responsiveness to a reporter gene coding for chloramphenicol acetyltransferase (CAT).

The identification of such ethanol-responsive elements and their cognate DNA-binding proteins might eventually allow us to determine more proximal sites of action by ethanol. For example, modulation of transcription by ethanol might occur through changes in phosphorylation of a particular DNA-binding protein. The kinase involved in phosphorylation of this DNA-binding protein could also be involved in other cellular actions of ethanol, such as modulation of ion channel activity. Thus, through study of the mechanisms underlying regulation of transcription by ethanol, we hope to elucidate a detailed understanding of cellular adaptation to ethanol.

Deletion Analysis of Tyrosine Hydroxylase Promoter

The data in Figure 5 confirms that an EtRE exists in the proximal 773 base pairs of the TH promoter. As shown in Figure 6, this region contains a number of known DNA-binding protein consensus recognition sites. Deletion of the TH promoter to -272 produced no significant change in ethanol responsiveness despite removing three copies of the consensus heat shock element (HSE, GAAnnTTCnn) found between -546 and -525 in the TH promoter. This corroborates our data on Hsc70 suggesting that ethanol, at the concentrations used in our experiments, is not simply inducing a classic heat shock response.

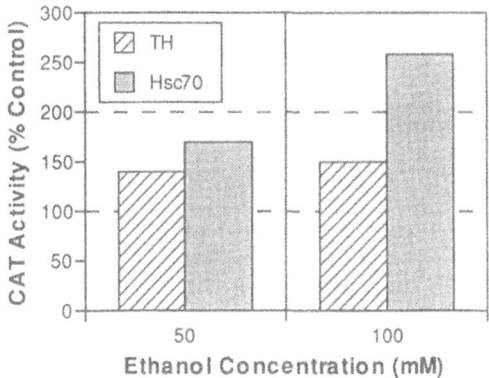

Figure 5. Ethanol-Responsive Promoter Regions in TH and Hsc70 Genes. Plasmid constructs containing 773 or 2500 bp, respectively, of the TH and Hsc70 promoters were coupled to a reporter gene coding for chloramphenicol acetyltransferase (CAT). Constructs were then co-transfected with pSV3Neo into N1E-115 (TH) or NG108-15 (Hsc70) neuroblastoma cells. Stable transfectants were isolated by G418 selection. Clonal isolates were then exposed to the indicated concentrations of ethanol for 24 hours and CAT activity assayed. Control cells were mock-treated. Results are representative of experiments repeated 4-6 times.

Further deletion of the TH promoter to position -41 did eliminate ethanol-responsiveness. This suggests that the EtRE should exist in the 231 bp between position -272 and -41. There is a CRE element located at the -41 position which is lost with this deletion, thus eliminating cAMP-responsiveness of the TH promoter (data not shown). To determine whether a CRE element could also confer ethanol-responsiveness, we studied the effect of ethanol on expression of a promoter construct (p25VIPCAT) containing the vasoactive intestinal protein (VIP) CRE element and TATA box (Montminy and

Bilezikjian, 1987). Although the p25VIPCAT construct showed an excellent induction by 0.5 mM 8-bromo-cyclic AMP (306% of mock-stimulated control), there was no response to 100 mM ethanol (107% of control). This data suggests that the CRE element is not sufficient for conferring ethanol-responsiveness to the TH promoter.

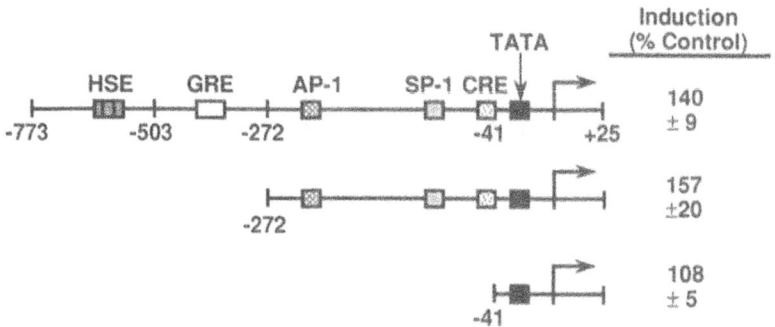

Figure 6. Deletion Analysis of TH Promoter - CAT Constructs. Mutations of the TH promoter were made by restriction enzyme digestion and coupled to CAT. Constructs were then electroporated into NE-115 cells and stable transfectants isolated. Cells were then analyzed for response to ethanol (24 hours, 100 mM). The position of possibly relevant promoter sequences are indicated. Results are the mean ± S.D. and are representative of experiments repeated 4-6 times..

Deletion Analysis of the Hsc70 Promoter

Figure 7 shows a similar deletion analysis of the Hsc70 promoter. Interestingly, deletion of 2400 bp of the Hsc70 promoter had no significant effect on ethanol-responsiveness in transient transfection assays. Thus, constructs with 2500 bp or 93 bp of Hsc70 promoter showed similar 2-fold inductions with 200 mM ethanol treatment.

When the proximal 93 bp of the Hsc70 promoter is compared with the -41 to -272 region of the TH promoter, there is very little homology between the two genes. The only known DNA-binding protein motif found in common between these regions is a consensus Sp1 site found at position -60 and -113 in the Hsc70 and TH promoters, respectively. We therefore constructed point mutations of the Sp1 site at -60 in the Hsc70 promoter. Figure 8 shows that a single point mutation in this region caused a dramatic decrease in ethanol-responsiveness. This strongly suggests that an Sp1 site is required for induction of Hsc70 gene expression by ethanol.

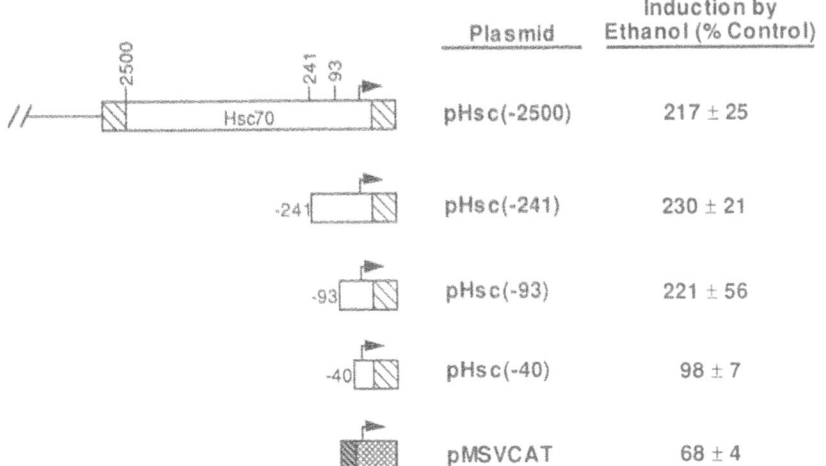

Plasmid	Induction by Ethanol (% Control)
pHsc(-2500)	217 ± 25
pHsc(-241)	230 ± 21
pHsc(-93)	221 ± 56
pHsc(-40)	98 ± 7
pMSVCAT	68 ± 4

Figure 7. Transient Transfection Analysis of Hsc70 Promoter Deletions. Deletion mutants of the Hsc70 promoter were coupled to CAT and used in transient transfection assays of NG108-15 cells. Twenty four hours after electroporation, cells were treated with ethanol (200 mM). Cells were lysed and CAT activity assayed after 24 hours of ethanol treatment. Control cultures were treated identically except for addition of ethanol. Control transfections with the viral promoter, pMSVCAT, showed a mild decrease with ethanol treatment (68% of control). Results are the mean ± S.D. and are representative of experiments repeated 2-16 times.

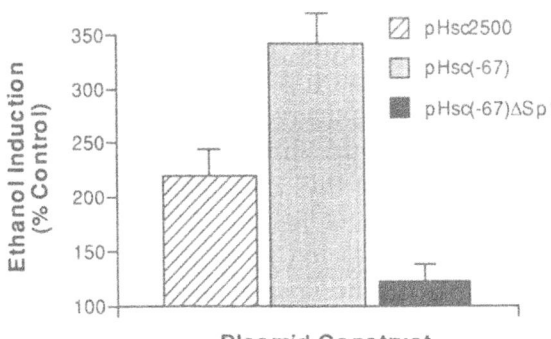

Figure 8. Point Mutation Analysis of Hsc70 Promoter. PCR was used to make CAT constructs containing the -67/+1 region of the Hsc70 promoter. Plasmid pHsc(-67ΔSP1)CAT contains a point mutant in the Sp1 site at -60 which markedly reduces the binding of Sp1 as determined by gel retardation analysis. Underlined bases denote copies of the consensus heat shock element (HSE) contained in both constructs. Transient transfection analysis was used to determine relative ethanol-responsiveness using conditions as described for Hsc70 in Fig. 7. Results are the mean ± S.D. and are representative of experiments repeated 4-5 times.

DISCUSSION

Our studies outlined here have produced the most detailed evidence to date supporting the hypothesis that changes in neuronal gene expression account, at least in part, for adaptation of the CNS to chronic ethanol exposure. The data from our 2-D gel studies and subtractive hybridization cloning suggest that ethanol induces coordinate changes in the expression of a substantial number of genes. As mentioned above, this strongly suggests a single or limited set of mechanisms underlying ethanol-responsive gene expression. The identification of such a mechanism(s) should produce a major advance in our understanding of the cellular responses to chronic ethanol exposure.

The results of our studies on TH and Hsc70 gene expression (see Fig. 5) show that ethanol-induced changes in mRNA abundance identified with the 2-D gel and subtractive hybridization studies are most likely secondary to alterations in gene transcription. Furthermore, our transfection analyses (Figs. 5-7) show that cis-acting DNA elements in the promoter regions of the Hsc70 and TH genes are able to confer ethanol-responsiveness on a heterologous reporter gene (CAT). This has previously been suggested to occur with the MHC-1 antigen gene (Parent, Ehrlich et al., 1987).

Deletion analyses of the TH and Hsc70 promoters (Figs. 6-7) suggest that the Sp1 motif contained in these promoters is required for ethanol-responsiveness. This is particularly compelling with the Hsc70 promoter where a single point mutation in the Sp1 site at -60 caused a virtual elimination of any ethanol-responsiveness (Fig. 8). Obviously, more detailed point mutation analysis of the TH promoter is needed to confirm the importance of the Sp1 site in the ethanol response of that promoter. Furthermore, one should see a return of ethanol-responsiveness in the mutated Hsc70 or TH promoters following the re-addition of functional Sp1 sites.

The identification of Sp1 as possibly conferring ethanol-responsiveness is an interesting observation since this DNA-binding protein motif is found in many promoters. It would appear difficult to explain any specificity of ethanol-induced changes in gene expression based solely upon an effect by Sp1. However, it should be noted that Tijan and co-workers have observed that binding of Sp1 to a promoter region can cause either increases or decreases in transcription depending on the "context" of the remainder of the promoter region (Briggs et al., 1986). Thus, in some promoters activation of Sp1 by ethanol could increase transcription while in others there could be a decrease or no change in transcription despite having identical Sp1 sites.

Recent evidence suggests other possible mechanisms that could confer specificity to an ethanol modulation of Sp1 activity. Several authors have shown that Sp1 likely interacts with other DNA-binding proteins to produce alterations in transcription (Pugh and Tijan, 1990; Li et al., 1991). Through interaction with other transcription factors, Sp1 may differentially regulate gene expression. Thus, ethanol could potentially affect the interactions between Sp1 and other transcription factors and produce varying effects on the expression of specific genes.

A final interesting aspect of our results to date concerns the question of how ethanol might modulate Sp1 activity. As suggested above, it may well be that ethanol does not directly alter the transcriptional activity of Sp1 but rather affects the interaction of other transcription factor(s) with Sp1. In this scenario Sp1 would be required but not sufficient for a gene to be ethanol-responsive. Other possible mechanisms include direct alterations in Sp1 abundance or transcriptional activity. In this regard, it is interesting to note that Sp1 becomes phosphorylated upon binding to DNA (Jackson et al., 1990). Although the function of Sp1 phosphorylation is currently unknown, it is a candidate mechanism for ethanol regulation since chronic ethanol exposure is known to modulate the activity of protein kinase cascades (Gordon, Collier et al., 1986; Diamond et al., 1987; Charness, Simon et al., 1989).

In conclusion, our studies have documented conclusively for the first time that chronic ethanol exposure can produce coordinate changes in gene expression in neural cells. Our studies on TH and Hsc70 have documented that ethanol appears to modulate gene expression at the level of transcription. Sp1 may be a required transcription factor for ethanol-responsive gene expression as shown by our point mutation analysis of Hsc70. Ongoing study in our laboratory on the mechanisms underlying regulation of EtRG expression will focus on three major questions: First, does a Sp1 recognition site mediate ethanol-responsiveness of EtRGs other than TH and Hsc70? Secondly, is Sp1 the only factor required for the ethanol response? Finally, how does chronic ethanol exposure modulate the transcriptional activity of Sp1 and/or other DNA-binding proteins required for EtRG regulation? The results of these studies should provide important new understanding of the molecular and cellular mechanisms underlying adaptation of the CNS to ethanol.

ACKNOWLEDGEMENTS

We thank Drs. I. Diamond, A. Gordon, D. Lowenstein, R. Messing and D. Mochly-Rosen for invaluable suggestions and discussions. J. Diaz, S. Barhite and W. Chin deserve high praise for providing excellent technical assistance. This work was supported by intramural funding from the Ernest Gallo Clinic and Research Center, National Institute on Alcohol Abuse and Alcoholism Research Scientist Development Award AA0018 and Research Grants AA07750 and AA08353, a grant from the Alcoholic Beverage Medical Research Foundation and Basil O'Connor Starter Scholar Research Award No. 5-746 from the March of Dimes Birth Defects Foundation (all to M.F.M.).

REFERENCES

Beckmann, R.P., Mizen, L.A. and Welch, W.J., 1990, Interaction of Hsp70 with newly synthesized proteins: Implications for protein folding and assembly., *Science* 248: 850.

Briggs, M.R., Kadonaga, J.T., Bell, S.P. and Tjian, R., 1986, Purification and biochemical characterization of the promoter-specific transcription factor, Sp1, *Science* 234: 47.

Chappell, T.G., Welch, W.J., Schlossman, D.M., Palter, K.B., Schlesinger, M.J. and Rothman, J.E., 1986, Uncoating ATPase is a member of the 70 kilodalton family of stress proteins, *Cell* 45: 3.

Charness, M.E., Gordon, A.S. and Diamond, I., 1983, Ethanol modulation of opiate receptors in cultured neural cells, *Science* 222: 1426.

Charness, M.E., Querimit, L.A. and Diamond, I., 1986, Ethanol increases the expression of functional delta-opioid receptors in neuroblastoma x glioma NG108-15 hybrid cells, *J. Biol. Chem.* 261: 3164.

Charness, M.E., Querimit, L.A. and Henteleff, M., 1988, Ethanol differentially regulates G proteins in neural cells., *Biochem. Biophys. Res. Commun.* 155: 138.

Charness, M.E., Simon, R.P. and Greenberg, D.A., 1989, Ethanol and the nervous system, *N. Engl. J. Med.* 321: 442.

Chirico, W.J., Waters, M.G. and Blobel, G., 1988, 70K heat shock related proteins stimulate protein translocation into microsomes, *Nature* 332: 805.

Dave, J.R., Eiden, L.E., Karaman, J.W. and Eskay, R.L., 1986, Ethanol exposure decreases pituitary corticotropin-releasing factor binding, adenylate cyclase activity, proopiomelanocortin biosynthesis and plasma β-endorphin levels in the rat., *Endocrinology* 118: 280.

Deshaies, R.J., Koch, B.D., Werner-Washburne, M., Craig, E.A. and Schekman, R., 1988, A subfamily of stress proteins facilitates translocation of secretory and mitochondrial precursor polypeptides., *Nature* 332: 800.

Di-Chiara, G. and Imperato, A., 1985, Ethanol preferentially stimulates dopamine release in the nucleus accumbens of freely moving rats, *Eur. J. Pharmacol.* 115: 131.

Diamond, I., Wrubel, B., Estrin, W. and Gordon, A., 1987, Basal and adenosine receptor-stimulated levels of cAMP are reduced in lymphocytes from alcoholic patients, *Proc. Natl. Acad. Sci. USA* 84: 1413.

Gayer, G.G., Gordon, A. and Miles, M.F., 1991, Ethanol increases tyrosine hydroxylase gene expression in N1E-115 neuroblastoma cells, *J. Biol. Chem.* 266: 22279.

Goldstein, D.B., 1973, Alcohol withdrawal reactions in mice: effects of drugs that modify neurotransmission, *J. Pharmacol. Exp. Ther.* 186(1): 1.

Gordon, A.S., Collier, K. and Diamond, I., 1986, Ethanol regulation of adenosine receptor-stimulated cAMP levels in a clonal neural cell line: An *in vitro* model of cellular tolerance to ethanol, *Proc. Natl. Acad. Sci. USA* 83: 2105.

Hightower, L.E., 1991, Heat shock, stress proteins, chaperones, and proteotoxicity, *Cell* 66: 191.

Jackson, S.P., MacDonald, J.J., Lees-Miller, S. and Tjian, R., 1990, GC box binding induces phosphorylation of Sp1 by a DNA-dependent protein kinase, *Cell* 63: 155.

Jaffe, J., 1985, Drug addiction and drug abuse, *in:* "The pharmacological basis of therapeutics," 7th, ed., A.S. Gilman et al., eds., Macmillan Publishing Company, New York, p 532.

Jungmann, R.A., Kelley, D.C., Miles, M.F. and Milkowski, D.M., 1983, Cyclic AMP regulation of lactate dehydrogenase, *J. Biol. Chem.* 258: 5312.

Kolber, M.A., Walls, R.M., Hinners, M.L. and Singer, D.S., 1988, Evidence of increased class I MHC expression on human peripheral blood lymphocytes during acute ethanol intoxication, *Alcoholism: Clin. Exptl. Res.* 12: 820.

Lewis, E.J., Harrington, C.A. and Chikaraishi, D.M., 1987, Transcriptional regulation of the tyrosine hydroxylase gene by glucocorticoid and cyclic AMP. , *Proc Natl Acad Sci U S A* 84(11): 3550.

Li, G.C., 1983, Induction of thermotolerance and enhanced heat shock protein synthesis in Chinese hamster fibroblasts by sodium arsenite and by ethanol, *J. Cell. Physiol.* 115: 116.

Li, G.C. and Hahn, G.M., 1978, Ethanol-induced tolerance to heat and to adriamycin, *Nature* 274: 699.

Li, G.C. and Werb, Z., 1982, Correlation between synthesis of heat shock proteins and development of thermotolerance in Chinese hamster fibroblasts, *Proc. Natl. Acad. Sci. USA* 79: 3218.

Li, R., Knight, J.D., Jackson, S.P., Tjian, R. and Botchan, M.R., 1991, Direct interaction between Sp1 and BPV enhancer E2 protein mediates synergistic activation of transcription, *Cell* 65: 493.

Lindblad, B. and Olsson, R., 1976, Unusually high levels of blood alcohol?, *J. Amer. Med. Assn.* 236: 1600.

Maurer, R.A., 1981, Transcriptional regulation of the prolactin gene by ergocryptine and cyclic AMP, *Nature* 294: 94.

Miles, M.F., Diaz, J.E. and DeGuzman, V., 1992, Ethanol-responsive genes in neural cell cultures., *Biochem. Biophys. Acta* 1138: 268.

Miles, M.F., Diaz, J.E. and DeGuzman, V.S., 1991, Mechanisms of neuronal adaptation to ethanol: Ethanol induces Hsc70 gene transcription in NG108-15 neuroblastom x glioma cells, *J. Biol. Chem.* 266: 2409.

Miles, M.F., Elliot, M., Tanner, W., Wilke, N. and Shah, S., 1992, Isolation of ethanol-responsive genes from neural cell cultures using subtractive hybridization, *submitted for publication* .

Miles, M.F., Hung, P. and Jungmann, R.A., 1981, Cyclic AMP regulation of lactate dehydrogenase, *J. Biol. Chem.* 256: 12545.

Mochly-Rosen, D., Chang, F.-H., Cheever, L., Kim, M., Diamond, I. and Gordon, A.S., 1988, Chronic ethanol causes heterologous desensitization by reducing α_s mRNA, *Nature* 333: 848.

Montminy, M.R. and Bilezikjian, L.M., 1987, Binding of a nuclear protein to the cyclic-AMP response element of the somatostatin gene, *Nature* 328: 175.

Montminy, M.R., Sevarino, K.A., Wagner, J.A., Mandel, G. and Goodman, R.H., 1986, Identification of a cyclic-AMP-responsive element within the somatostatin gene., *Proc. Natl. Acad. Sci. USA* 83: 6682.

Nagatsu, T., Levitt, M. and Udenfriend, S., 1964, Conversion of L-tyrosine to 3,4-dihydroxyphenylalanine by cell-free preparations of brain and sympathetically innervated tissues. , *Biochem Biophys Res Commun* 14: 543.

Parent, L.J., Ehrlich, R., Matis, L. and Singer, D.S., 1987, Ethanol: an enhancer of major histocompatibility complex antigen expression, *FASEB J.* 1: 469.

Pugh, B.F. and Tijan, R., 1990, Mechanism of transcriptional activation by Sp1: Evidence for cofactors, *Cell* 61: 1187.

Ritzmann, R.F. and Tabakoff, B., 1976, Dissociation of alcohol tolerance and dependence. , *Nature* 263(5576): 418.

Schlesinger, M.J., 1990, Heat shock proteins, *J. Biol. Chem.* 265: 12111.

Schmid, S.L. and Rothman, J.E., 1985, Enzymatic dissociation of clathrin cages in a two-stage process, *J. Biol. Chem.* 260: 10044.

Travis, G.H. and Sutcliffe, J.G., 1988, Phenol emulsion-enhanced DNA-driven subtractive cDNA cloning: Isolation of low-abundance monkey cortex-specific mRNAs, *Proc. Natl. Acad. Sci. USA* 85: 1696.

Urso, T., Gavaler, J.S. and Van Thiel, D.H., 1981, Blood ethanol levels in sober alcohol users seen in an emergency room, *Life Sci.* 28: 1053.

Waterman, M., Murdoch, G.H., Evans, R.M. and Rosenfeld, M.G., 1985, Cyclic AMP regulation of eukaryotic gene transcription by two discrete molecular mechanisms, *Science* 229: 267.

Welch, W.J., 1987, The mammalian heat shock (or stress) response: a cellular defense mechanism, *in*: "Immunobiology of proteins and peptides," M.Z. Atassi, ed., New York, Plenum Publishing Corp., p 287.

SUPER-INDUCTION OF C-FOS-LIKE PROTEIN IN BOVINE ADRENAL CHROMAFFIN CELLS ASSOCIATED WITH ETHANOL WITHDRAWAL

Ouahiba Bouchenafa and John Littleton

Division of Biomedical Science
Kings College
Mansera Rd. Chelsea
London SW3

INTRODUCTION

We have previously described changes that occur in bovine adrenal cells cultured in medium containing ethanol that mimic several of the changes observed in central neurones in ethanol dependence in vivo (Harper & Littleton, 1990; 1991). These changes generally reflect the development of tolerance to the inhibitory effects of ethanol while it is present, and cellular hyperexcitability on removal of the drug. One sign of cellular excitability that we did not measure in these earlier papers is the induction of c-fos proteins. If our cell culture model does accurately reflect changes that occur in vivo then increased expression of the *c-fos* gene is predicted since this has been shown in mouse brain during ethanol withdrawal (Jitenda et al, 1989). Increased expression of *c-fos* has also been shown in the central nervous system in vivo associated with seizures caused by electrical stimulation (Cole et al, 1990), kindling (Dragunow & Robertson, 1987) and convulsant drugs (Morgan et al 1987).

Increased expression of the *c-fos* gene may be particularly important in some of the long-term consequences of alcohol withdrawal since the *c-fos* gene encodes transcription regulatory factors which mediate long-term adaptive responses to transynaptic signals (Sheng & Greenberg, 1990). In most cell types the basal level of *c-fos* expression is low but it increases rapidly and transiently on stimulation of the cell, *c-fos* is one of the "immediate-early" genes and the factors it encodes are the first in a cascade of reactions controlling cellular gene expression. Induction of c-fos in ethanol withdrawal could therefore have consequences far beyond the duration of the induction.

In the central nervous system *c-fos* expression seems to be invariably associated with seizures (see refs above) and we and others have previously provided a wealth of evidence that ethanol withdrawal seizures are at least partly a consequence of up-regulation of dihydropyridine (DHP)-sensitive voltage-operated calcium channels in brain (Dolin et al, 1987; Brennan et al, 1990). The chronic exposure of adrenal-derived cells, either PC12 (Messing et al, 1986) or bovine adrenal chromaffin cells (Harper et al, 1989), to ethanol also causes an apparent up-regulation of DHP-sensitive calcium channels. Since the expression of *c-fos* in PC12 cells seems to be regulated by Ca^{2+} influx through DHP-sensitive calcium channels (Morgan & Curran, 1986; Greenberg et al, 1986) and since Ca^{2+} influx through these channels is increased in ethanol withdrawal (Greenberg et al, 1987) it seemed very likely indeed that *c-fos* expression would also be increased. We therefore estimated *c-fos* expression by immunocytochemistry for the c-fos protein during ethanol withdrawal from adrenal chromaffin cells. These experiments are part of an ongoing attempt to relate changes in cells in culture to those that underlie drug tolerance and dependence in the central nervous system in vivo.

METHODS and MATERIALS

Cell cultures

Bovine adrenal chromaffin cells (BACC) were prepared as described previously (Harper & Littleton 1990). Isolated BACC were suspended in Dulbecco's modification of Eagle's medium (DMEM) supplemented with 10% calf serum. They were plated (0.3ml/chamber, 0.5 x 10^6 cells/ml) in multichamber slides (Gibco) previously coated with poly-L-lysine (30,000-70,000MW, Sigma). They were left in an incubator (37°C, 5% CO_2) for 72 hours to allow cell attachment. They were then fed with DMEM medium containing 200mM ethanol in closed containers (to prevent ethanol evaporation). The cells were used on the 9th day after plating i.e. after 6 days of ethanol treatment. Control cells were treated in exactly the same way except that ethanol was not added to the medium at any time.

Immunocytochemistry

The culture medium was removed from the culture plates and BACC were pre-incubated for 15min with Locke's solution in which the $CaCl_2$ was replaced with $MgCl_2$ (NaCl 150mM, KCl 5.4mM, HEPES 10mM, NaOH 5mM, $MgCl_2$ 2.5mM, Glucose 2g/l). The preincubation solution was then removed and cells were exposed for various times to either Locke's solution as above or to Locke's solution containing 2.5mM $CaCl_2$. At the end of the incubation BACC were fixed for 10min in 4% paraformaldehyde solution (pH 7.4 with 0.1M PBS:sodium phosphate buffered saline), washed twice for 10min with PBS plus 0.3% Triton X-100 then incubated with methanol/hydrogen peroxide (19:1 vol/vol) for 30min. After further washing (2 x 10min in PBS plus Triton as before) the cells were incubated in normal goat serum (vector) for 30min, the serum was then poured off and replaced with polyclonal rabbit c-fos antibody (anti c-fos 456, MEDAG, Germany) at a 1:2000 dilution and left for 48 hours at 4°C. The bound antibody was detected by biotin-conjugated secondary antiserum subsequently incubated with avidin peroxidase complex (ABC kit, Vector laboratory, UK). These steps were separated by two washes of 10min each and finally the cells were incubated with diaminobenzidine (0.5mg/ml) and 0.01% hydrogen peroxide for 5min. All steps were carried out at room temperature unless otherwise stated.

RESULTS

Staining representing c-fos like immunoreactivity was seen after exposure of BACC to Locke's solution containing calcium for 15 minutes (Fig.1). The staining is seen as a darkening of the cell nucleus against a light grey background, the nucleoli do not appear to stain. In control cells c-fos like immunoreactivity was seen in rather few cells and the intensity of staining was not great (Fig.1a). In contrast both the number of cells showing immunoreactivity and the intensity of staining was greater in cells withdrawn from ethanol (Fig.1b). The enhanced staining was not however seen in BACC that had been withdrawn from ethanol 24 hours before exposure to Locke's solution containing calcium (Fig.1c). The absolute requirement for external Ca^{2+} is shown in Fig.1d, neither ethanol-treated, nor control cells (not shown), showed immunoreactivity on exposure to Locke's solution in the absence of added calcium. Despite this requirement

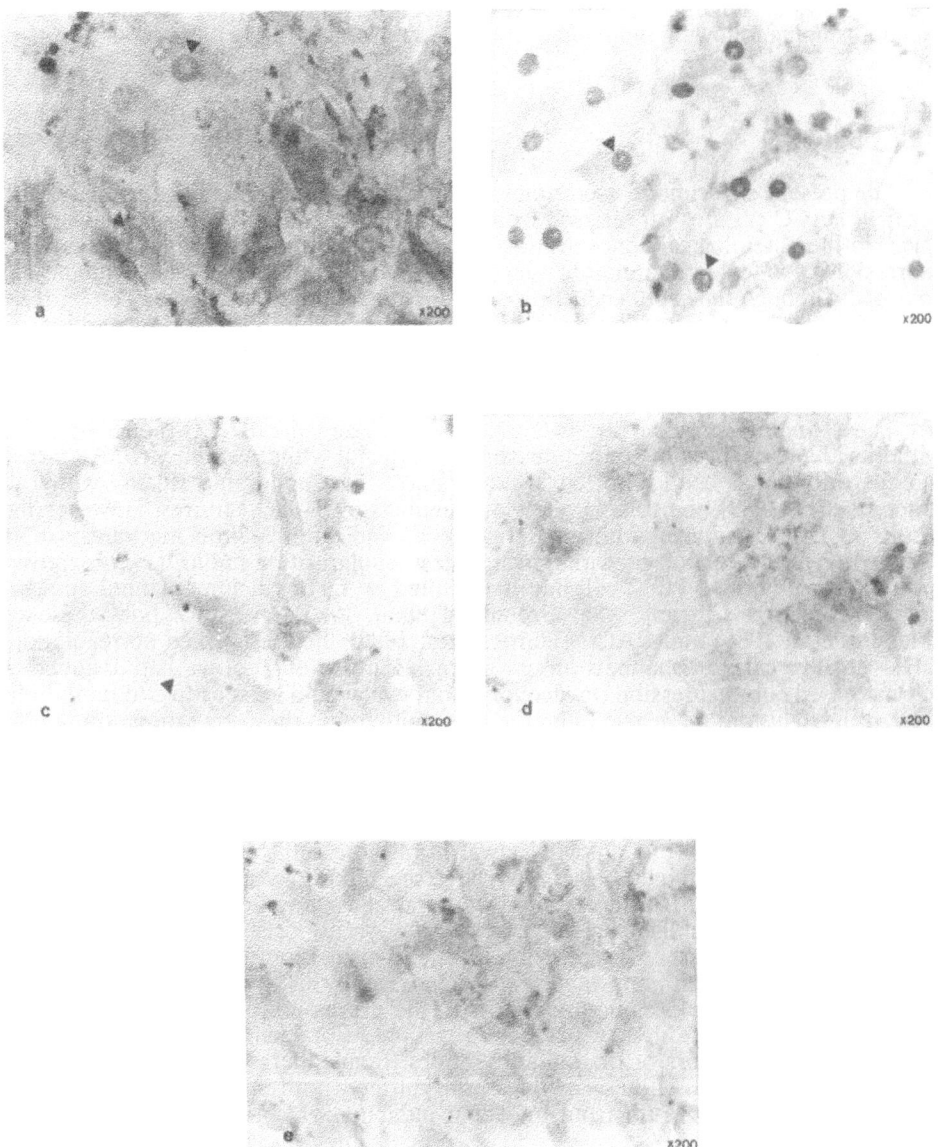

Figure 1. Shows photomicrographs of bovine adrenal cell cultures stained for *c-fos* like protein. (1a) shows the weak nuclear immunostaining in control cells exposed to Locke's solution containing 2.5mM $CaCl_2$ for 15 minutes. (1b) Shows more intense staining in adrenal cells removed from ethanol for 15 minutes prior to exposure to Ca^{2+} in the same way as controls. (1c) shows that this effect is lost if the cells are removed from ethanol 24 hours before exposure to calcium-containing Locke's solution. (1d) illustrates that staining for *c-fos* in ethanol withdrawal is dependent on external calcium, it shows cells withdrawn from ethanol for 15 minutes before exposure to calcium-free Locke's solution for 15 minutes. (1e) shows that omission of the *c-fos* anti-serum abolishes the staining associated with exposure to calcium-containing medium.

301

for calcium neither nifedipine (1μM) nor cadmium chloride (25μM) appeared to affect the immunoreactivity shown by ethanol withdrawn or control BACC (not shown). The specificity of the immunochemical reaction for c-fos is shown by the absence of nuclear immunostaining following omission of the c-fos antiserum (Fig.1e).

DISCUSSION

The present experiments were undertaken to characterise further the changes that occur in BACC cultures on removal from ethanol and exposure to solutions containing Ca^{2+} in the physiological concentration range. BACC cultures show an increase of the expression of c-fos like immunoreactivity in ethanol withdrawn cells compared to controls. Although the super-induction of *c-fos* is dependent on external Ca^{2+}, it does not appear to be inhibited by the dihydropyridine calcium channel antagonists. The increase in *c-fos* expression was also unaffected by the presence of 25μM cadmium, a concentration that almost completely abolished the enhanced catecholamine release from ethanol withdrawn cells (Harper & Littleton, 1991).

These findings provide further similarities between the BACC model of ethanol withdrawal and changes in central neurones in ethanol withdrawal in vivo. An increase in c-fos mRNA has been reported in the brain of mice undering ethanol withdrawal (Jitenda et al, 1989) and this parallels our findings in BACC cultures. However these results in BACC cultures unfortunately give no indication of the mechanism of the change in *c-fos* expression on withdrawal. Since depolarisation-induced *c-fos* expression in the adrenal-derived PC12 cell line is inhibited by DHP calcium channel antagonist drugs (Morgan & Curran, 1986; Greenberg et al, 1986) and since both these cells (Messing et al 1986) and BACC (Harper et al 1989) show a marked up-regulation in DHP-sensitive calcium channels on chronic exposure to ethanol we had assumed that any increased *c-fos* expression on alcohol withdrawal would be sensitive to inhibition by DHP drugs. This did not prove to be the case neither was the *c-fos* expression inhibited by the inorganic calcium channel antagonist cadmium. Both of these agents block calcium entry through the "L-type" of voltage-operated calcium channel and since the expression of *c-fos* was dependent on external calcium we can only conclude that calcium entry through some other route (e.g. the "N-type" of voltage-operated channel) is also enhanced during ethanol withdrawal from these cells. The situation may not be identical in the intact animal because we have recently shown that the parttern of *c-fos* expression in rat brain during ethanol withdrawal is modified by administration of dihydropyridines (unpublished results).

Although the apparent increase in *c-fos* gene expression on ethanol withdrawal is transient it may have an importance extending beyond this time frame. The *c-fos* gene products are themselves regulatory factors in the expression of other important genes and represent the first in a cascade of reactions. For example c-fos regulates the expression of pre-cursors of the opioid peptides (Sonnenberg et al 1989) and so long-term alterations in the synthesis of these transmitters (both in the brain in vivo and in BACC in vitro) might result from the super-induction of *c-fos* that occurs in ethanol withdrawal.

The results provide further evidence of an increase in cellular excitability associated with withdrawal of ethanol from BACC cultures. This increased excitability extends to the Ca^{2+}-dependent induction of c-fos-like protein expression but the mechanism remains obscure.

ACKNOWLEDGEMENT

These experiments were supported by Lipha Pharmaceuticals.

REFERENCES

Brennan CH, Crabbe J & Littleton JM.(1990) Genetic regulation of dihydropyridine sensitive calcium channels in brain may determine sensitivity to ethanol physical dependence. Neuropharmacol 29:429-432.

Cole AJ, Abu-Shakra S, Saffen DW, Baraban JM & Worley PF. (1990) Rapid rise in transcription factor mRNA in rat brain after electroshock induced seizures. J Neurochem. 55: 1920-1927.

Dolin SJ, Hudspith MJ, Pagonis C, Little HJ & Littleton JM. (1987) Increased dihydropyridine-sensitive calcium channels in rat brain may underlie ethanol physical dependence. Neuropharmacol 26: 275-279.

Dragunow M & Robertson HA.(1987) Kindling stimulation induces *c-fos* protein(s) in granule cells of the rat dentate gyrus. Nature 329: 441-444.

Greenberg DA, Carpenter CL & Messing RO.(1987) Ethanol-induced component of $^{45}Ca^{2+}$ uptake in PC12 cells is sensitive to calcium channel modulating drugs. Brain Res 410: 143-146.

Greenberg ME, Ziff EB & Greene LA.(1986) Stimulation of neuronal acetylcholine receptors induces rapid gene transcription. Science 234: 80-83.

Harper JC, Brennan CH & Littleton JM.(1989) Genetic up-regulation of calcium channels in a cellular model of ethanol dependence. Neuropharmacol 28: 1299-1302.

Harper JC & Littleton JM.(1990) Development of tolerance to ethanol in bovine adrenal chromaffin cells in culture. Alcoholism: Clin. Exp. Res.14: 508-512.

Harper JC & Littleton JM.(1991) Characteristics of catecholamine release from adrenal chromaffin cells cultured in ethanol. I Spontaneous and K+ -induced release. Alcohol & Alcoholism 26: 25-32.

Jitenda RD, Tabakoff B & Hoffman PL.(1989) Ethanol withdrawal seizures produce increased c-fos m RNA in mouse brain. Mol. Pharmacol. 37:367-371.

Messing RO, Carpenter CL, Diamond I & Greenberg DA (1986). Ethanol regulates calcium channels in clonal neural cells. Proc Nat Acad Sci USA 83: 6213-6215.

Morgan JI, Cohen DR, Hempstead JL & Curran T.(1987) Mapping patterns of *c-fos* expression in the central nervous system after seizures. Science 237: 192-197.

Morgan JT & Curran T.(1986) Role of ion-flux in the control of *c-fos* expression. Nature 322: 552-555.

Sheng M & Greenberg ME.(1990) The regulation and function of *c-fos* and other immediate-early genes in the nervous system. Neuron 4: 477-485.

Sonnenberg DL, Rausher FJ, Morgan JI & Curran T.(1989) Regulation of proenkephalin by *fos* and *jun*. Science 246: 1622-1625.

REFERENCES

EFFECT OF ACUTE AND CHRONIC ADMINISTRATION OF ETHANOL ON *c-fos* EXPRESSION IN BRAIN

Fei Le, Peter Wilce, David Hume and Brian Shanley

Alcohol Research Unit
Department of Biochemistry
University of Queensland
Australia 4072

INTRODUCTION

Trans-synaptic activation of neurons occurs over a time frame ranging from milliseconds to hours. Recently, it has been shown that the slower long-term responses, which are thought to underlie neuronal plasticity, are associated with induction of gene expression. The genes involved fall into two major groups: the cellular immediate early genes (IEGs), where transcription is activated rapidly and transiently within minutes of stimulation (Greenberg et al, 1985; Morgan and Curran, 1986) and the late response genes, where expression is induced more slowly over a period of hours (Merlie et al., 1984; Castelluci et al., 1988).

Among the neuronal IEGs one of the first to be discovered was the proto-oncogene, *c-fos* (Greenberg et al., 1985, 1986; Morgan and Curran, 1986). Basal levels of *c-fos* expression are very low in neurons. However, induction occurs rapidly and transiently *in vivo* following generalized seizures produced by convulsants, such as pentylenetetrazole (PTZ) (Morgan et al., 1987; Dragunow and Robertson, 1988), thereby providing a biochemical marker for mapping of activated neurons (Sagar et al., 1988; Dragunow and Faull, 1989).

The starting point for the present study was the finding that acute administration of ethanol can inhibit PTZ-induced induction of *c-fos* in rat brain (Le et al., 1990). The aims were to investigate the receptor systems and brain areas involved in mediating both the acute and chronic effects of ethanol.

Alcohol, Cell Membranes, and Signal Transduction in Brain
Edited by C. Alling *et al.*, Plenum Press, New York, 1993

MATERIALS AND METHODS

Materials

Adult male Wistar rats weighing 200-230 g were obtained from the Central Animal Breeding House of the University of Queensland. All pharmacological agents were purchased from Sigma (St. Louis, MO) except for (5R,10S) - (+)-5-methyl-10,11-dihydro-5-H-dibenzo[a,d]cyclo-hepten-5,10-imine hydrogen maleate (MK-801) which was obtained from Research Biochemicals Incorporated (Natick, MA).

Chemicals were dissolved in 0.9% NaCl, except for clonazepam, picrotoxin and ethyl-8-azido-5,6-dihydro-5-methyl-6-oxo-4-H-imidazo[1,5α],[1,4]benzodiazepine-3-carboxylate (Ro15-4513), which were dissolved in dimethyl sulphoxide (DMSO).

Acute Experiments

In acute experiments, the antagonist (ethanol, clonazepam or MK-801) was injected intraperitoneally 10 minutes prior to treatment with agonist (PTZ, picrotoxin, N-methyl-D-aspartic acid (NMDA), kainic acid or caffeine). Ro 15-4513 was administered 10 minutes prior to injection of ethanol. Doses used were as follows: PTZ (35 mg/kg or when used in combination with Ro15-4513, 30 mg/kg), picrotoxin (5 mg/kg), clonazepam (10 mg/kg), Ro15-4513 (4 mg/kg), NMDA (125 mg/kg), MK-801 (7.5 mg/kg), kainic acid (35 mg/kg), caffeine (200 mg/kg) and ethanol (1.5 g/kg). Blood ethanol levels were determined using the method of Lundquist (1959). Animals were killed by decapitation 30 minutes after treatment with PTZ, picrotoxin, NMDA, kainic acid or caffeine, except as otherwise indicated.

Chronic Experiments

Ethanol was administered chronically, either in a liquid diet for 4 weeks (Ward, 1987), or by inhalation for 3 weeks (Hillmann et al., 1988). Both regimes produce tolerance and physical dependence. In withdrawal studies, ethanol administration was discontinued and animals were sacrificed at 4, 8, 10, 12 or 18 h thereafter. Studies were also carried out using the abovementioned agonists and antagonists at the doses stated. Naloxone (10 mg/kg) was injected subcutaneously.

Northern Analysis

Brains were rapidly removed, immediately frozen in liquid nitrogen and stored at -70°C until RNA preparation. Northern Analysis was carried out as described previously (Le et al., 1992).

In Situ Hybridization

Animals were anaesthetized 12 h after ethanol withdrawal and then perfused with 4% (w/v) paraformaldehyde. Following postfixation, the brains were frozen and stored at -70°C prior to performance of *in situ* hybridization (Wilson & Higgins, 1989).

Statistics

Data were obtained from 3 independent experiments for each study. Results show typical Northern Blots. Membranes were also scanned using an AMBIS radioanalyser and results calculated as mean cpm ± S.E.M. (Fig 2B). Statistical significance was calculated, where necessary, using Student's t-test.

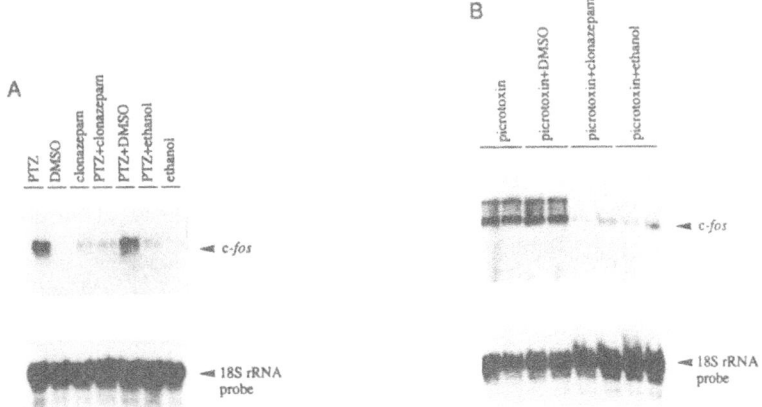

Figure 1.   Effect of ethanol and clonazepam on induction of *c-fos* in rat brain. Representative Northern blots showing effect of A; PTZ and B: picrotoxin.

RESULTS

Acute Experiments

Fig. 1A confirms that synthesis of *c-fos* mRNA was rapidly induced in rat brain following intraperitoneal injection of PTZ. Both ethanol and clonazepam completely blocked induction of *c-fos* by PTZ, while neither ethanol nor clonazepam alone had any effect on *c-fos* mRNA synthesis. Similarly, DMSO, the carrier solvent for several drugs used in this study, had no effect on basal or PTZ-induced levels of *c-fos* mRNA in brain. PTZ-induced seizures were prevented by ethanol or clonazepam at the doses effective in inhibiting *c-fos* induction.

Fig. 1B shows that both ethanol and clonazepam were equally effective in inhibiting picrotoxin-induced *c-fos* mRNA synthesis. This experiment was carried out because PTZ is believed to act at the picrotoxin binding site on the gamma-aminobutyric acid ($GABA_A$) receptor (Squires et al., 1984). It is noteworthy that picrotoxin caused induction of at least

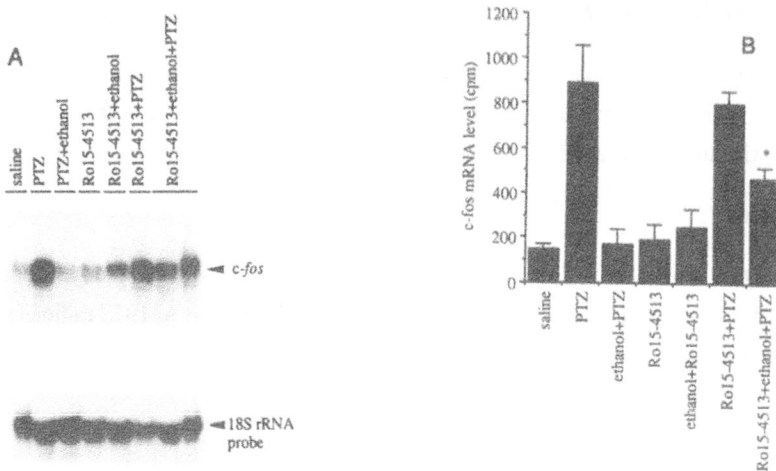

Figure 2. Partial reversal by Ro15-4513 of the inhibitory effect of ethanol on induction of *c-fos* in rat brain by PTZ. A: Representative Northern blot. B: Scans of Northern Blots using AMBIS radioanalyser, expressed as cpm ± S.E.M. (n=3). *p< 0.05 compared with Ro15-4513 + PTZ treatment.

one additional mRNA species of higher molecular weight than the usual one observed with other convulsants. The identity of the extra band is unknown.

Fig. 2 illustrates the effect of Ro15-4513, alone and together with ethanol and PTZ on brain *c-fos* mRNA synthesis. Ro15-4513 is an azido analogue of the classical GABA$_A$ receptor antagonist, Ro15-1788 (Möhler et al., 1984; Sieghart et al., 1987). It has been shown to block certain behavioural effects of ethanol selectively (Suzdak et al., 1986a) and to inhibit augmentation by ethanol of GABA-induced ^{36}Cl$^-$ flux in synaptoneurosomes (Harris et al., 1988). This drug partially antagonized the inhibitory effect of ethanol on PTZ-induced *c-fos* mRNA synthesis, but had no effect on *c-fos* induction when administered alone.

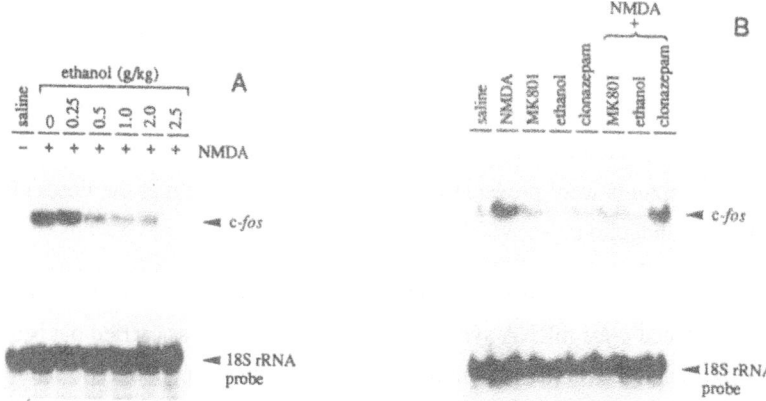

Figure 3. Effect of ethanol, clonazepam and MK-801 on induction of *c-fos* in rat brain by NMDA. Representative Northern blots showing A: dose-dependent inhibition of NMDA induction of *c-fos*. B: inhibition of inducing effect of NMDA by ethanol or MK-801, but not clonazepan.

PTZ-induced brain mRNA synthesis can be partially inhibited by MK-801, a non-competitive antagonist of the NMDA receptor (Wong et al., 1986). This suggests involvement of the NMDA receptor in the action of PTZ on *c-fos* induction. Hence experiments were conducted to examine the effect of ethanol on induction of *c-fos* in brain by NMDA. Fig. 3A shows that ethanol opposed the effect of NMDA in a dose-dependent manner. Fig. 3B indicates that ethanol and MK-801 produced a similar degree of inhibition, whereas clonazepam had no effect on NMDA-induced *c-fos* mRNA synthesis.

An obvious criticism of the results described thus far, is that ethanol may exert a global, inhibitory effect on *c-fos* induction in brain in response to any sufficiently powerful stimulus. This led to the experiments illustrated in Fig. 4. While ethanol again blocked. *c-fos* induction by NMDA, it was ineffectual in opposing the action of kainic acid which acts through the kainate type of glutamate receptor (Fig. 4A). Similarly, ethanol had no effect on the action of caffeine which is thought to induce *c-fos* by competitive antagonism at adenosine receptors (Marangos and Boulenger, 1985).

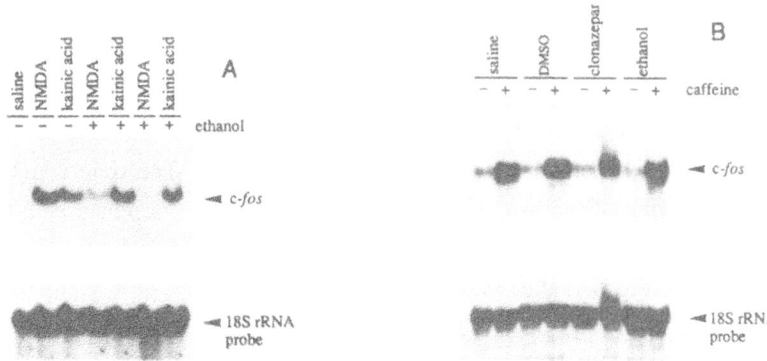

Figure 4.   Lack of effect of ethanol on induction of *c-fos* in rat brain. Representative Northern blots showing effect of A: kainic acid and B: caffeine.

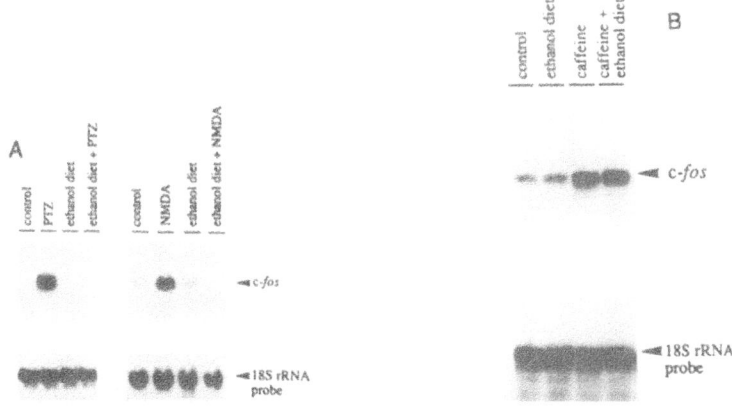

Figure 5.   Effect of chronic ethanol administration on induction of *c-fos* in rat brain. Representative Northern blots showing effect of A: PTZ or NMDA, and B: caffeine. Animals were maintained on a liquid diet for 4 weeks before acute administration of PTZ, NMDA or caffeine.

Chronic Experiments

Fig. 5A shows that chronic administration of ethanol, with accompanying development of physical dependence, did not result in tolerance with respect to the inhibitory effect of ethanol on induction of brain *c-fos* by PTZ or NMDA. The mean blood ethanol concentration in these animals was 110 mg/dl (Fig 7A.) Similarily, chronic administration of ethanol did not change the response to an acute dose of caffeine (Fig 5B).

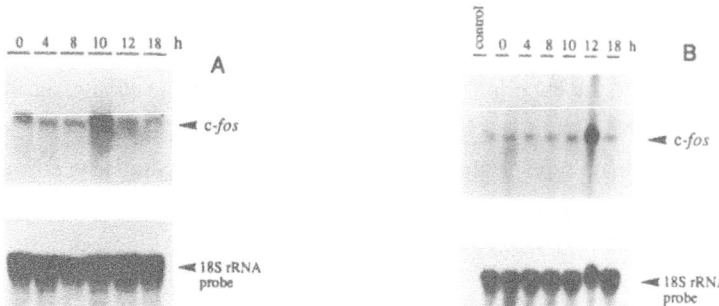

Figure 6.   Time course of ethanol withdrawal and induction of *c-fos* in rat brain. Representative Northern blots showing results for   A: liquid diet treatment and B: ethanol inhalation treatment. Animals were withdrawn after 3-4 weeks treatment and sacrificed at the times indicated.

Following cessation of ethanol administration, rats showed signs of withdrawal, including tremor, hyperactivity and spontaneous seizures. The time course of *c-fos* induction following ethanol withdrawal is shown in Fig. 6. A sharp burst of *c-fos* mRNA synthesis was observed in both groups of animals. In those treated with a liquid diet this occurred at 10 h (Fig. 6A) after withdrawal, while in the vapour-treated group the peak was observed after 12 h (Fig. 6B). The different time courses are probably related to the difference in blood ethanol levels at the time of withdrawal in the two groups and the consequent difference in the rate of ethanol elimination (Fig. 7A).

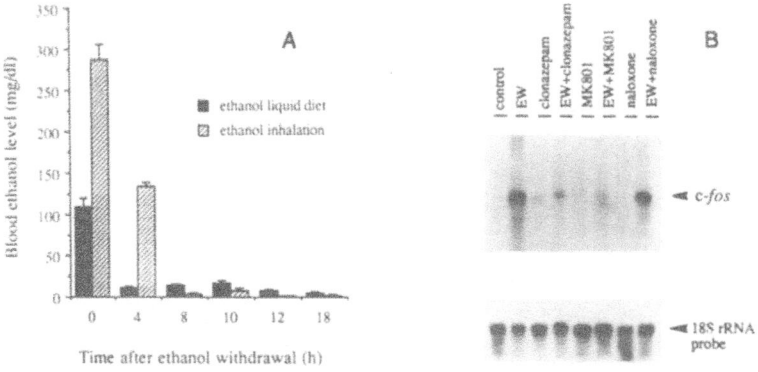

Figure 7.   Pharmacokinetics and pharmacology of ethanol withdrawal. A: time course of blood ethanol levels in rats withdrawn from liquid diet and ethanol vapour treatments. B: representative Northern blots showing effect of clonazepam, MK-801 and naloxone on induction of *c-fos* in rat brain following withdrawal from ethanol vapour treatment. The drugs were administered at 4, 8 and 10 h and animals sacrificed at 12 h after withdrawal.

The induction of *c-fos* following ethanol withdrawal was found to be inhibited by administration of the GABA$_A$ receptor agonist, clonazepam, and the NMDA receptor antagonist, MK-801. By contrast, the opiate receptor antagonist, naloxone, had no effect (Fig. 7B).

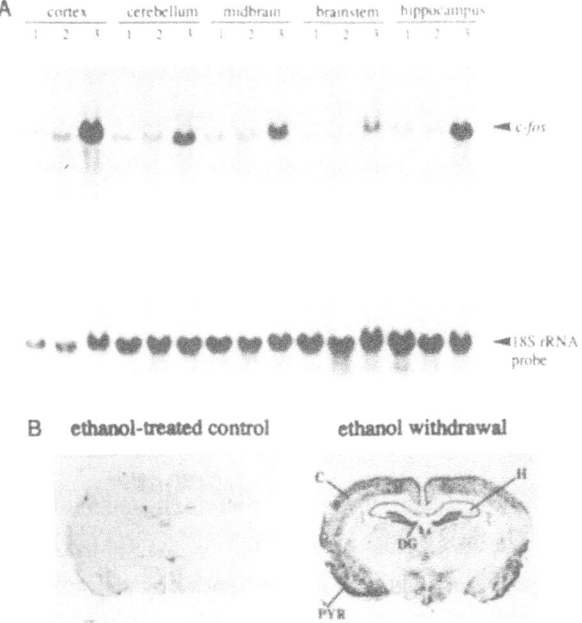

Figure 8. Regional distribution of *c-fos* induction in brains of rats following ethanol withdrawal Animals were treated with ethanol vapour, withdrawn from ethanol and sacrificed 12 h thereafter. A: representative Northern blots showing regional distribution of *c-fos* induction determined by dissection followed by mRNA extraction and Northern analysis. 1 = control (no treatment); 2 = ethanol-treated (not withdrawn); 3 = ethanol-treated (withdrawn). B: regional distribution of *c-fos* induction demonstrated by *in situ* hybridization using coronal section. C = cerebral cortex; DG = dentate gyrus; H = hippocampus; PYR = pyriform cortex.

Fig. 8A shows that significant induction of *c-fos* was found in several brain regions of rats undergoing withdrawal from ethanol. Induction was most marked in the cerebral cortex and hippocampus. However, significant increases were also observed in the cerebellum and midbrain. Studies using *in situ* hybridization (Fig. 8B) confirmed the marked induction in the cerebral cortex and hippocampus, particularly the dentate gyrus.

DISCUSSION

The results of the acute studies confirmed our previous finding that administration of PTZ causes rapid and transient induction of *c-fos* in rat brain, a phenomenon which can be inhibited by prior injection of ethanol (Le et al., 1990). Whether this action of ethanol is nonspecific, or is mediated via a specific receptor(s) is of considerable interest. The major objective of the acute studies was to attempt to answer this question.

Induction of *c-fos* in brain is a function of neuronal excitation (Morgan and Curran, 1987). Hence, it can be argued that any drug which is sufficiently sedative should be capable of preventing *c-fos* mRNA expression caused by stimuli such as PTZ-induced seizures, simply on the basis of the balance between excitatory and inhibitory signals impinging on a particular neuron or group of neurons. While this is the most obvious explanation for ethanol's inhibitory effect on induction of *c-fos* by PTZ, our findings suggest some degree of specificity of action.

PTZ is thought to exert its excitatory effects on the central nervous system partly through interaction with the picrotoxin binding site on the $GABA_A$ receptor (Squires et al., 1984). Ethanol has pharmacological properties similar to GABA, including anxiolytic, anticonvulsant and muscle relaxant actions. Interaction between ethanol and the $GABA_A$ receptor is supported by the demonstration that it potentiates GABA-induced ^{36}Cl flux in cultured spinal cord neurons (Ticku et al., 1986) and synaptoneurosomes (Suzdak et al., 1986b). Therefore, it is logical to suggest that ethanol's action in antagonizing *c-fos* induction in brain by PTZ may be mediated via the $GABA_A$ receptor.

The results of the acute study with picrotoxin, ethanol and clonazepam are compatible with an action of ethanol mediated by the $GABA_A$ receptor but are not conclusive. The induction of additional mRNA species by picrotoxin was unexpected but probably represents hybridization of the *c-fos* probe to related mRNAs encoding Fos B and/or Fos-related antigen (Fra-1). Why these mRNAs were not observed with the other inducing agents remains to be investigated.

Ro15-4513, a partial inverse agonist of the $GABA_A$ receptor (Nutt and Lister, 1987) antagonizes the behavioural (Suzdak et al., 1986a) electrophysiological (Palmer et al. 1988) and neurochemical (Harris et al., 1988) effects of ethanol on GABAergic function. Consequently, it is of particular interest that, when Ro15-4513 was administered in combination with PTZ and ethanol, it caused partial reversal of the inhibitory effect of ethanol on PTZ-induced *c-fos* mRNA synthesis, but had no effect on *c-fos* induction when administered alone. It is tempting to conclude that this is further evidence of ethanol's action via the $GABA_A$ receptor.

However, the fact that MK-801 is also capable of antagonizing PTZ- and picrotoxin-induced increases in *c-fos* mRNA synthesis casts doubt on this as the only interpretation. MK-801 is a non-competitive antagonist of the NMDA receptor, suggesting involvement of this receptor in the stimulatory effects of PTZ on *c-fos* mRNA synthesis. Ethanol was found to have a pronounced dose-dependent inhibitory effect on induction of *c-fos* in brain by NMDA. It is noteworthy that the inducing action of NMDA was inhibited be more important than the GABA$_A$ receptor in mediating the inhibitory effect of ethanol on PTZ- and NMDA-stimulated *c-fos* mRNA synthesis.

The results of the experiments, showing that ethanol was not capable of inhibiting *c-fos* induction in brain by kainic acid or caffeine are additional evidence that ethanol is unlikely to be acting simply as a nonspecific, global inhibitor of stimulus-induced *c-fos* mRNA synthesis.

Chronic administration of ethanol to animals and humans is well known to lead to the development of tolerance and physical dependence. The regimes used in the present study were shown to be effective in this regard, as evidenced by the appearance of withdrawal phenomena when administration was stopped. Thereafter, it is surprising that tolerance toward the inhibitory effect of ethanol on induction of *c-fos* in brain by PTZ or NMDA was not observed. Further work is required to explain this apparent inconsistency. The fact that induction of *c-fos* by caffeine was unaffected by chronic ethanol administration is predictable, in view of the lack of effect of acute dosage with ethanol on the action of this drug.

Withdrawal phenomena, including tremor, hyperactivity and seizures occurred in animals taken off ethanol after 3-4 weeks of chronic administration. The development of seizures co-incided with a burst of brain *c-fos* induction at 10 h and 12 h after withdrawal of ethanol in the liquid diet and inhalation treatment groups, respectively. This difference in time course was consistent and appears to be due to the differences in blood-ethanol levels at the time of withdrawal and the consequent difference in the rate of decline in blood ethanol concentration. These findings are in accordance with those of Dave et al. (1990) and Morgan et al (1992). However, in the latter study induction of brain *c-fos* was found to commence 2-4 h after cessation of ethanol vapour treatment and to peak at about 8 h. Induction of *c-fos* was also found by these authors to be independent of the development of seizures. A similar observation was reported for NMDA-induced *c-fos* mRNA synthesis in brain (Morgan and Linnoila, 1991). These findings add further weight to the argument that ethanol's acute effect in inhibiting *c-fos* induction by PTZ or NMDA is not simply the action of a nonspecific, anticonvulsant drug.

Induction of brain *c-fos* following ethanol withdrawal was preventable by administration of clonazepam or MK-801 but not naloxone, suggesting that the GABA$_A$ and NMDA receptors may be involved in mediating this phenomenon. Decreased sensitivity of the GABA$_A$ receptor (Morrow et al., 1988) and increased sensitivity of the NMDA receptor (Grant et al., 1990) are thought to be adaptive responses to chronic ethanol administration. Therefore, it is likely that the excitatory phenomena observed during withdrawal are a reflection of increased glutamatergic transmission via NMDA receptors, assisted by decreased GABAergic activity. This concept is supported by studies on regional distribution of *c-fos* induction, discussed below, and by the recent observation that intrahippocampal injection of NMDA leads to increased cellular damage in animals undergoing ethanol withdrawal (Davidson et al., 1992).

Regional dissection of brains of rats withdrawn from chronic ethanol treatment (12 h) showed widespread induction of *c-fos* in the cerebral cortex, hippocampus and midbrain (Fig. 8A). This was confirmed using *in situ* hybridization, showing maximal induction in the dentate gyrus and pyriform cortex (Fig. 8B). The pattern resembles that observed with NMDA-induced *c-fos* mRNA synthesis and can be blocked stereospecifically by MK-801 (Morgan et al., 1992).

In conclusion, acute administration of ethanol leads to inhibition of induction of *c-fos* in brain by PTZ and NMDA, but not by other excitatory agents, such as kainic acid or caffeine. This suggests that ethanol is acting via a specific receptor(s). There is evidence indicating possible involvement of the $GABA_A$ and NMDA receptors, particularly the latter. Chronic administration of ethanol does not alter the abovementioned responses. This is unexpected in view of the development of tolerance and physical dependence under these conditions. Withdrawal following chronic ethanol administration leads to a predictable, marked but transient increase in brain *c-fos* mRNA synthesis which can be inhibited by administration of $GABA_A$ agonists or NMDA antagonists. The regional distribution of *c-fos* induction following withdrawal resembles the pattern seen after administration of NMDA, suggesting that the role of the NMDA receptor may be dominant under these conditions. The study of *c-fos* induction and possibly that of other IEGs promises to be a valuable tool in the further elucidation of the neurochemical processes associated with the acute and chronic effects of ethanol.

REFERENCES

Castelluci, V.F., Kennedy, T.E., Kandel, E.R., and Goelet, P., 1988, A quantitative analysis of 2-D gels identifies proteins in which labelling is increased following long-term sensitization in Aplysia, *Neuron* 1:321.

Dave, J.I., Tabakoff, B., and Hoffman, P.L., 1990, Ethanol withdrawal seizures produce increased *c-fos* mRNA in mouse brain, *Mol Pharmacol*, 37:367.

Davidson, M., Wilce, P.A., and Shanley, B.C., 1992, Increased sensitivity of the hippocampus in ethanol-dependent rats to the toxic effect of N-methyl-D-aspartic acid *in vivo, Brain Res.* in press.

Dragunow, M., and Faull, R., 1989, The use of *c-fos* as a metabolic marker in neuronal pathway tracing, *J Neurosci Meth.* 29:261.

Dragunow, M., and Robertson, H.A., 1988, Localization and induction of *c-fos* protein-like immunoreactive material in the nuclei of adult mammalian neurons, *Brain Res.* 440:252.

Grant, K.A., Valverius, P., Hudspith, M., and Tabakoff, B., 1990, Ethanol withdrawal seizures and the NMDA receptor complex, *Eur J Pharmacol.* 176:289.

Greenberg, M.E., Greene, L.A., and Ziff, E.B., 1985, Nerve growth factor and epidermal growth factor induce rapid transient changes in proto-oncogene transcription in PC 12 cells, *J Biol Chem.* 260:14101.

Greenberg, M.E., Ziff, E.B., and Greene, L.A., 1986, Stimulation of neuronal acetylcholine receptors induces rapid gene transcription, *Science* 234:80.

Harris, R.A., Allan, A.M., Daniell, L.C., and Nixon, C., 1988, Antagonism of ethanol and pentobarbital actions by benzodiazepine inverse agonists: neurochemical studies, *J Pharmacol Exp Ther.* 247: 1012.

Hillmann, M., Wilce, P.A., and Shanley, B.C., 1988, Effects of chronic ethanol exposure on the GABA-benzodiazepine receptor complex in rat brain, *Neurochem Int.* 13:69.

Le, F., Wilce, P., Cassady, I., Hume, D., and Shanley, B., 1990, Acute administration of ethanol suppresses pentylenetetrazole-induced *c-fos* expression in rat brain, *Neurosci Lett.* 120:271.

Le, F., Wilce, P.A., Hume, D.A., and Shanley, B.C., 1992, Involvement of γ-aminobutryic acid and N-methyl-D-aspartate receptors in the inhibitory effect of ethanol on pentylenetetrazole-induced *c-fos* expression in rat brain, *J Neurochem.* in press.

Lundquist, F., 1959, The determination of ethyl alcohol in blood and tissues, *Meth Biochem Anal.* 7:215

Marangos, P.J., and Boulenger, J.P., 1985, Basic and clinical aspects of adenosinergic neuromodulation, *Neurosci Behav Rev.* 9:421.

Merlie, J.P., Isenberg, K.E., Russell, S.D. and Sanes, J.R., 1984, Denervation supersensitivity in skeletal muscle: analysis with a cloned cDNA probe, *J Cell Biol.* 99:332.

Möhler, H., Sieghart, W., Richards, J.G., and Hunkelev, W., 1984, Photo-affinity labelling of benzodiazepine receptors with a partial inverse agonist, *Eur J Pharmacol.* 102:181.

Morgan, J.I., Cohen, D.R., Hempstead, J.L. and Curran, T., 1987, Mapping patterns of *c-fos* expression in the central nervous system after seizure, *Science* 237:192.

Morgan, J.I. and Curran, T., 1986, Role of ion flux in the control of *c-fos* expression, *Nature* 332:552.

Morgan, P.F. and Linnoila, M., 1991, Regional induction of *c-fos* mRNA by NMDA: a quantitative *in situ* hybridization study, *NeuroReport* 2:251.

Morgan, P.F., Nadi, N.S., Karanian, J. and Linnoila, M., 1992, Mapping rat brain structures activated during ethanol withdrawal: role of glutamate and NMDA receptors, *Eur J Pharmacol.* 255:217.

Morrow, A.L., Suzdak, P.D., Karanian, J.W. and Paul, S.M., 1988, Chronic ethanol administration alters GABA, pentobarbital and ethanol mediated ^{36}Cl$^-$ uptake in cerebral cortical synaptoneurosomes, *J Pharm Exp Ther.* 246:158.

Nutt, D.J., and Lister, R.G., 1987, The effect of the imidobenzodiazepine, Ro15-4513, on the anticonvulsant effects of diazepam, sodium pentobarbital and ethanol, *Brain Res.* 413:193.

Palmer, M.R., Van Horne, C.G., Harlan, J.T., and Moore, E.A., 1988, Antagonism of ethanol effects on cerebellar Purkinje neurons by the benzodiazepine agonists, Ro15-4513 and FG 7142: electrophysiological studies, *J Pharmacol Exp Ther.* 247:1018.

Sagar, S.M., Sharp, F.R., and Curran, T., 1989, Expression of *c-fos* protein in brain: metabolic mapping at the cellular level, *Science* 240:1328-1331.

Sieghart, W., Eichinger, A., Richards, J.G., and Möhler, J., 1987, Photo-affinity labelling of benzodiazepine receptor proteins with the partial inverse agonist [^{3}H]Ro15-4513: a biochemical and autoradiographic study, *J Neurochem.* 48:46.

Squires, R.F., Saederup, E., Crawley, J.N., Skolnick, P., and Paul, S.M., 1984, Convulsant potencies of tetrazoles are highly correlated with actions on GABA/benzodiazepine/picrotoxin receptor complexes in brain, *Life Sci.* 35:1439.

Suzdak, P.D., Glowa, J.R., Crawley, J.N., Schwartz, R.D., Skolnick, P., and Paul, S.M., 1986a, A selective imidobenzodiazepine antagonist of ethanol in the rat, *Science* 234:1243.

Suzdak, P.D., Schwartz, R.D., Skolnick, P., and Paul, S.M., 1986b, Ethanol stimulates gamma-aminobutyric acid receptor-mediated chloride transport in rat brain synaptoneurosomes, *Proc Natl Acad Sci.* 83:4071.

Ticku, M.K., Lowrimore, P., and Lehoullier, P., 1986, Ethanol enhances GABA-induced ^{36}Cl$^-$ influx in primary spinal cord cultured neurons, *Brain Res Bull.* 17:123.

Ward, L.C., 1987, Animal models of chronic alcohol ingestion: the liquid diet, *Drug Alc Depend.* 19:333.

Wilson, M.C. and Higgins, G.A., 1989, *In Situ* Hybridisation, *in* "Neuromethods: Molecular Neurobiological Techniques", Vol 16, A.A. Boulton, G.B. Baker and A.T. Campagnoni, eds., Humana Press, New Jersey.

Wong, E.H.F., Kemp, J.A., Priestly, T., Knight, A.R., Woodruff, G.N., and Iversen, L.L., 1986, The anticonvulsant, MK-801, is a potent N-methyl-D-aspartate antagonist, *Proc Natl Acad Sci.* 83:7104.

INDEX

The manufacturer's authorised representative in the EU is Springer
Nature Customer Service Centre GmbH, Europaplatz 3, 69115 Heidelberg,
Germany. If you have any concerns regarding our products, please
contact ProductSafety@springernature.com

Printed and bound by CPI Group (UK) Ltd, Croydon, CR0 4YY
23/04/2026
02095629-0016